Advances in Japanese Business and Economics

Volume 25

Advances in Japanese Business and Economics (AJBE) showcases the work of Japanese and non-Japanese scholars researching the Japanese economy and Japanese businesses. Published in English, the series highlights for a global readership the unique perspectives of Japan's most distinguished and emerging scholars of business and economics. It covers research of either theoretical or empirical nature, in both authored and edited volumes, regardless of the sub-discipline or geographical coverage, including, but not limited to, such topics as macroeconomics, microeconomics, industrial relations, innovation, regional development, entrepreneurship, international trade, globalization, financial markets, technology management, and business strategy. At the same time, as a series of volumes written by Japanese and non-Japanese scholars studying Japan, it includes research on the issues of the Japanese economy, industry, management practice, and policy, such as the economic policies and business innovations before and after the Japanese "bubble" burst in the 1990s.

AJBE endeavors to overcome a historical deficit in the dissemination of Japanese economic theory, research methodology, and analysis. The volumes in the series contribute not only to a deeper understanding of Japanese business and economics but to revealing underlying universal principles.

Overseen by a panel of renowned scholars led by Editor-in-Chief Professor Ryuzo Sato, AJBE employs a single-blind review process in which the Editor-in-Chief, together with the Managing Editors and specialized scholars designated by the Editor-in-Chief or Managing Editors, rigorously reviews each proposal and manuscript to ensure that every submission is a valuable contribution to the global scholarly readership.

All books and chapters in AJBE are indexed in Scopus.

More information about this series at http://www.springer.com/series/11682

Minoru Shimamoto

National Project Management

The Sunshine Project and the Rise of the Japanese Solar Industry

Springer

Minoru Shimamoto
Graduate School of Commerce
and Management
Hitotsubashi University
Kunitachi, Tokyo, Japan

ISSN 2197-8859 ISSN 2197-8867 (electronic)
Advances in Japanese Business and Economics
ISBN 978-981-15-3179-8 ISBN 978-981-15-3180-4 (eBook)
https://doi.org/10.1007/978-981-15-3180-4

This Springer imprint is published by the registered company Springer Nature Singapore Pte Ltd.
The registered company address is: 152 Beach Road, #21-01/04 Gateway East, Singapore 189721, Singapore

Preface

National projects refer to social planning spearheaded by the government. This book is mainly focused on discussing cases where industry, government, and academia have collaborated on research into and development of technologies to achieve policy goals. Finding solutions for environmental energy problems and developing new business and new industry are some of the innovations that have emerged out of national projects. At the same time, the innovation process provides new ideas for finding common ground between public administration and business administration.

Background to the Research

It has been several decades since my research into national projects was first launched. I finally arrived at this research theme after a long time and many twists and turns. When studying for my master's degree, my interests revolved around industrial policy and management development, more specifically clarifying the impact of industrial policy on corporate management. My research topic was to shed light on the impact of the wartime economic system on corporate management based on the thoughts and actions of reform bureaucrats during the 1930s and 1940s. The idea that the Japanese style of business administration was rooted in wartime controls was attracting attention at the time, and the policies behind the attempts to revolutionize the Japanese economic system through a planned economy were considered extremely interesting.

By extension, my research interests then shifted to the industrial rationalization movement at the Ministry of Commerce and Industry (subsequently the Ministry of International Trade and Industry (MITI); currently, the Ministry of Economy, Trade and Industry (METI)). Later on, I became increasingly interested in shedding light on development processes in peacetime instead of the special conditions of wartime, so I shifted the focus of my research to postwar industrial policy.

Building on these ideas for my doctoral degree, I started to research the transition to domestic production of computers and aircraft as an example of industrial policy at MITI during the period of rapid economic growth in the 1950s and 1960s. Together with Professor Yonekura Seiichirō, I co-authored an article on the domestic production of computers, which was published in *Nihonteki keiei no seisei to hatten: Kēsubukku Nihon kigyō no keiei kōdō* (Yuhikaku 1998). One of the main protagonists in the article was Hiramatsu Morihiko (then Oita governor after a period at MITI) who had participated in formulating industrial policy for computers. I will never forget receiving a postcard from Mr. Hiramatsu in which he described his impressions of the article, which he had read by chance. In addition, an article about the domestic production of aircraft was published in the *Hitotsubashi Review* (vol. 121, no. 5, 1999). Both articles concluded that the crux of industrial policy by government is where to draw the line between imposing planning and where to defer to corporate independence concerning competition in a seminal industry and that such skills determine the effectiveness of policies. This is the source of the viewpoint that looks at the conflict between emergence and planning where policy and management are involved.

Nearer to the present, the industrial policy of MITI became increasingly focused on technology and innovation policies through national projects as well as the rescue of ailing industries and protection of infant industries. As the focus of my interests and research themes shifted to the later period, I quite naturally turned to technology policy by the government and strategies for research and development of corporations. For my doctoral work, I carried out case study research appertaining to innovation policies and management in the hi-tech technologies. My themes were the new energy technologies of the 1970s, which became the prototype for this book, and the fine ceramics and biotechnology industries of the 1980s. As a result, it was not long before my observations could no longer be contained within the framework of management history. Meanwhile, the need arose to analyze decision-making processes in the government, corporations, and other large organizations, so I turned my attention to the study of the conceptual frameworks for organizational theories of management and case study techniques.

Looking back, I realize how unsophisticated we were in graduate school. It was extremely optimistic to think that positive outcomes in line with the expectations of policymakers were a given where technology policy was concerned. Thinking about it now, it was extremely naïve to assume that supporting industry with huge budgets would necessarily have the desired effect on industrial development. It was only later on, when studying a selection of cases, that I noticed the Unintended Consequences of a policy process that tries to deliver industrial development by artificial means.

The argument is extremely simple as long as we only consider the impact of government policy on corporations. However, in reality, corporations also respond to government policy based on their own management strategies. Both government and corporations are players who advance the game one move at a time while trying to predict what the other protagonist is thinking. The relationship between the two protagonists is not determined in a single move but involves complicated interaction. Therefore, government policy is not the key condition that determines policy outcome. Rather, it may well be the response from corporations pretending to listen to directives. Gradually, I started to sense the boundaries of a style of research that tries to extract outcomes from the history of policy, so I proceeded to venture into the area of the business activities of private enterprise, moving toward observations of how they interact with policy interventions by the government.

Based on this awareness of the issues, I then selected new energy development in the 1970s and 1980s for my next research topic. It was the very moment in the late 1990s when global environmental issues were being much discussed in Japan. Set in this context, the timing seemed to be right for researching the relationship between government and corporations in regard to developing technology in the field of environmental energy. Public interest was high, industrialization was difficult, and the future market was huge, so developing technologies for new energy would require innovation in the public and private sectors. Simply clarifying the static structures of the system was inadequate as a method of analysis. Rather, it was crucial to delineate the dynamic interaction between the players in industry, government, and academia. This research theme may ultimately enable us to map out the historical transformations in the national innovation system in Japan. Such were my expectations when I started to research the national project for new energy, which is the central topic of this book.

Origins of the Sunshine Project

Does the Sunshine Project ring a bell? It has nothing to do with the renewable energy boom after the Great East Japan Earthquake in 2011. Rather, the Sunshine Project was a national project for developing new energy that was launched more than 45 years ago at the time of the first oil crisis. It may surprise some readers that Japan has had a program for researching new energy technology since then.

The two oil crises that hit Japan in the early 1970s when the period of rapid economic growth was drawing to a close were critical events for a country with limited energy resources. These events served to remind government and corporations alike that any delay to the stable supply of oil would strike a great blow at economic activities in Japan.

Research into and development of new energy technologies to replace oil was one of the countermeasures implemented by the government. New energy refers to solar energy and other natural energy sources, so it is more or less what we call renewable energy today. For present purposes, it can be defined as eco-friendly clean energy subject to policy development. As of the 1970s, MITI greatly expanded its policies for industrial technology to focus on energy-related development in step with the fully fledged launch of new energy research and development. The Sunshine Project, the focus of this book, was the government's national project for developing new energy technologies. It was a national project involving the entire country on a large scale and over the ultra-long term. Industry, government, and academia mobilized all available resources to advance the research into and development of new energy technologies, setting the grand goal of liberating resource-poor Japan from the problems of resource scarcity.

By the 1990s, people began to recognize that the project would not only contribute to solving the energy problems, but also to the global environmental problems. For example, after the Kyoto Protocol, it was emphasized that the development of new energy would also provide a solution to global warming by reducing carbon dioxide emissions. This is how the (New) Sunshine Project was given new meaning as a project that would save the global environment by developing clean energy technologies. But is the idea of finding solutions to environmental energy problems by developing new energy technologies under the auspices of a national project anything other than a grand concept and an impossible dream?

An overview of the situation suggests that the late 1990s were a time when photovoltaic energy systems spread rapidly in Japan. Solar cells made by Sharp, Kyocera, Sanyo, and other companies were appearing in all sorts of places. It was a time when solar cells were installed at public facilities, and when we started to see people having them installed on the roofs of their homes. For example, at the time, I did a stakeholder survey at Kyoto station where the New Energy and Industrial Technology Development Organization (NEDO) had installed a large-scale photovoltaic energy system as a field test. Solar panels also glittered proudly on the roofs of the private homes of stakeholders in the solar energy corporations when I visited their homes for interviews. Photovoltaic energy was the symbol of clean energy, and the fact that these things were visible in all sorts of places seemed to prove that, a quarter century after its launch, the Sunshine Project was finally bearing fruit.

However, the situation in the second half of the 1990s was in no way equal to what it is now in the 2010s. Much time has passed between then and now. Aside from the people who were acutely aware of environmental issues, systems for generating photovoltaic power were not then something that ordinary people bought on the spot, due to the high price and the poor conversion efficiency. In the 1990s, a 3 kW power generation system for a residential home cost more than six million yen, which was far too expensive to install on the roof of your house if you were looking for financial advantage. In those days, photovoltaic energy was still only a hobby for some wealthy people. The energy buyback system whereby power companies buy power generated in private homes had barely got off the ground at the time.

However, the technology for photovoltaic energy developed very quickly and when it was coupled with subsidies from government and local municipalities it became very popular. At the same time, people grew increasingly aware of global warming and other issues with an impact on the global environment. In the early twenty-first century, by the tail end of these global changes, the time to reap the rewards of the long-term efforts of policymakers, researchers, engineers, and business operators had at long last arrived. The recent popularity of photovoltaic energy is amazing, with solar panels now extensively available at local electronics retail stores. It takes the breath away to see this happening.

Legacy of the Sunshine Project

As originally planned, the Sunshine Project (as of 1993, the New Sunshine Project) came to a close in the early 2000s. A special symposium commemorating the 40th anniversary of the Sunshine Project was held at the Pacifico Yokohama convention complex on September 18, 2014. It was an anniversary event held to coincide with the annual briefing session on new energy achievements sponsored by the NEDO. It was a magnificent event. As someone who has studied the history of renewable energy policies and the history of renewable energy administration for many years, I felt extremely pleased to attend this gathering. At the same time, I was aware once again of the great significance of leaving matters not talked about on a ceremonial occasion like this on record in the manner of books like this one.

The event praised the significance of the Sunshine Project and its contributions throughout, including the panel discussion gracefully presided over by Araki Yukiko, the former chief of new energy measures at MITI, who is currently the manager of an environment-related division at Hitachi.

Sakaiya Taichi, former comprehensive research and development officer in charge of the Sunshine Project, who is also renowned as the author of *Yudan!* [*Negligence!*] and *Dankai no sedai* [*Baby boomers*], was invited to deliver a commemorative speech at the symposium. In the keynote lecture that preceded Sakaiya's speech, Kuwano Yukinori, President of the Photovoltaic Power Generation Technology Research Association (PVTEC), spoke on the history of solar cells. Kurokawa Kōsuke, specially appointed professor at the Tokyo Institute of Technology, and other experts discussed the course of technological development through the project on the platform based on their respective research themes (solar power, wind power, coal, and hydrogen) in a panel discussion that followed.

In his commemorative speech, Sakaiya shared memories from his research and development officer days. Sakaiya said that MITI's Sunshine Project was totally played down in those days compared with a nuclear fusion project of the Science and Technology Agency, that his efforts to set up NEDO were challenging because one corporation with special semigovernmental status had to be dissolved to set up another, and that NEDO's establishment was his final work at MITI. In his speech, Sakaiya emphasized that amusing explanations are vital for a policy if it is to win the support of the public and receive funding. This point left a strong impression on me. As readers of this book will become aware, MITI craved the establishment of NEDO as an organization for executing its projects, and MITI officials in charge of policies aggressively promoted new energy development to the public.

In his keynote lecture, Kuwano recounted the history of amorphous solar cells, which became the source of the HIT system that boasts the highest sunlight conversion efficiency of 25.6%, and the efforts made to popularize them. *Taiyo denchi wa donoyōni hatsumei sare seichōshitanoka: Taiyō denchi kaihatsu no Rekishi* [*How solar cells were invented and how they grew*: *The history of solar cell development*] (Ohmsha 2011) written by Kuwano is the most important piece of literature and the best textbook for people studying the history of solar cells. I met Kuwano for the first time in a room near Umeda Station in Osaka. Kuwano was the managing director of Sanyo Electric in those days. Despite his position, which kept him busy, Kuwano kindly spoke with me for a long time. I left the interview with a pocket calculator equipped with a solar cell, which he gave me as a gift.

Kurokawa is the originator of grid-connected, reverse-flow solar power generation systems in Japan. The current spread of solar power generation systems for home use would not have been possible without power transmission studies by Kurokawa. He appears frequently in this book as well, at times when the Sunshine Project reached important stages, such as the development of solar cell manufacturing plants. In the panel discussion, Kurokawa shared the story of when Horigome Takashi of the Electrotechnical Laboratory took the trouble to come to his house to ask for his cooperation in solar energy studies. I met Kurokawa for the first time when I visited him at his laboratory at the Tokyo University of Agriculture and Technology. I remember a three-dimensional artwork depicting the sun that covered an entire laboratory wall. Kurokawa had a shining necktie with a design based on the sun on his chest on the day of the symposium as well.

Hirano Katsumi, a professor at Nihon University, who was in charge of coal liquefaction studies, said that the suspension of the studies was regrettable in view of the survival of sunlight, wind power, and hydrogen as themes for development today. With the preliminary remark that it may be rude to make such a statement in the presence of Furukawa Kazuo, Chairman of NEDO, and Chūbachi Ryōji, President of the National Institute of Advanced Industrial Science and Technology (AIST), Hirano mentioned the absence of central government support for coal energy in the stage of practical application and socialization as the reason for the discontinuation of the studies. I found this remark extremely impressive as well. Readers of this book will note the huge budget allocated by the Sunshine Project to coal studies. The amount of government subsidies is not the only problem. Unlike solar power generation, no company had confidence in the practical application of developed coal liquefaction technologies in a manner that was commercially profitable. Even if achievements come from a national project, they do not spread throughout society unless companies can turn them into a business in a manner that is commercially profitable. Corporate eagerness for commercialization is the final key to the diffusion of project achievements in society. Well-known and experienced business managers, such as Furukawa, the former president of Hitachi, Ltd., and Chūbachi, the former president of Sony Corporation, know this as a matter of course (come to think of it, Watamori Tsutomu, the first NEDO chairman, also came from Hitachi, Ltd.). I believe that Furukawa and Chūbachi were in their current positions as the chairman of NEDO and the president of AIST for that reason.

At this event, I learned that Horigome Takashi, who had been actively involved in the Electrotechnical Laboratory and NEDO, and who had gained the nickname "Mr. Sunshine," had passed away in 2013. In addition to Horigome, Suzuki Ken, a research and development officer who first drafted the Sunshine Project at the AIST, and Kimura Kenjirō, a contributor to the solar cell business from its initial stage who moved from Sharp to Kyocera by way of Japan Solar Energy, left us in 2010 and 2005, respectively. Thinking about their departures, I feel very apologetic toward everyone who provided me with their assistance for having taken such a long time to write this book.

Successor Projects and the Future

Today, the baton for developing technologies has been passed to several successor projects. Times are changing much more rapidly than we ever imagined. Since the end of the Sunshine Project, there has been one unexpected event after another, including the escalation of global warming, the nuclear power accident and energy policy chaos stemming from the Great East Japan Earthquake, and the pressure on electronics companies exerted by manufacturers of solar panels in China, Taiwan, and Korea. Ironically, people are now expecting more from renewable energy than ever before. However, whether Japanese corporations will continue to be leading product suppliers is currently hanging in the balance. Much is expected of both the public and private sectors, including innovation based on technology development, support policies and market development, and energy system designs that consumers find more convenient. Looking back at the history of a project that was launched as early as the 1970s to put the challenges and outcomes of a national project on the record will no doubt provide important knowledge for overcoming the environmental energy problems and for developing industry in Japan. Initiatives designed to find solutions to the global environmental problems and the energy problems are vital in order for resource-poor Japan to achieve sustained economic development. Technology is key to moving forward by developing the compatibility of energy, ecology, and economy. To do so, what kinds of systems, organizations, and management styles do we need, and how should industry, government, and academia approach collaboration?

The conditions surrounding solar power generation have changed significantly in recent years. Even the final chapter of this book may consist of stories that readers sensitive to the trends find outdated. After all, success did not last long for Suntech Power in China and Q Cells in Germany, both of which had risen quickly to the top of the industry and prospered for a time. As explained in the final chapter, the two companies expanded their operations by serving markets that grew rapidly with the

FIT system in the middle of the 2000s. However, a sudden business downturn after the global financial crisis acted as a strong headwind for them. They both became insolvent as a result of the fierce price competition into which the entire industry plunged. Jiangsu Shunfeng Photovoltaic Technology took over Wuxi Suntech Power, a Suntech Power subsidiary. The Hanwha Group of South Korea acquired Q Cells in the summer of 2012.

These companies boldly raise funds by making efficient use of markets, engage in aggressive investments by watching out for the right time, and procure manufacturing technologies available for purchase from markets. Management resources such as manpower, materials, and money keep increasing to satisfy emergent demand by changing their forms, even if other companies absorb failed ones and the owners of individual companies change. The speed of change in this industry is amazing indeed.

In this environment, Sharp, Kyocera, Panasonic (formerly known as Sanyo Electric), and Mitsubishi Electric, all of which appeared to have entered the global competition late, are working to develop a profitable business model based on their respective original solar power generation plans. Japanese companies tend to work on maintaining the way they should be as communities called enterprises and display their individual strengths within those ways under capitalism. Whether the application of a market mechanism or accumulation in an organization is more beneficial when they try to maintain competitive advantages is a question of efficiency. At the same time, it is a question that is closely associated with a way of working that makes life happy for each and every one of us as workers.

The expansion of the solar power generation market itself shows great promise, even though market growth may become sluggish temporarily. In these circumstances, Japanese companies appear to be taking aim at the forthcoming century of green innovations by deciding on their respective fields of expertise. Global environmental and energy problems will inevitably remain serious issues from this point on. We cannot backtrack toward renewable energy expansion. There are still many topics remaining for research in this field. I shall further discuss future developments elsewhere.

Issues Dealt with by This Book

With this book, and now that such a long time has passed, I would like to shed light on the origins of the Sunshine Project, how it continued, and whether it produced any results. I would also like to consider the reasons why policies were successful in some areas but did not have the intended effect in other areas. By tracing the history of the project, we may learn how technology innovation was employed to achieve policy goals with a high degree of public interest, such as technology research and development, or finding solutions to environmental energy problems. Researching the history of the Sunshine Project is not the only aim of this book. Rather, if we can present suggestions for how to structure national projects, it may also be possible to identify ways for industry, government, and academia to put their collective heads together to find solutions not only to environmental energy problems, but also to social problems such as the food crisis, social disparity and poverty, and the aging population and the decline in the birthrate. Therein lies the goal of this book.

Although the development of new energy is the subject of this book, I am interested in broader perspectives on the national project and the series of processes formulated and implemented by industry, government, and academia to research and develop technologies. In this regard, the areas I cover are: (1) the process of formulating government policy, (2) systems for cooperation between industry, government, and academia, (3) corporate technology strategies. I believe that the outcome of the national project was strongly influenced by these factors.

The first issue is the governmental decision-making process involved in planning policy. By aiming the inquiry at the government and tracing the processes for formulating and implementing the national project, a true picture emerges of what decision-making by what participating agent led to the launch of the project. In general, there is a tendency to assume that capable government workers draft outstanding projects but, if you look into the history of actual projects, you soon find out that it is not at all a given that an omnipotent government formulates projects under its own steam from beginning to end. The support of corporations, universities, and technical experts at national research institutes is essential when formulating projects. The support of politicians and the nation is also necessary to have projects approved as policy. For these reasons, it is a given that several agents have to participate in the process of formulating a project. What is the process like when such large numbers of people are brought into a project and participate in its formulation? How are national projects created? Studying the case of new energy will likely unearth suggestions that will guide our understanding of the general processes of formulating national projects and industrial policy.

The second issue relates to the creative process and the design of systems of collaboration between industry, government, and academia. The term "industry–government–academia collaboration" has been particularly popular since the early 1980s, giving people the impression that research into and development of advanced technologies has led to one success after another as a result of such collaboration. In fact, there are cases as early as in the 1970s of rival corporations joining forces to research and develop technologies, producing outstanding results, such as the joint research and development of VLSI. In these cases, organizational theory research has focused on how the parties coordinated competition and cooperation. On the other hand, the increase in semigovernmental corporations and public-service corporations, the *amakudari* custom of high-ranking governmental officials taking up corporate positions after retirement, and other dysfunctions stemming from management organizations controlled by the government have long been identified as problematic. What kind of effective organizations and inter-organizational relations are required for collaboration between industry, government, and academia to be successful? How should we implement projects to develop technologies in the name of collaboration between industry, government, and academia? What kinds of organizations implement national projects? Where these issues are concerned, we can expect valuable insights from the organizational changes that took place in the case of new energy.

The third issue for a national project is the process of selecting the targets for technology development and the corporate strategic decision-making around technology development. The parties identify long-term goals and select the required technology development themes for projects, but how do they select promising themes for technology development with high potential for success? Where a long-term project is concerned, technology may progress and markets may change in unexpected ways during the lifetime of the project. As a result, there is no avoiding periodic reviews of the targets in relation to the prospects and feasibility of technology development. However, in these situations, it is not normally so easy to tell which themes and targets for technology development are the promising ones. There will be tension between corporations about the timing of decisions to discontinue themes that are not producing results, or how to evaluate the feasibility of emerging themes. Therefore, the parties involved in research and development make efforts to communicate the superiority of the technologies they are pushing, investing their energy into persuasive forecasts targeted at a broader audience. The success of such persuasion is linked to increased budgets for the projects, and budget increases raise the probability of successfully developing the targeted technology. Consequently, there is already intense competition at the stage of selecting the focused target. How do you identify guidelines for technologies and markets with uncertain futures? Studying cases of competition around targeting specific technologies for a project will shed light on the aspects of delineating future concepts for the project.

The issue of technology assessment is strongly linked to overall technology strategies at companies participating in a project. Governments get involved in a project to achieve policy goals where there is a high level of public interest and where the successful development of technologies will have a ripple effect on other industries. However, when corporations are involved in a project, the most important issue is using the technologies to develop products for successful commercialization. Corporations are able to link research and development of technologies to their own businesses as soon as profitable products using the technology start to emerge. When the interests of the government and the corporations overlap, cooperation between the public and private sectors when developing technologies is successful; but there are negative effects if the expectations of the government and the corporations are out of sync, or if corporations perceive participation in a project as simple association, or if the government for some reason refuses corporate lobbying for promising technologies. In particular, when the technologies targeted for project research and development are pivotal to the overall strategies of corporations participating in a project, they will be fully committed to development, even spending their own funds. Conversely, even if the government is considering commissioning major corporations with strong technological competence to carry out research and development, it is only natural that such corporations will have less appetite for research and development if they do not plan to use the technologies to develop their own businesses. When considering the development of technologies for promising projects, it is not possible to ignore overall corporate strategies. Studying cases where corporations have developed commercial versions of new energy products will shed light on how consistency between projects and corporate strategy influences the outcomes.

Structure of the Book and Recommendations for Readers

The main part of this book consists of three case studies interspersed with two reflective chapters. The first case study describes the Sunshine Project from the perspective of project management. Based on the perspective of government, the chapter investigates methods for managing projects effectively and efficiently to achieve the goals. The focal points in the chapter are environmental awareness, technology forecasts, organizational structure, budget allocation, and outcome assessment. The history of the Sunshine Project is mainly described from the perspective of project management based on the official history of the project.

The second case study is a detailed examination of the routines in all organizations, whether industry, government, or academia, and of the autonomy of the project organization. In the first case, the gaze is directed at policymakers, their environmental awareness, and the management of technology selection on the tacit assumption that the organizations contributing to the project generally follow the directives of the policymakers. However, in actual projects, organizations do not necessarily operate in line with the expectations of policymakers. Corporations and universities participating in projects act in line with their own missions and government agencies do not necessarily exert full control. Each organization carries out its day-to-day business operations in line with its own routines with all sorts of unintended consequences for the project as a whole. It is essential for the success of a project to deepen understanding of the mechanisms behind routines and their unintended consequences.

The third case study increases the degree of detail to focus on the smallest unit of analysis, the intentions and motivations of key individuals participating in the project. As a result, we begin to realize that the rational decision-making of government observed in the first case study, and the autonomous routines at organizations observed in the second case study are based on different intentions, expectations, hopes, and conjectures when viewed from the perspectives of the individuals concerned. What emerges is competition to mobilize knowledge, capital, and other resources from other parties, to set out a vision of the future, and to promote ideas in regard to an uncertain future. Even if the initial ideas are unsubstantiated, mobilizing resources may lead to implementation along the lines of the idea. It would seem that when someone predicts something, the mere existence of a prediction activates a mechanism for mobilizing the resources to realize the prediction. This is a characteristic shared among entrepreneurs who launch innovations. For a project to succeed, project management must move forward on the assumption that the predictions will fulfill themselves.

The book makes the following assumptions about its readers.

(1) *To readers with an interest in environmental energy issues*: Anyone with an interest in environmental energy issues may have an interest in the measures implemented by the Japanese government and corporations to counter the deteriorating global environment and the depletion of fossil fuels. The 1970s was a time when pollution, the energy crisis, and other negative aspects of rapid economic growth made themselves known in the shape of air pollution, water contamination, scarce power supply, and the two oil shocks. As a result, many people began to realize that rapid economic growth also had negative effects and that the supply of energy, taken for granted in the past, was actually not inexhaustible.

Under these circumstances, the government attempted to bring these problems under control by introducing a range of measures, such as calling for limits on energy usage, developing new energy and energy conservation technologies, and developing technologies to control pollution. One of these measures was to formulate and implement the Sunshine Project. At first, the aim of the Sunshine Project was to solve the energy problem by researching and developing technologies, but by the early 1990s the emphasis shifted to how the project might contribute to solving environmental problems. The research into and development of new energy technologies attracted attention as a measure to counter the global environmental issues.

Focusing on the Sunshine Project and photovoltaic energy, this book is an attempt to shed light on the outcomes of initiatives introduced in the past by government, prominent manufacturers, and power companies in light of the history of environmental energy policy in Japan and recent efforts to introduce photovoltaic energy. The study is expected to provide effective suggestions when considering measures to counter environmental energy problems.

(2) *To readers with an interest in innovation policy and management of technology (MOT)*: Energy conservation and new energy technology research and development were at the center of technology policy in Japan from the late 1970s to the early 1980s. In a sense, it was also a question of industrial technology policy and innovation policy. During this time, the former Agency of Industrial Science and Technology (AIST) spent at least 70% of its technology policy budget on new energy and energy conservation.

As suggested by the name of the project, solar energy research was the leading program at the Sunshine Project. The research and development of technologies was tirelessly promoted with the focus on solar thermal energy at first, before switching to photovoltaic energy at a later date. The latter, in particular, led to much innovation by national research institutes, universities, and private enterprise including the development of new systems, new manufacturing methods, and improved performance. Concerning project planning, management, and assessment of research and development of photovoltaic energy technologies, this book is an attempt to shed light on the implementation structures at the MITI, the AIST, the NEDO, and the PVTEC, as well as the true nature of the positioning of photovoltaic energy within companywide strategies and technology development at major manufacturers of solar cells. Such knowledge will benefit any consideration of policies and management to promote innovation.

(3) *To readers with an interest in management studies and organizational theory*: The aim of this book is primarily to seek out models for rational planning and implementation methods with a focus on national projects. However, from time to time, projects do not deliver the expected outcomes as a result of circumstances that were not at all foreseen in the planning stages. It may be a matter of something completely unexpected in the early stages, such as fluctuations in the price of oil or other drastic changes in the external environment, or bureaucratic dysfunction where organizations end up going in a different direction than first intended. The former can be countered by raising awareness of the environment, but the latter—where unintended consequences are the result of interaction—is more difficult to deal with because methods alone will not suffice.

National projects generally involve top-down planning, but new technologies and methods may emerge from bottom up during the lifetime of a project. In such cases, what standards does the organization apply to adapt to the new environment and the unexpected outcomes?

Any consideration of these issues requires not only a viewpoint on rational project planning and organizational design, but also a theoretical viewpoint to explain the deeply interesting organizational phenomena that emerge as a project moves forward.

Considering that these problems may also occur in social phenomena, this is not simply a preoccupation with theories of engineering project management. Rather, there is a need to consider the mechanism of unintended consequences during a project while referring to prior research in management studies and organizational theory. We also need to understand the thought world of the protagonists to understand why such mechanisms emerge.

As already mentioned, this book aims to bridge management history (historical research) with organizational theory (theoretical research). In addition to historical descriptions from specific perspectives, any facts that cannot be clarified using such frameworks will be verified by applying different frameworks. If such frameworks still shed no light on the phenomena, the book reorganizes them from yet another perspective. As a result, the book sheds light on diverse aspects of historical phenomena that do not come to light when viewed from a single viewpoint. The primary aim of this book is to describe historical phenomena from various theoretical viewpoints.

Only read by researchers in specialist fields, academic research is thought inaccessible to the general reader. Such high thresholds are often viewed as a matter of course by the academic world. In fact, this difficulty is not limited to students, business practitioners, and other general readers, but researchers with other fields of expertise also find it difficult to understand aims, awareness of the issues, research methods, terminology, and key points when reading authors in other fields. This tendency is particularly strong in fragmented specialist fields where only a small population shares the "general knowledge" of the field. However, it is still possible to make the descriptions accessible to the general reader. It has been my longstanding wish to write a book about research that contributes to broadening outlooks in addition to presenting historic case studies and their fascinating developments. This book is the first step.

Kunitachi, Japan
July 2020

Minoru Shimamoto

Acknowledgements

The Japan Broadcasting Corporation (NHK) reportedly interviewed many individuals concerned with the Sunshine Project, including Horigome, Suzuki Ken, and Kimura, to produce an episode of the *Project X* television documentary series focused on solar power generation in the middle of the 2000s, when Japan's solar power generation industry ruled the world. I assisted with the episode's production by telling an NHK reporter who visited my university laboratory stories that could help him as clues for gathering data. However, the episode broadcast was a tale of Kyocera, with Inamori Kazuo in the leading role. As I detail in this book, Inamori certainly played an important role in the history of solar power generation. That is obvious and unmistakable. However, the individuals concerned who cooperated with NHK's coverage but made no appearance in the episode at all must have felt disappointed when they saw it broadcast. The storyline and the connection of events are oversimplified when clarity for viewers is emphasized within the limited broadcast time for the program. NHK gathered a massive amount of data by relying on the goodwill of the people concerned. However, the greater part of this information never saw the light of day. I heard that interviewees criticized NHK's production policy for the *Project X* program series in many ways. I suspect that similar things have happened in other places. It would be my pleasure if this book conveyed the history of solar power generation from a broader perspective, however small the difference may be.

Numerous people have provided me with their warm support through the course of my activities up to this point. Words cannot describe how much Yonekura Seiichirō helped me as my advisor in my graduate school days. Yonekura was genuinely worried about me when I lost a job because of a course completion error at the time of graduation from university. He advised me to study the history of business administration at a graduate school of commercial science. That advice changed my life completely. Yonekura gave me a push in the back with a smile whenever I fell into a mental slump after entering graduate school two years after my classmates to study abroad and repeat a year. I would like to become a researcher like Yonekura, who goes abroad with great ambition, communicates the importance of entrepreneurship, and makes many people happy.

Numagami Tsuyoshi taught me the right attitude toward my studies and how I should feel about learning. Numagami was extremely strict, both in lectures and elsewhere. His strictness caused me to stay up all night, time after time. At the end of those trials, however, I reached the self-awareness that humans grow through studies. For me, that was an experience that nothing else could replace. If I were to add rigorous footnotes to each of the ideas on theoretical considerations presented in the chapters of this book, they would be filled with important quotes from Numagami's *Kōi no keieigaku* [*Toward an action system theory of management*] (Hakuto-Shobo 2000).

This book owes a lot to Numagami's opinions to that extent. I would like to remain a researcher like Numagami, who has a strong will and carves out a field of original thought based on profound contemplation.

I can still recall vividly the instructions I received during my graduate school days from Nonaka Ikujirō, Itami Hiroyuki, and Suzuki Yoshitaka, including the expressions on their faces on those occasions. In many cases, I realized later, sometimes many years later, what they were trying to teach me. Every time I recall those instructions, a sense of gratitude for the warmth of these mentors who patiently guided this idiot and a feeling of apology for my lack of understanding fill my heart.

Kikkawa Takeo allowed me to attend his postgraduate seminars as an instructor when he took up his new post at Hitotsubashi University some time later. His permission made me extremely happy. Kikkawa shared his extensive knowledge of the history of business administration in his seminars. In addition, he developed appropriate reasoning by freely combining this knowledge and sought to answer problems in society through solutions that resulted from this reasoning. Kikkawa always overwhelmed me with his precise application of the history of business administration. The way he taught and guided undergraduate and graduate students in his seminars impressed me as an ideal image of a university faculty member. I gained an appreciation of how delightful it was to meet a great scholar on those occasions.

I spent the end of the 1990s as a graduate student. It was a fortunate time, with many outstanding seniors and colleagues surrounding me. During this time, I met current Hitotsubashi University faculty members Tanaka Kazuhiro, Katō Toshihiko, Karube Masaru, Fukukawa Hironori, Matsui Takeshi, Kagaya Tetsuyuki, and Shimizu Hiroshi. We underwent hardships together during our studies. We are still able to work together now, which is like a dream. In those days, scholars in the position to shoulder the responsibility for business administration studies in Japan, namely Inayama Kenji, Kawai Kazuhisa, Senō Dai, Terahata Masahide, Nukada Haruka, Taotao Bi-Matsui, Fukushima Eishi, Fukushima Michi, and Yanagida Takuji, gathered in a corner of a building for postgraduate studies at Hitotsubashi University's Graduate School of Commerce and Management. They complained when things were difficult and shared their joy when things were fun.

They were precious people. They left for various parts of Japan after completing their postgraduate programs, just as I went to Aichi to start a new job.

Many people assisted with my research during the period until the completion of this book. This book could not have existed without the support of people involved in the Sunshine Project, business managers and engineers at private companies, and senior researchers at national institutes and universities. I would like to take this opportunity to express my heartfelt appreciation for the cooperation extended by many individuals, including Kuwano, Kurokawa, Tanaka Kazunobu, Suzuki Norio, Murozono Mikio, Tani Tatsuo, Sawada Shinji, Saitō Tadashi, Watanabe Chihiro, and Konagai Makoto, to name just a few.

At Aichi Gakuin University, my first place of employment, Naitō Isao took care of me more than just a little. Naitō hosted extremely lively study meetings by inviting superior graduate students from Nagoya University. Kobashi Tsutomu, Iwata Ittetsu, and Furusawa Kazuyuki, who studied with me at these meetings, all became university faculty members later, which made me extremely happy. After the study meetings were over, I discussed research with Wakuta Yukihiro of Nagoya University over drinks. From time to time, I drank too much and stayed over at Wakuta's place. That is just one of the many good memories from those days. I returned to Hitotsubashi University in 2004. At that point, I regretted the loss of opportunities to attend study meetings with Naitō, Wakuta, and Hayashi Tōru.

Regarding everyday academic affairs at Hitotsubashi University, I would like to thank my assistants Ōwada Keiko, Hasebe Michiko, Takagi Ritsuko, and Arakane Mayumi for their thoughtful and warm attention to detail.

I received assistance from the Takeyama Fund for the publication of this book. Anonymous referees gave me many useful comments when the Fund examined my project. Based on their comments, I touched up the draft for this book as much as time permitted me, but what I did was far from perfect. I would like to keep studying diligently so that I can resolve the many unsatisfactory points in this book by deepening my thoughts.

I made many more unreasonable requests of Tokuchi Michiyo of Yuhikaku Publishing Co., Ltd. compared with the time when her company published my earlier work, *Idemitsu Kōsan no Jiko Kakushin* [*The renewal of Idemitsu Kōsan*] (Yuhikaku 2012). The publication of this book would have been impossible without the assistance that Tokuchi provided to me in many ways. I thank and respect Tokuchi from the bottom of my heart for her dedicated and sincere work as a professional book editor.

Finally, I would like to ask readers to allow me to share my memories of Amano Tomofumi. He entered the graduate school at Hitotsubashi University two years after me. We studied in the same laboratory for graduate students. Amano was a junior with outstanding abilities, who studied under Itami Hiroyuki. Furthermore, Amano was a man of great caliber with intellectual, mental, and physical stamina who was able to undertake steady research tirelessly. Compared with ordinary graduate students who had anxieties about their future, Amano was naturally cool and collected. Thinking about it now, in his youth he already reflected the character of a great figure through his words and manner.

Amano helped me proofread my doctoral dissertation in the laboratory for graduate students during an increasingly cold early winter. He was busy himself, but he provided me with many constructive comments. The memory of this causes tears to well up in my eyes. Amano left this world due to a sudden illness after taking up the post of associate professor at the University of Tokyo's Faculty of Economics. Just before his untimely death, I met him by chance at Narita Airport. We happened to board the same flight to Britain. I cannot help feeling that a strange stroke of fortune enabled me to speak with him on that final occasion.

I still keep the draft of the dissertation as my personal treasure. It contains red handwriting that points out omissions, errors, and logical inconsistencies found in its contents. That handwriting belongs to Amano. This book is the final version of that draft, which I was able to publish after repeated refinements.

Kunitachi, Japan Minoru Shimamoto
July 2020

Contents

About the Author

Minoru Shimamoto is a professor at the Graduate School of Business Administration of Hitotsubashi University, where he received his Ph.D. in commerce and management in 1999. His research focuses on business history, innovation, and sustainability. He has won the Nikkei Prize for Best Books in Economics and the Takamiya Award from the Academic Association for Organizational Science (AAOS).

Chapter 1
Defining the Problem: Solutions Based on Innovative Answers to Social Problems

1.1 Unintended Consequences of National Projects

The shift in Japan's energy policy following the Great East Japan Earthquake has led to a rising interest in renewable energy. As the answer to the sustainable energy problem, much is expected of the introduction and spread of renewable energy. The aim of this book is to identify tactics that are useful for promoting policies for renewable energy development.

There were already reasons to be concerned about the future of energy in Japan before the earthquake. Specifically, the problems were, firstly, the post-Kyoto Protocol issue of global warming and, secondly, the need to deal with the steep increases in the price of crude oil as of the early 2000s. Involving the environment and resources, these issues forced a reconsideration of the nature of energy in Japan, a country conventionally reliant on fossil fuel. The government recognized the difficulties inherent in continuing to maintain an oil-dependent economy in the long term. Seen in that light, Japan was already under pressure to take a more assertive policy approach to the environmental and resource aspects of energy before the great earthquake struck.

However, I would also like to remind the reader of another matter. There were some tacit assumptions made about energy policy in Japan before the Great East Japan Earthquake. Before the disaster, nuclear power was considered the most beneficial formula for countering the energy problem. As a contributing factor to greenhouse gas reduction, nuclear power was considered the trump card as far as global environmental problems were concerned. Nuclear power generation was even the premise for calculating emissions targets under the Kyoto Protocol. Thinking back on it now, the Democratic Party of Japan, the ruling administration at the time, made very confident declarations to the international community about achieving emissions targets. One might even say that they brought on a situation where there was no choice but to rely on nuclear power to meet the targets. At that time, the vulnerability of nuclear power to natural disaster, spent nuclear fuel

M. Shimamoto, *National Project Management*,
Advances in Japanese Business and Economics 25,
https://doi.org/10.1007/978-981-15-3180-4_1

disposal, and decommissioning issues were still relatively neglected. In addition to the environmental and resource perspectives, nuclear power also held out bright prospects for industrial development in Japan through the export of plants to other countries.

However, the accident at the Fukushima Daiichi Nuclear Power Plant in the wake of the Great East Japan Earthquake was a critical moment for reexamining such optimistic plans to rely on nuclear power. In addition to the unresolved matter of radioactive waste treatment, the high overcall costs associated with nuclear power if decommissioning was included were once again recognized as the enormous scale of the damage caused by the disaster began to sink in. This encouraged reconsideration of the conventional energy policy premised on the use of nuclear power. However, reverting to oil is also associated with major problems. Conventional fossil fuel is a cause of global warming and comes with a risk of resource depletion. Excessive imports are a drain on the national wealth of Japan and cause the trade balance to deteriorate, factors that are directly linked to a crisis of energy security for Japan. Faced with this situation, the spotlight fell on one particular option that invited more expectation than ever before. That option was renewable energy.

As a parallel measure alongside nuclear power for resolving the environmental and energy issues, renewable energy already attracted attention before the earthquake disaster. Even when setting targets for greenhouse gas emissions, it was taken for granted that renewable energy products would become widespread in the near future when technical innovation was achieved. Policy support for developing, introducing, and installing the technology had been launched before the earthquake and, according to preliminary calculations, renewable energy would cover a substantial amount of energy need by 2020 or 2030. Then, when the approach to nuclear power was revised following the earthquake, renewable energy once again invited great expectations.

Looking back at history, this is not the first time that renewable energy has all of a sudden attracted similar levels of attention. This is one point I want to raise in this book. We should remember that something very similar happened already, at the time of the oil crises. By studying the examples of the past, we can avoid repeating failures caused by ignorance. Reflecting on the past sequence of events sets the course toward future success. This is the essence of studying history.

The development of renewable energy has a long history spanning several decades. It began as early as the first oil crisis in 1973. At that time, "renewable energy" was not as common a term as it is today. Energy derived from natural resources to replace oil was referred to as "new energy"—because it was something new. Below, this book uses the term "new energy" when discussing historical events and the term "renewable energy" as a more general designation.

When the oil crises jeopardized the stable supply of energy to Japan, the government announced a scheme to develop new energy technologies, where the public and private sectors would join forces to develop technologies that would find solutions to the energy problem in Japan. This became the starting point for the Sunshine Project, which is the topic of this book. Focusing on the Sunshine Project, this book looks back on the history of the government-led national project for new energy, which has been in place since the time of the oil crises. This exercise should provide thought-provoking suggestions for the future of energy policy in Japan.

With the exception of the outbreak of the Gulf War, oil prices have fortunately been stable since the two oil crises in the 1970s. Therefore, the development of new energy has lost some of its initial sense of urgency and the name of the Sunshine Project has been forgotten.

However, in recent years the low-price stability has once again crumbled. Specifically, this change is due to unforeseen fluctuations in the price of crude, natural disasters such as the Great East Japan Earthquake, increased awareness of global environmental problems, and tension in the international situation surrounding Japan following the rise of China. For Japan, energy policy is the bedrock of national security and stable economic growth. A stable energy supply based on international cooperation and self-sufficiency to the maximum possible extent laid the foundation for the sound development of politics and economy in Japan, while failure would lead to a life or death situation for the nation. This is no different now than it was at the time of the oil crises.

In the meantime, the development and spread of renewable energy has surfaced as the trump card in terms of measures to counter the energy problem. For the future of Japan, it is crucial that the stance in favor of developing the technology, promoting installation, and supporting growth for the industry remain the same, and that the government leads by setting effective policy for industry and technology.

However, it is not as straightforward as expecting development and installation to progress more or less automatically as long there is policy support. Now is the time for learning from history. A careful study of examples from the past will show numerous unintended consequences of national projects that are based on collaboration between university, industry, and government. In this regard, the central issue of this book is historical analysis from multiple perspectives, which aims to clarify the mechanisms concerned and to provide a formula for reflection and improvement.

1.2 A Peek Inside Projects Based on Industry–Government–Academia Collaboration

Resolving social problems is not the task of government alone. Rather, it is essential to have the cooperation of corporations and research institutes. Society today is beset with a range of problems including global warming, acid rain, and other environmental problems, as well as issues with the energy and food supply, poverty, and social welfare. With respect to these issues, the government employs policy measures such as financial assistance or tax breaks, denoting the preferred policy based on public perspectives. Seeking new business opportunities, the private sector will without doubt be motivated to participate as long as the government makes a serious effort. This provides corporations with the opportunity to exercise their social responsibility as expected by the nation and to commercialize technologies that serve the public interest while at the same time safeguarding the profit-seeking principle. In their role as educational institutions, universities and other research institutes can cultivate human resources with knowledge and skills while also supporting corporate research and development by making research outcomes publicly available, in particular where scientific and technological aspects are concerned. Aiming for successful technical research and development, industry, government, and academia each bring their own perspectives to the table. Finding solutions to social problems is not only a matter of government policy. Rather, research institutions at corporations, universities, and national research laboratories should also participate and team up with the government to tackle the issues. This is where the issue of inter-organizational management emerges.

The efforts of diverse organizations are required to deal with the problems of environmental energy, the topic of this book. To briefly summarize, it is vital (1) that the government promotes technology development and supports market development with the nation's understanding; (2) that corporations develop and commercialize new products, and that conscious consumers use environmental energy products; (3) that universities and other research institutes advance science and technology research while educating the next generation of specialist human resources. The key to resolving social issues is in harmonious collaboration between industry, government, and academia.

However, it is not necessarily self-evident that the policies considered effective from the individual viewpoints of organizations in industry, government, or academia are actually beneficial for the whole. We need to pay careful attention to this point because we must avoid situations where localized gain brings overall disadvantages. When considering this matter, we need to pay attention to the interaction between each organization. To put it simply, we need to undertake a detailed examination of organizational phenomena and what effect interaction between policy-making, business strategy, and research policies has on reality.

For example, the government promotes research into and development of renewable energy by providing subsidies or trust money to corporations or universities. Corporations receive financial assistance from the government and use the outcome of university research to develop, manufacture, and sell products. Government-supported universities cooperate with corporations on technology, supporting development and evaluating technologies. Universities also equip the next generation of human resources with vital knowledge and skills. Each organization considers measures for dealing with the problem of environmental energy from their own viewpoints, contributing by the means available to them. However, organizations where industry, government, and academia collaborate are not single-mindedly committed to such work. The government deals with other important policy issues than renewable energy development. Corporations cannot continue to develop, manufacture, and sell unprofitable products in the long term. Universities are under increasingly strong pressure from a society committed to successful research. Government, corporations, and universities guarantee the survival of their own organizations by playing their expected roles from their own perspectives. However, where overall social challenges are concerned, undesirable and inconvenient phenomena may arise when these organizational tasks work against each other. Each organization endeavors to fulfill their own mission based on their own norms, which may cause a mechanism that brings overall disadvantage from the viewpoint of meeting policy challenges.

In order to prevent such situations, there should be some kind of overall coordination of the individual endeavors of each organization. If this works well, there is no doubt that synergy between the activities at each organization will improve the results. Therefore, we could say that the topic of this book is how to facilitate coordination between organizations where industry, government, and academia collaborate. The motives of industry, government, and academia are not necessarily always in agreement. In such cases, what are the processes that promote collaboration between multiple organizations? Specifically, we will look at actual case studies to see who takes the initiative and under what motivation. Clarifying these points is one of the necessary conditions for the successful development of renewable energy.

As seen from the above, research involving technology innovation is both an engineering topic geared toward technology and a management topic aimed at organizations. The topics concern the natural sciences as well as the social sciences. Naturally, technology development will not succeed if it bypasses the logic of the natural sciences. However, technology development is a joint task between organizations where many people participate and, wherever large numbers of people are required to make decisions in order to accomplish a goal, there is also a task for the social sciences. For this reason, agreement on the selection of correct natural scientific methods also falls with the domain of social science research topics.

1.3 Solutions Based on Innovative Answers to Social Problems

It is good news when systematic technology development leads to successful new technologies that bring solutions to a range of social issues. For example, accomplishing new energy technologies to replace fossil fuels would relieve the profound crisis of energy depletion facing humankind. Successful technology development would also have an impact on a wide range of associated industries, leading to an array of new products and indirectly promoting economic and industry growth. Therefore, this book highlights the importance of innovation through collaboration between multiple organizations as a strategy for finding solutions to energy and environmental issues. To extract the conditions for successful technology development through collaboration between industry, government, and academia, it is necessary to identify the management approaches that bring results and to clarify the mechanism of interaction among the various entities involved.

There is, of course, no reason why a company in the private sector could not systematically develop technologies on its own. Even without the support of the government or universities, corporations can take on the challenge of finding solutions to issues of public interest. At present, it is also possible to swiftly raise large-scale funding for venture businesses thanks to the development of the stock market. In reality, aside from planned national projects led by the government, recent theories on industrial policy and technology policy have ignited the debate in light of corporate entrepreneurial initiatives to create new business by leveraging venture support and industry clusters. In particular, from the mid-1980s to the 1990s, it was not necessarily clear even to policymakers which new technologies and new industries to encourage. As a result, interest turned to flexible networks that could respond to technology changes instead of big projects with pre-specified grand objectives. In fact, depending on the industry, it is often small-scale, emerging venture companies that achieve important innovation. If this is the premise, the role of government, in the views of some, should be to create the structures that promote new business development or to develop the infrastructure that supports business. After all, it is a principle of capitalist society that corporate entrepreneurs mobilize funds, human resources, equipment, and other management resources based on their own business judgment, developing new technologies and new business through capital investment or investment on research and development.

However, it is also a fact that this is not as easy as it sounds. Even if a particular technology is useful from a public interest perspective, it is difficult for private sector corporations to develop such technologies on their own where the potential for successful development is either remarkably low, or the required management resources are exceedingly expensive. Recent pioneering technologies require funds far in excess of what a single company can spend on development, or extremely long time scales. Therefore, it is difficult to successfully develop advanced technologies where the degree of difficulty is high by simply leaving research and development or capital investment to corporate volunteers.

In such cases, support from the government, universities, or other public organizations in the form of funds, intellectual resources, or technology constitutes a significant measure of support until new ventures based on the technology become financially viable as economic activities in the private sector in the future. Even when the aim for the future is profitable economic activity by corporations, the examination of effective approaches has practical significance since support from the government and academia is effective in the run-up to getting projects on track.

There are all kinds of policy objectives where the level of public interest is high. For example, some typical domains are national defense, medical care, and environmental energy where some aspects of public interest would not necessarily be implemented if we simply rely on market mechanisms for research and development. Therefore, the government has tried to achieve these policy goals by legal and economic means.

To take a broader approach, development takes on public interest to the extent that advanced technologies become the underlying technologies for a broad range of industries. This is because broad areas of industry benefit from technology spillovers as a result of the development of the underlying technologies. Consequently, where such advanced technologies are concerned, the government organizes projects to commission research and development, or to provide subsidies to multiple corporations, justifying its participation from a public interest perspective.

This book attempts to clarify the effects of government schemes to devise solutions to energy and environmental problems by promoting innovation in renewable energy. If the conditions for such policy successes are identified, the tactics for achieving objectives that are in the public interest will also become clear, making it possible to reproduce policy successes by artificially complying with the conditions.

However, caution must be exercised in these cases. Policies made by the government or other planning authorities are not always carried out as originally intended. In particular, in case of collaborative projects between industry, government, and academia, the projects are executed in participation with corporations, universities, and other organizations, so it is only natural that government directives do not necessarily determine the behavior of other organizations. As a result, there emerges an inter-organizational dynamic created by the entity where the intention resides. How do we go about controlling such interactive structures between several entities? Is it even possible to control them in the first place?

Some would argue that it is possible if the instructions and directives from the top are followed to a tee. Others believe that even when there is no control from the top, organizational capability will be demonstrated if good routines for member conduct have been developed in keeping with the norms of individual organizations. Naturally, such ideas are extremely important to the practical application of policy and management, and this book will be especially careful when considering these measures.

However, to anticipate the final conclusion of this book, there are dynamic constructions where individuals gamble their own existence to generate meaning and to create history together and it is, in fact, not possible to make rigorous distinctions between those who control and those who are controlled.

This book will employ multiple case studies to convey this message to its readership.

1.4 Twists and Turns in the Development of Renewable Energy

Below, I will look into the history of the development of renewable energy in Japan. Cumulative installed capacity and technical capability indicate that Japan has been successful with solar cells and photovoltaic energy systems. Even today, installations of photovoltaic energy systems in residential homes and public facilities in Japan are rising steadily from year to year. As for the prospects for the future, photovoltaic energy is expected to expand as an industry with a vast international market due to rapid market expansion in Europe and other regions. The technology contributes to energy self-sufficiency all over the world by taking advantage of local climate characteristics. It is a particularly effective method of generating power in regions where the development of the power infrastructure is lagging behind. As mentioned above, photovoltaic energy is a textbook example showing how to use innovation to overcome environmental energy problems.

This is not something that Japanese corporations accomplished in a day. There is a history of tenacity and persistent effort on the part of industry, government, and academia to develop renewable energy before getting to the point where photovoltaic energy became reality.

In retrospect, moves to introduce renewable energy have always been at the mercy of the ups and downs of oil prices. The two oil shocks in the 1970s exposed the fragility of economies that were dependent on oil. Japan and other advanced countries were forced to increase the efficiency of the existing sources of energy and to develop new energies to replace oil as the basis for sustainable economic development. At the time of the first oil crisis when the price of oil resources rose steeply, corporations were researching and developing energy-saving technologies to reduce the amount of oil used. A range of technical innovations emerged out of these efforts and Japan made good progress with energy-saving technologies. As a result, the second oil shock had a relatively slight impact on Japan compared to many other countries. In the 1980s, the Japanese electrical appliance and automobile industries were then able to make a strong showing on the international market.

It was around this time that corporations began full-scale development and commercialization of the technologies for photovoltaic energy and other renewable energy. This was a useful tactic with regard to the environmental and energy problems. But getting results required more time than anticipated because the price of crude dropped in the late 1980s resulting in less incentive to develop new energy. The drop in the price of crude was good news for the country as a whole, but it also worked strongly against the research into and development of renewable energy.

By the early 1990s the situation changed once again. At this time, measures to counter environmental issues were tacked on to the development of renewable energy, which, once again, gained prominence as an eco-friendly energy source. By the late 1990s, the Japanese economy had entered a prolonged slump after the collapse of the bubble economy. However, by this time, little by little, development had started to show definite results and capital investment by corporations coupled with efforts by the government to develop markets and systems led to an increase in production and installations in some photovoltaic energy fields. By the 2000s, the interest in renewable energy was revived due to resource issues, including another rise in the price of crude, and the worsening environmental crisis in the wake of global economic development. In this way, the development of renewable energy has been greatly influenced by resource energy issues (steep increases in oil prices) and environmental issues (controls on carbon dioxide emissions). Fortunately,

Japan has been able to continue the long-term development of energy technologies at the level of government, corporations, universities, and national research institutions. The setting has been industry, government, and university projects for researching and developing renewable energy technologies. Collaborative efforts by industry, government, and academia to develop technologies, markets, and other forms of industrialization have had a major impact on the rapid breakthroughs in the photovoltaic energy industry in Japan today.

Photovoltaic energy is one example of how remarkably successful the new energy business has been for Japan. With Sharp, Kyocera, Sanyo Electric, and other corporations at the top of the world rankings, Japan has so far been the world's largest producer of solar cells. However, in recent years, Japanese companies have dropped down the rankings, under extreme pressure from corporations in the emerging nations. The decline in the share of the international market indicates the need for further strategic measures to maintain Japan's competitive strength in this remarkable growth industry.

In Japan, the history of researching and developing technologies for photovoltaic energy through collaboration between industry, government, and academia already spans several decades. The national project format for researching and developing technology based on government-led technology policy has had a great impact on technology research and development as well as the development of practical applications. Taking photovoltaic energy as the topic, this book aims to use case studies to clarify the conditions for successful innovation by means of projects run by industry, government, and academia.

1.5 Research Angles on National Projects

In general, a national project refers to a large-scale plan for industrial development, social infrastructure, or research and development initiated by the government and carried out by the private sector.[1] In each case, the objectives are strongly influenced by public interest and state intervention is generally considered to be significant for the discipline in question. Strictly speaking, these three configurations are not completely separate. For example, there are interconnected cases where research and development establishes social infrastructure, which then leads to industrial development. This book specifically focuses on large-scale research and development that uses the facilities and human resources at companies in the private sector and the research facilities at universities funded with trust money or subsidies for research from the public purse in order to deliver specific policy objectives set by the government.

It is not in Japan, but in the United States where we find numerous textbook examples envisioned as national projects. In the United States, large-scale research

[1]Kanamori et al. (1990, p. 552).

projects were implemented approximately every ten years in the twentieth century. Projects such as the Tennessee Valley Authority in the 1930s, the Manhattan project in the 1940s, missile development in the 1950s, the Apollo project in the 1960s, energy independence in the 1970s, and the Strategic Defense Initiative in the 1980s were initiated by the Department of Defense, NASA, or the Department of Energy, mainly targeting military technology or space development. In addition to projects with such direct national objectives, research has also been commissioned on a wide variety of topics. In particular the Defense Advanced Research Projects Agency (DARPA) at the Department of Defense is planning industrialization projects that are focused on the field of electronics. In a broad sense, such endeavors can also be understood as national projects for technology policy as defined in this book.

Unlike the United States, the development of military technology is a subject with strong constitutional restrictions in Japan. Since space development is not significant enough to add to national prestige, the implication of national projects as a means of achieving national objectives is weak. However, launched in the mid-1970s, the Sunshine Project met the national objective of guaranteeing energy security. In the view of the Ministry of International Trade and Industry (MITI), which had already participated in relatively diverse small-scale technology development, it was the perfect theme for developing policy for a major project since it was also linked to the promotion of a new industry.

According to Yoshikai Masanori, a former official at MITI who later took up a post as professor at Saitama University, technology policy (= industrial technology policy) at MITI formed a part of industrial policy together with industrial structural policy and industrial organizational policy. Technology policy was defined as something that "optimizes resource allocation, selects the format for policy participation, extracts and evaluates the technologies required to implement policy with a connection to industry".[2] Aspects of this definition highlight the identity of MITI as it tries to strike a balance with the ministries and agencies in charge of technology policy. In Japan, science and technology policy in its broader application was the responsibility of the Science and Technology Agency while the Ministry of Education was responsible for science education and research policy at universities. In this context, MITI felt the need to assert the reasons for participating in technology policy by labeling it "industrial" technology policy.[3] The Sunshine Project was a matter of both industrial policy and technology policy. Since the project was also committed to delivering a national objective, technology policy at MITI placed the most emphasis on this policy program as of the 1980s.

National projects are implemented to deliver industrial development, environmental protection, stable energy supplies, or other policy objectives. However, if a technology policy is premised on developing a self-reliant business in the future,

---

[2]Yoshikai (1985, p. 8).

[3]For the proportion of each ministry and agency in the budget for science and technology, see Imai (1984, p. 187).

another approach is to produce results by finding a match with corporate management strategy or university research strategy. When considering technology policy based on such premises, it is necessary to study the relationship between policy and management by getting to the core of the corporation.

National projects are not only determined top down by the planners. This fact makes it difficult to understand the true situation. Some projects have been created by engineers, researchers, or managers at the frontlines of research and development who found something feasible. In other words, some projects exist because of the bottom-up efforts of individuals. Consequently, whether proposed by an individual or a small group of people, the initial draft plan is not realized as intended when formulating and executing projects. The shape of the project undergoes dynamic changes depending on a series of actions by the participating entities. One could say that a project consists of the trajectory of such a sequence of actions. Seen in this light, it is not the prior draft plan and its direct path to success that we should consider, but the management that guided the project through a process of learning and improvement. When considering effective projects, we must develop an understanding of the conflict between the autonomous nature of a project that causes a sequence of actions by multiple entities, and the intent of the entities trying to exert some control in line with their own principles of behavior.

Since there is currently an insufficient store of prior research concerning what kinds of policies are linked to success, it should be possible to reflect knowledge obtained by analyzing the mechanisms that link policy and success in future policy. In conventional theories on industrial policy, policy methods have been directly linked to research and development success or the outcome of industrial development.[4] Researchers in various countries have overestimated the authority of MITI based on such perspectives. However, private sector corporations and universities do not simply obey the directives of the government for technology policy programs. Assuming that corporations and universities behave like independent social entities, we need to consider collaboration between industry, government, and academia as a form of resonance between multiple systems.

Seen in this light, it is necessary to plan a structure that facilitates organic collaboration yet maintains the unique character of each system in industry, government, and academia in order for collaboration between the entities to have an effect. The reason being that there is a management logic to corporations and an academic logic to universities in the same way that there is an administrative logic to government. The three entities define themselves and fulfill their social functions in line with their own unique logic. For collaboration between industry, government, and academia to succeed, we have to consider how these three systems, with their different functions, will collaborate and sympathize with each other in line with their own individual and specific truths. Assuming that there is some public policy significance in nominally retaining overall profits and the shortsighted economic rationality of individual entities, to curb initial profits for the sake of

[4]See Anchordoguy (1989), Callon (1995), and Fransman (1999).

overall profits at a later stage is a collective action based on agreement. This is the point where it is necessary to temporarily shelve the economic rationality of the individual entities. In other words, one might say that the issue is to establish agreement in relation to public interest. Such angles are indispensable when discussing national projects.

Regarding national projects, the government commissions research and development for specific policy objectives from the private sector, from universities, or from other research institutes. If successful, delivering results of public interest is linked to national well-being. This also motivates universities and corporations to cooperate with the government because by participating in a state project universities can produce scientific results and corporations can produce economic results. Work that aims to deliver policy objectives based on cooperation between several organizations including industry, government agencies, and universities is at the heart of collaboration between industry, government, and academia.

Of course, whether universities and corporations will always cooperate in ways expected by the government is an unknown. Various factors add to the uncertainty. For example, university researchers will hesitate if research and development is not linked to academic performance. Undoubtedly, corporations will also think twice in cases where it is difficult to commercialize the outcomes of research and development. To avoid such situations, national projects need effective management in order to fulfill the government's policy objectives with the help of industry, government, and academia. This book aims to throw some light on these matters.

1.6 The History of National Projects

The history of large-scale projects is a long one. If we look back in time, the construction of the pyramids, sphinxes, and other huge monuments in ancient Egypt are also large-scale projects. We can even say that the history of flood control, city-building, and other projects is as old as human civilization. In modern times, urban planning and other public works, even the pursuit of war on a national scale, are also a kind of public or national project.

The existing research on conventional large-scale projects has mainly focused on internal management mechanisms. In Miller and Lessard (2000), project management research is mainly focused on engineering assessments, technology choices, outcomes, and the management of independent contractors.[5]

In *The Anatomy of Major Projects*, Morris and Hough analyze eight case studies: the Channel Tunnel, Concorde, the Advanced Passenger Train, the Thames Barrier, the Heysham 2 nuclear power station, the Fulmar North Sea oil field, the computerization of the PAYE system, and Project Giotto. According to Morris and Hough (1987), a project is an undertaking to achieve a specified objective, defined

[5]Miller et al. (2000).

in terms of technical performance, budget, and schedule.[6] Project management originated in the chemical industry just before the Second World War and developed in the defense and petrochemical industries in the 1950s. Since then, research has developed into an important management discipline. There are basically two kinds of projects: the ones that are complete in themselves, such as an oil platform or a tunnel, and those that represent a series of products or projects, such as an aircraft or an aid program. In either case, projects are accomplished according to a common lifecycle. Every project, regardless of its kind or duration, essentially follows the activity sequence of prefeasibility and feasibility, design and contract negotiation, implementation, hand-over, and in-service support. In many fundamental aspects, the skill of the project manager depends on an innate appreciation of the requirements for advancing the project through its lifecycle. In the process, many issues arise that are common to all kinds of projects, for example leadership and organization, financing, planning and control, and contracting of third parties. A project has an end objective. It follows a lifecycle, it requires financial analysis and funding, it needs planning during which trade-offs are made between objectives. Project management originated out of these needs and has developed into the study of management techniques.[7] As the starting point for the discussion, the present book adheres to such perspectives on projects in the prior research.

In Japan, national projects have mainly been implemented as industrial policy. New industry and the new technologies that serve as its foundation were extremely important policy issues in Japan as the country modernized its industry following the precedents set by Europe and North America. This situation remains unchanged today. In modern society, where the international competitiveness of corporations is largely dictated by the implementation and commercialization of sophisticated technologies, it is increasingly important to clarify the effectiveness of policy. There are generally two reasons for the government to intervene in industry by means of policy—firstly, to regulate a declining industry and, secondly, to protect a fledgling industry. The government may try to avoid a situation where an industry goes into decline shortly after it has been established if the potential for growth is nipped in the bud through exposure to harsh competition, or there is a situation where corporate bankruptcy may create mass unemployment or have a negative effect on related industries.

In recent years, the government has directed its efforts at support for technology development in addition to opening up markets or expanding public-sector demand and introducing preferential taxation when protecting fledgling industries. In the past, the Ministry of Economy, Trade and Industry (METI) referred to this as industrial technology policy, but in recent years this approach is often referred to as innovation policy since it also covers the commercial development of new technologies. With the sophistication of industrial technology and industrial structures becoming more and more hi-tech, innovation policy is increasingly important as a

[6]Morris and Hough (1987).

[7]Morris and Hough (1987).

policy package of direct and indirect government support ranging from technology development to commercialization. Developing new technology is the foundation for new industry and, for a resource-poor country like Japan, the role of research into and development of scientific techniques and industrial technology is vital for the nation.

1.7 The History of Industrial Policy in Japan

I would like to start with an overview of the history of industrial policy in Japan based on Nishida (2000) before discussing specific national projects below.[8] From the 1950s to the 1970s, Japan adopted policy development measures for a range of industries under the guise of industrial policy. Specifically, by introducing sophisticated technologies and preferential allocation of the foreign currency required to import equipment for the steel industry, the shipbuilding industry, the machine industry, the synthetic fiber industry, the oil industry, and the petrochemical industry, the aim was rapid growth across these industries on a national scale.

The purpose of industrial policy is to protect and promote new industries that aim to develop the national economy. Tsuruta (1982) claims that the model for postwar industrial policy in Japan was created in the 1950s and systematized in the 1960s.[9] Once a national economy has achieved a certain level of development, such policies gradually lose their relative importance as the emphasis tends to shift toward promoting science and technology research, developing intellectual property rights, establishing professional education and vocational training systems, and preparing the competitive environment for small and medium-sized enterprises rather than industrial protection.[10] In fact, the terms “industrial technology policy” or “innovation policy” have often been used to designate industrial policy at MITI since the 1980s.

The history of industrial policy in Japan dates back to the Meiji policy of promoting new industry. The Japanese government adopted the advanced technologies of the major Western powers to promote the modernization of industry in Japan and to assert its independence as a nation. The systematic introduction of technology and industrial stimulation measures by a centralized and authoritarian bureaucratic nation is a pattern of behavior characteristic of latecomers to capitalism. By the late Meiji and early Taisho period, the achievements of the Industrial Revolution had also been introduced to Japan, where capitalism took root with a focus on the cotton spinning industry. Later, in step with the growing economic and military power of Japan, the international conflict over economic interests on the

---

[8]Nishida (2000, p. 159).

[9]Tsuruta (1982).

[10]Nishida (2000, p. 157).

Chinese mainland intensified. Japan established a wartime economic structure to prepare for the expected all-out war and corporations came under the control of the military authorities and the civil service. The government implemented strong controls of the entire economy to get ready to conduct a war.

When the Second World War ended with the defeat of Japan, a materials mobilization plan, referred to as the priority production system, was launched in 1947 to start the reconstruction of Japan. Focused on productive resources such as coal, power, and steel, the system attempted to revive the rotation of investment and output between multiple industrial departments. Financing for the key industries came from the Reconstruction Finance Bank, which was established to provide financial support. The policy during the postwar reconstruction period inherited the methods of the wartime-controlled economy.

However, enforcing investment in the key industries during a time of extreme shortages of goods led to steep inflation. When Joseph Dodge, President of the Detroit Bank, was posted to Japan as economic advisor in 1949, he adopted the so-called Dodge Line, a policy of inflation controls based on austere fiscal measures and industry rationalization to stabilize the Japanese economy. The aim of the deflationary policy was to put an end to the controlled economy and to restore the functionality of the free market economy.

When the Korean War broke out in 1950, the American policy toward occupied Japan changed from the demilitarization and permanent weakening of Japan to supporting the economic growth of Japan as a member of the free world. To do so, the United States and other Allied countries supported the recovery and development of the Japanese economy by providing the essentials for economic development: export markets, technology, and industrial financing.

Such is the context for the strong industrial policies implemented in Japan from the 1950s to the 1970s. The steel, shipbuilding, machine, synthetic fiber, oil, petrochemical, and other industries were targeted for strategic growth and the foreign currency needed to introduce advanced technologies and import production facilities from the West was preferentially allocated to these industries. The measures included financing for industry through the Japan Development Bank to stimulate investment, industrial location development and other industrial infrastructure programs, as well as an accelerated depreciation system and other tax breaks for capital investment.[11]

The following were the basic guidelines for industrial policy in the 1950s and 1960s. Firstly, the government applied restrictions on imports and foreign capital to protect domestic industry. Secondly, the government formulated industrial rationalization plans, aiming to rationalize and develop specific industries. Thirdly, in addition to special taxation measures and subsidies, the government launched the Fiscal Investment and Loan Program in 1953 as a way of systematically allocating industrial financing. Under this program, funds deposited by households into the Postal Savings System and postal life insurance trust funds were allocated to

[11]Nishida (2000, p. 160).

industry and public enterprise. Fourthly, the government recognized depressed industry cartels and implemented management stabilization. The government also regulated capital investment competition for large-scale development by means of administrative guidance.

The basic principle of the postwar economic system was free trade supported by a framework that included the General Agreement on Tariffs and Trade (GATT), the International Monetary Fund (IMF), the World Bank, and the Organisation for Economic Co-operation and Development (OECD). After the defeat in the war, Japan also participated in these frameworks, gaining entry to the IMF in 1952 and the GATT in 1955. However, Japan was given a grace period before fully accepting free trade since developing domestic industry was considered difficult if free trade was immediately implemented at a time when the economic potential of Japan was weak.

In the 1960s, industrial reorganization moved forward in policy terms to expand corporate scale with the aim of strengthening the competitive clout of domestic industry. In 1964, the Industrial Rationalization Council was reorganized as the Industrial Structure Council, which aimed to use its findings to pursue a more sophisticated industrial structure. The specific measures employed were a system of permits and licenses, special taxation measures, subsidies, policy lending by monetary institutions, the Fiscal Investment and Loan Program, technology promotion, upgrading industrial infrastructure, restructuring industrial organizations, and stabilization by means of cartels. The Industrial Policy Bureau was established at MITI to implement policy across industry.[12]

Industrial policy in the 1970s was a transition period when technology policy was preeminent and policy objectives were implemented through large-scale national projects. The Manhattan Project, the Apollo Project, and other large-scale projects served as models. Typically, the image of a national project was of a government-centered initiative to achieve some grand policy objective. However, by the early 1980s Japan had become an international leading player in the technology field and policies were formulated with the aim of promoting and upgrading industrial science and technology in Japan instead of protecting and developing specific fledgling industries. Parallel with the growth of the Japanese economy, the significance of industrial policy was questioned anew. Already in the 1970s, there were arguments in favor of a retreat from aspects of managing and regulating industrial policy and a shift toward policies that would guarantee more complete functionality than the market mechanism.[13] In addition to MITI, the Ministry of Education (the Ministry of Education and the Science and Technology Agency) formulated its own technology policy in the name of science and technology at the time, as did the Ministry of Posts and Telecommunications, the Ministry of Transport, the Ministry of Agriculture and Forestry, and the Ministry of Health and Welfare in their respective policy fields.

---

[12]Ono (1999).

[13]Sangyō kōzō shingikai (1971).

In addition to a licensing system for industrial policy, MITI directly promoted technology research at research institutes associated with the Agency of Industrial Science and Technology (Kōgyō Gijutsuin: hereafter AIST) while simultaneously providing indirect support for technology research and development at private sector corporations under the pretext of industrial technology policy. In fact, by the late 1970s and early 1980s, technology promotion and research and development were cited as unique policy goals. For example, the VLSI Technology Research Association was set up in 1976 based on the Act on Research and Development Partnership concerning Mining and Manufacturing Technology enacted in 1961. The Technopolis policy was developed based on the 1983 Law for Accelerating Regional Development based on High-Technology Industrial Complexes.[14] These policies aimed to promote competition rather than to protect industry and this period was a turning point for industrial policy at MITI. In the 1980s, fine ceramics and other new materials, micro electronics at the leading edge of computer and semiconductor technologies, and biotechnologies that made use of genetic engineering raised expectations. But, in fact, one large-scale technology idea after another appeared without anyone knowing when they would be commercialized and implemented. Since the scope of application was unclear, research and development strongly tended toward basic research, promoting technology development within a long range.

As described above, there was a strong shift from emulating other countries to developing own technologies as time passed. Industrial policy had started with the protection of fledgling industry, but the focus had shifted to innovation policy for advanced technologies.

1.8 Technology Policy Tools at the MITI AIST

The following is a look at the particulars of industrial policy, in particular technology policy. Immediately after the war, MITI established a structure for technology policy. The Industrial Technology Agency, the predecessor to the AIST, was established as early as 1948.

Amid the devastation immediately after the war, the Industrial Technology Agency was set up within MITI as the agency for promoting the development of technologies that would become fundamental to the industrial recovery of Japan. The experimental research organizations at the Ministry of Commerce and Industry (the predecessor of MITI) were consolidated as the Industrial Technology Agency. The Ministry of Commerce and Industry had been vertically organized by field of industry, which is why all experimental research organizations were affiliated with the general affairs sections at each bureau. Therefore, when the administrative mechanism was restructured after the war, the Industrial Technology Agency was

[14] Itō (1998).

established as an independent agency with jurisdiction over industrial technology—the idea being to adjust for the structural deficiencies and consolidate the experimental research institutes.[15] In the 1952 reforms, which aimed to reduce the administrative bodies, the Industrial Technology Agency became the AIST when it was reorganized as a body affiliated with MITI rather than an external agency. Still today, the AIST has jurisdiction over many extremely important matters relating to industrial recovery including standards, weights and measures, support for technology introduction, and subsidies for technology development.

The initial task of the research institutes affiliated with the AIST was to undertake experimental research that would aid the understanding and absorption of sophisticated technologies introduced from abroad. The Agency supervised the national research institutes and was responsible for commissions and subsidies to private corporations. The Sunshine Project was one of the policy programs at the Agency.

In broad terms, the technology policy tools at the AIST can be divided into two categories. One category refers to collaboration between the government and the private sector to promote research and development. Specifically, the projects are implemented based on themes selected by the government from the perspectives of public interest and policy goals. The second one aims to promote private sector technology—specifically, to assist corporations taking part in the development of specified technologies by providing subsidies for corporate research and development, special taxation measures, or low-interest financing. To take an example from later years, the Large-Scale Project System and the Sunshine Project are representative of the former, while the latter includes the subsidized Mining and Industry Technology Research Association and the Japan Key Technology Center.

Introduced immediately after the AIST was set up in 1952, the system of subsidies for the Mining and Industry Technology Research Association became a pioneering technology policy at the Agency. The system subsidized the development of important technologies necessary for postwar recovery and retained its importance to corporate research and development until the mid-1960s.[16] These subsidies were invested in the transistor and the prototype for the YS-11 passenger aircraft, with subsidies to research and development projects accounting for 15–40% of total research funding in Japan.[17] By the early 1960s, the need for large-scale technology development increased again and, based on the findings of the Industrial Structure Research Committee set up in 1961, a system for funding commissioned experimental research in industrial technology was established in fiscal year 1964. This was expanded into the Large-Scale Industrial Technology Research and Development System (the Large-Scale Project System) in 1966.

---

[15]MITI Kōgyō gijutsuin Kenkyū kaihatsukan shitsu [MITI AIST, Office for Research and Development] (1987, p. 18).

[16]Gotō and Wakasugi (1984, p. 171).

[17]Gotō and Wakasugi (1984, p. 172). Also, MITI Kōgyō gijutsuin, ed., *Wagakuni no sangyō gijutsu seisaku no enkaku* [*History of the Industrial Technology Policy of Japan*] (self-pub., n.d.), p. 2.

The Large-Scale Project System was significant for upgrading the industrial structure, strengthening international competitiveness, developing natural resources, preventing industrial pollution, and other public works. The system also mobilized industry, government, and academia to research and develop technologies that private sector corporations could not undertake on their own due to the vast funding and long-term research periods required. In its first fiscal year (1966), ultra high-performance electronic calculators, MHD power generation, and desulfurization technologies were among the themes selected for large-scale projects. Development of these technologies began with a total budget of 730 million yen. The number of themes handled under the Large-Scale Project System increased from year to year with eight project systems set up by 1973. Initially, the system was covered by the general account budget but, following a reorganization of the tax system in 1979, there was an increase in special account budgets. Under the system, several themes were developed in parallel as the technology for a theme was developed over a span of five to ten years and, as soon as one theme finished, a new one started up (Table 1.1).

In general, MITI industrial policy addressed technically difficult themes, but the focus was on developing important new industries that would have a ripple effect on peripheral industry. The objective was to develop the petrochemical industry, aircrafts, and computers, already available in the industrialized Western countries. The technology policy was positioned as supporting the introduction and development of the underlying technologies for industries requiring advanced technologies. Therefore, in the period from 1966 to the 1970s, the Large-Scale Project System allocated large-scale financing to themes that supported key industries, such as ultra high-performance electronic calculators (10 billion yen total development funding during the period), pattern information processing system (22 billion yen, as above), and aircraft jet engines (13 billion yen, as above).[18] This trend becomes evident when drawing comparisons with other projects in the same period, such as the seawater desalination project (6.7 billion yen), electric automobiles (5.7 billion yen), and desulfurization technology (2.6 billion yen).

In addition to the Large-Scale Project System in 1966, the Sunshine Project in 1974 (new energy technologies), and the Moonlight Project in 1978 (energy-saving technologies), other policy programs at the AIST included commissioned research into and development of medical and welfare equipment in 1976, and commissioned research and development of next-generation industrial key technologies in 1981.

Started in 1976, the Medical and Welfare Equipment Technology Project aimed to deliver an advanced welfare society. The goal of the project was to overcome problems with medical and welfare services by developing the related equipment and technologies. Despite the urgency of developing technology for the medical and welfare services field, it is difficult to enter the market due to the special characteristics of the equipment, which requires advanced technologies. Under this

[18]MITI Kōgyō gijutsuin Kenkyū kaihatsukan shitsu (1987, p. 29).

Table 1.1 Large-scale project themes (unit: million yen)

Project Name	Total R&D expenses	Year																											
		66	67	68	69	70	71	72	73	74	75	76	77	78	79	80	81	82	83	84	85	86	87	88	89	90	91	92	93
Ultra high-performance electronic calculators	10000																												
Desulfurization technology	2600																												
Olefin manufacturing methods	1100																												
Seawater desalination and use of by-products	6700																												
Remote-controlled deep-sea bottom oil drilling equipment	4500																												
Electric automobiles	5700																												
Pattern information processing system	22000																												
Aircraft jet engines	13000																												
Traffic control systems	7300																												
Direct steelmaking using high-temperate reductive gas	13700																												
Olefin production from heavy oil	13800																												
Resources recycling technologies	11300																												
Ultra high-performance laser-based flexible manufacturing system	13700																												
Sea-bottom oil production system	15000																												
Optical measurement and control system	15700																												
Production of basic chemicals from carbon monoxide	10500																												
Manganese nodule mining system	20000																												
High-speed calculation system for science and technology	23000																												
Automatic sewing systems	10000																												
Robots for work in extreme conditions	20000																												
Resource exploration systems	23000																												
Water recycling systems	11800																												
Interoperable database system	15000																												
Ultra-precision mechanical processing	1500																												

Source MITI Kōgyō gijutsuin Kenkyū kaihatsukan shitsu [MITI AIST, Office for Research and Development] (1987, p. 30)

project, the government provided support for the field. The project had a significance of its own and continued into the 1980s.

Launched in 1981, the Next Generation Project was set up with the aim of implementing research and development of innovative basic technologies with a closer focus on the underlying technologies than the Large-Scale Project System. The context for the project was the recognition that in order to become a leader in technology, Japan needed to accumulate not only applied research, but also basic research to rival the advanced West. Super-conductivity, biotechnology, and other technologies were developed under the project.

Incidentally, in the budget for fiscal year 1985, 41.2 billion yen was allocated to the Sunshine Project, 9.9 billion yen to the Moonlight Project, 12.4 billion yen to the Large-Scale Project System, 500 million yen to the Next Generation Project, and 260 million yen to the Medical and Welfare Equipment Technology Project (and international joint development expenses). In the 1980s, the significance of the latter two programs was relatively small when compared to the Sunshine, Moonlight, and Large-Scale Projects.[19]

In 1993, the Large-Scale Project System, the Next Generation Project, and the Medical and Welfare Equipment Technology Project were consolidated as the Industrial Science and Technology Frontier Program. For the remainder of the 1990s, this program and the New Sunshine Program were positioned as the two major technology policy projects at the AIST.

In 1985, the AIST also set up the Japan Key Technology Center, a semi-governmental corporation, with the aim of creating an environment where private corporations could channel their energies into the development of key technologies. The center provided private sector corporations with financing and conditional interest-free loans to develop key technologies. Projects funded in fiscal year 1985 included non-oxide glass research and development, protein engineering, and natural language processing for developing electronic dictionaries.[20] The approach to technology policy at the center was not only to commission projects that strongly reflected the policy objectives of the government, but the center also aimed to encourage any budding industries that emerged out of the corporations.

The AIST used these multiple policy tools to support key technologies and technologies with a degree of public interest. However, it would be fair to say that relative importance was placed on developing new energy technologies, in other words the Sunshine Project from the late 1970s through the 1980s. In this period, the Sunshine Project was the main policy program at the AIST at MITI.

The subject of this book is the history of national projects, which were intended to develop important technologies. It is an attempt to gain useful knowledge about which policies, management, and research formats are effective for reaching public

---

[19]The English designations, the "Sunshine Project" and the "New Sunshine Program", suggest that the Sunshine Project was a single project and not necessarily a program at the outset. However, in light of the subsequent increase in new energy technology targets, the project functioned as a program in all but name.

[20]MITI Kōgyō Gijutsuin (1985, p. 58).

goals. The book focuses on the Sunshine Project (the New Sunshine Program after 1993), which is a national project implemented in the early 1970s with the goal of developing solar, geothermal, coal liquefaction and gasification, hydrogen energy, and other new energy technologies. This book will consider the processes of (1) system design, (2) implementation structure, and (3) program utilization by industry, government, and academia. In short, it is the aim of this book to trace a detailed outline of the launch of a national project, the formal implementation system, and the use of corporations and universities to promote research and development. The case studies for these three phases will provide much food for thought regarding effective approaches to resolving environmental energy issues and other important policy issues. What should we do to achieve policy goals and to bring together the human resources, facilities, financing, and information for specific technology development? What are the conditions for successful technology research and development by means of organic collaboration between industry management, academic research, and government policy?

1.9 The Sunshine Project

The AIST at MITI played a central role in the Sunshine Project, a national project launched in 1974. Its goal was to develop renewable energy technologies to replace oil. Specifically, to substantially reduce Japan's dependence on oil for its primary energy supply by the year 2000, replacing oil with solar, geothermal, and other clean new energies.[21] The development office of the AIST studied the matter in 1973, ahead of the first oil crisis, and launched the project in the following year amid favorable public opinion due to the oil crisis. Initially, there were four development themes: solar energy, geothermal energy, coal gasification and liquefaction, and hydrogen energy, but the themes changed in step with subsequent advances in technology development.

Initially, the AIST handled corporate commissions directly, but once the New Energy Development Organization (NEDO), a semi-governmental corporation, was set up in the 1980s, it fell to NEDO to commission services and to manage the implementation cycle. In 1993, the Sunshine Project and the research and development project for energy conservation technologies (the Moonlight Project) were reorganized as the New Sunshine Program, which continued until fiscal year 2000 (a part of the program continued until fiscal 2002) when it was ended by linking it to a successor project.

This book reviews the history of the Sunshine Project to clarify the outcomes of the project and the nature of the organizational processes. The aim is to suggest successful approaches to national projects and approaches that will prevent failure.

---

[21]*The Nikkei/Nihon Keizai Shimbun*, January 25, 1974.

Focusing on the perspective of the government, the book scrutinizes the ministries and agencies, semi-governmental corporations, private sector corporations, and other individual organizations as well as the individuals who worked at these institutions and advanced the project. These organizations and individuals contributed to the progress of the Sunshine Project and influenced the decision-making about the direction of the project. The survey of the actions of these protagonists (or the accumulation of the actions of organizations) focuses on the following three points.

1. From the perspective of the government, what were the processes for determining policy objectives within the government and for executing the project?
2. How was the semi-governmental organization that implemented the project set up? What role did it fulfill positioned as it was between the corporations, government, academia, and national research institutes?
3. From the perspective of the private sector, to what degree did corporations act autonomously with regard to the government projects? What about the decision-making involving project development or communication between the government and corporations during project implementation? These three questions will shed light on the process of developing the Sunshine Project.

Since an extremely wide range of technologies was targeted for development under the Sunshine Project, describing every detail would require a massive effort. Therefore, this book will focus on solar energy, specifically, photovoltaic energy. This is a theme that cannot be omitted from any discussion of the history of the Sunshine Project and it is possible to understand the essence of the project by analyzing this theme. The reasons for selecting this theme as the research subject are outlined in the following three points.

Firstly, this theme was representative of the Sunshine Project from the start. The project name "Sunshine" evokes images of clean energy and derives from the development of solar energy technologies. Consequently, without this theme, the project would likely never have existed.

Secondly, this theme was around from the start of the Sunshine Project and it was a part of the project until its termination. Therefore, it is possible to gain an overview of the whole project, from establishment to termination, by focusing on this theme.

Thirdly, the market for photovoltaic power generation exists and, of all the themes, it is the one that comes closest to an autonomous industry. Consequently, the theme facilitates observations of outcomes and tangible achievements, which is in line with the essence of this book: to clarify the mechanisms that advance national project.

For the reasons stated above, this book investigates the progress of the Sunshine Project with the focus on photovoltaic power generation.

References

Anchordoguy, M. (1989). *Computers Inc.: Japan's challenge to IBM*. Cambridge, Mass.: Council on East Asian Studies, Harvard University Press.

Callon, S. (1995). *Divided Sun: MITI and the breakdown of Japanese High-tech industrial policy, 1975–1993*. Stanford, Calif.: Stanford University Press.

Fransman, M. (1999). *Visions of innovation: The firm and Japan*. New York: Clarendon Press.

Gotō, A., & Wakasugi, R. (1984). "Gijutsu seisaku [Technology policy]. In R. Komiya, M. Okuno, & K. Suzumura (Eds.), *Nihon no sangyō seisaku* [*Industrial policy of Japan*]. Tokyo: University of Tokyo Press.

Imai, K. (1984). Gijutsu kakushin kara mita saikin no sangyō seisaku. In R. Komiya, M. Okuno, & K. Suzumura (Eds.), *Nihon no sangyō seisaku* [*Industrial policy of Japan*]. Tokyo: University of Tokyo Press.

Itō, T. (1998). *Tekunoporisu seisaku no kenkyū* [*The study of technopolis policy in Japan*]. Tokyo: Nippon Hyōron sha.

Kanamori, H., Ara, K., & Moriguch, C. (Eds.). (1990). *Keizai jiten* [*Yuhikaku dictionary of economic terms*]. Tokyo: Yuhikaku.

Miller, R., Lessard, D. R., & IMEC Research Group. (2000). *The strategic management of large engineering projects: Shaping institutions, risks, and governance*. Cambridge, Mass.: MIT Press.

MITI Kōgyō Gijutsuin. (Ed.). (1985). *Kōgyō gijutsuin nenpō*. Tokyo: MITI Kōgyō Gijutsuin.

MITI Kōgyō gijutsuin Kenkyū kaihatsukan shitsu [MITI AIST, Office for Research and Development]. (Ed.). (1987). *Ōgata Purojekuto nijūnen no ayumi: Wagakuni sangyō gijutsu no ishizue o kizuku* [*The course taken by large projects in the last 20 years: Laying foundations for industrial technologies in Japan*]. Tokyo: Tsūshō Sangyō Chōsakai.

Morris, P. W. G., & Hough, G. H. (1987). *The anatomy of major projects: A study of the reality of project management*. Chichester; New York: Wiley.

Nishida, M. (2000). *Inobēshon to keizai seisaku* [*Innovation and economic policy*]. Tokyo: Yachiyo Shuppan.

Ono, G. (1999). *Gendai Nihon no sangyō seisaku: Dankaibetsu seisaku kettei no mekanizumu* [*Industrial policy of contemporary Japan*]. Tokyo: Nihon Keizai Shimbunsha.

Sangyō kōzō shingikai. (1971). *70-nendai no tsūshō sangyō seisaku: Sangyō Kōzō Shingikai chūkantōshin* [*International trade and industry policy in the 1970s*]. Tokyo: Ōkurashō Insatsukyoku.

Tsuruta, T. (1982). *Sengo nihon no sangyō seisaku* [*Industrial policies in postwar Japan*]. Tokyo: Nihon Keizai Shimbunsha.

Yoshikai, M. (1985). *Nihon no sangyō gijutsu seisaku* [*Industrial technology policy of Japan*]. Tokyo: Tōyō Keizai Inc.

Chapter 2
What is the Sunshine Project: Overview of the Project

2.1 Introduction

Before launching into the case studies, this chapter will look at budgets and other investments in the Sunshine Project and data on the output of results to shed some light on the real contributions made by the project.

At the outset, the immediate goal of the Sunshine Project was to successfully develop new energy technology as a matter of technology policy. If this goal were achieved, Japan could meet a substantial portion of its domestic demand for energy with new energy instead of oil, thereby ensuring energy security for Japan and putting a stop to emergency imports of oil. In addition, setting such a goal would have ripple effects for the future when the new energy technologies were industrialized and market-based, resulting in new employment and technology ripples for the Japanese economy.[1] But what were the outcomes of the Sunshine Project in view of the dual perspectives of achieving policy objectives through innovation and establishing new industries?

In terms of outcomes, the Sunshine Project did not achieve the goal of introducing new energy because the sense of urgency around developing new energy lost intensity when the prices of crude oil fell in the 1980s. In fact, new energy as a ratio of primary energy increased hardly at all after the late 1970s. Early on, much had been expected of large-scale power stations running on new energy but, when they proved unprofitable, no further progress was made after the experimental plants of the early 1980s. Even though the potential was maintained in case of emergencies, subsequent work went no further than accumulating research and development outcomes. So, the original goal of the Sunshine Project, to replace a substantial portion of oil with new energy, was not achieved

---

[1]A third goal, measures for dealing with global environmental problems, was added in the 1990s. MITI (1993, p. 38).

M. Shimamoto, *National Project Management*,
Advances in Japanese Business and Economics 25,
https://doi.org/10.1007/978-981-15-3180-4_2

However, so far the other objective of establishing new industry is concerned, photovoltaic energy and fuel cell technologies were the most successful of all the new energy-related technologies, making remarkable progress in terms of developing the technologies and delivering an industry that was next to self-reliant. Although not true of all the research themes included in the Sunshine Project, one could make the claim that it is possible to reach a stage where major industrial developments can be expected when industry and government collaborate on developing technologies over a long period of time based on a long-term and consistent project.[2] In fact, as will be shown below, there was a sharp increase in production output of Japanese solar cells in the late 1990s and, for a long time, Japanese solar cells were the best in the world. This was not only due to the successful development of technologies under the Sunshine Project. Other factors include effectively functioning government support for infrastructure development, such as subsidies for setting up grid-connected photovoltaic energy systems and monitoring systems for their owners.

In retrospect, when the Sunshine Project began in the mid-1970s, there were limitations on the practical applications of photovoltaic energy due to the cost, and it was generally thought that development by private enterprise based on market mechanisms was the only solution.[3] Photovoltaic energy was not a highly profitable business for most corporations and they would hardly have given it any weight in their corporate technology strategies had the government not provided political leadership on developing the technology. In this sense, it becomes clear that, in one way or another, the Sunshine Project ultimately contributed a great deal to the emergence of the photovoltaic energy industry.

The Sunshine Project was a national project, but what were the intentions behind it and who were the actors that moved the project forward? What impact did it have on the emergence of new industries? Below, I will first look at some concrete numbers before attempting an overview of the project.

2.2 Project Budget

In this section, I would like to capture an overview of the Sunshine Project by surveying the numerical data.

The government is the party responsible for implementing national projects and taxation is the source of funding. The goal of a national project is to develop technologies that are not market-based, so the cost of development is shared by the whole nation in one form or another. Therefore, I would like to start with an

[2]Fuel cells and photovoltaic energy were both earmarked for accelerated promotion under the New Sunshine Project. MITI (1993, p. 64).

[3]Taiyōkō Hatsuden Gijutsu Kenkyū Kumiai [Photovoltaic Power Generation Technology Research Association] (1998, p. 31).

overview of the budgets invested in the Sunshine Project. Below, I will identify individual policy programs funded by the Agency of Industrial Science and Technology (AIST) at the Ministry of International Trade and Industry (MITI) and provide detailed breakdowns of the budgets.

The policy goals and the scale of the budgets were changed as of fiscal year 1993 when the Sunshine Project was integrated with other projects, including the Moonlight Project, and reorganized as the New Sunshine Project. Therefore, this chapter will focus primarily on data for the period prior to the reorganization at the end of fiscal year 1992. Where statistics are available, the chapter will also trace the New Sunshine Project to the point of termination and, where available, statistics for the 2000s will be provided as reference material. Some important statistics, including figures dating to the period after the Sunshine Project was terminated, will be reproduced at the end of this chapter and in the case studies.

Basically, the budget for the Sunshine Project came out of a technology policy program at the AIST at MITI.[4] Figure 2.1 illustrates the breakdown of the budget for the AIST during the implementation period for the Sunshine Project (Fig. 2.1).

As indicated by Fig. 2.1, the Sunshine Project was a policy program that accounted for nearly half of the AIST budget throughout the 1980s. If we add in the project to develop technologies for energy conservation (the Moonlight Project), which was launched in 1979, the AIST spent more than 50% of its budget for technology policy on energy-related projects in this period. As already mentioned, when viewed from a budget perspective, the development of energy-related technologies was without doubt the central issue at the MITI AIST throughout the 1980s.

How does the amount of budget spent on developing new energy technologies compare to other countries? Figure 2.2 compares government expenditure on renewable energy in several countries.

Although government expenditure on new energy (renewable energy) in Japan lagged far behind that of the United States, it was generally speaking higher than European countries. The reason for the conspicuously high figures of the United States in the late 1970s is that this was a period when renewable energy research was given huge amounts of funding in the name of energy independence. The existence of a similar national project in the United States proved very useful to the Sunshine Project when asserting the significance of the project. However, the United States significantly reduced the budget in later years. As mentioned below, the context was the fall in the prices of crude oil and a reduced budget for developing energy conservation and new energy technologies in the wake of the shift to the Strategic Defense Initiative (SDI) under the Reagan administration. Compared to data for several overseas countries, we see that the budget for

[4]The Agency of Industrial Science and Technology budget was composed of the national research institute budget and the AIST budget. Here, reference is made to the share of the AIST budget allocated to the Sunshine Project. As of the late 1980s, subsidies for installing photovoltaic energy systems were made available under the Agency of Natural Resources and Energy budget for energy policy.

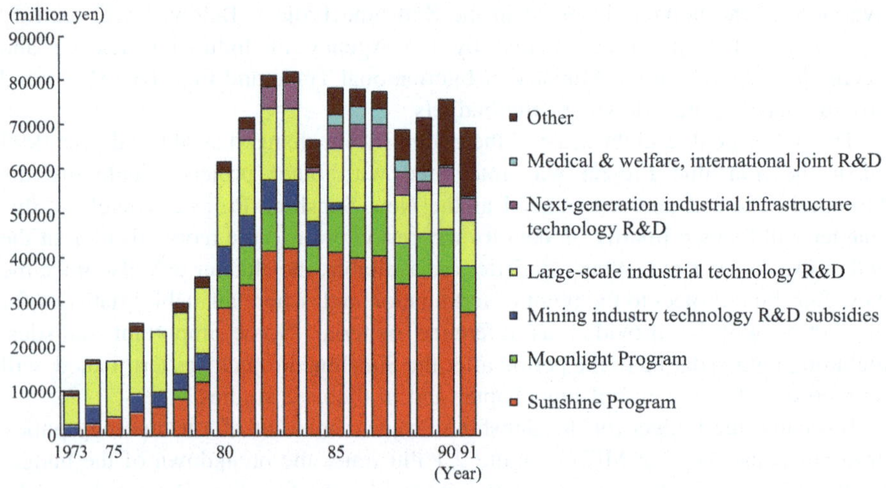

Fig. 2.1 Budget for the agency of industrial science and technology broken down by program. *Note* The figures are actual values for each fiscal year. *Source* MITI Kōgyō gijutsuin, ed., *Kōgyō gijutsuin shōkai* [Introducing the AIST], editions for the respective years

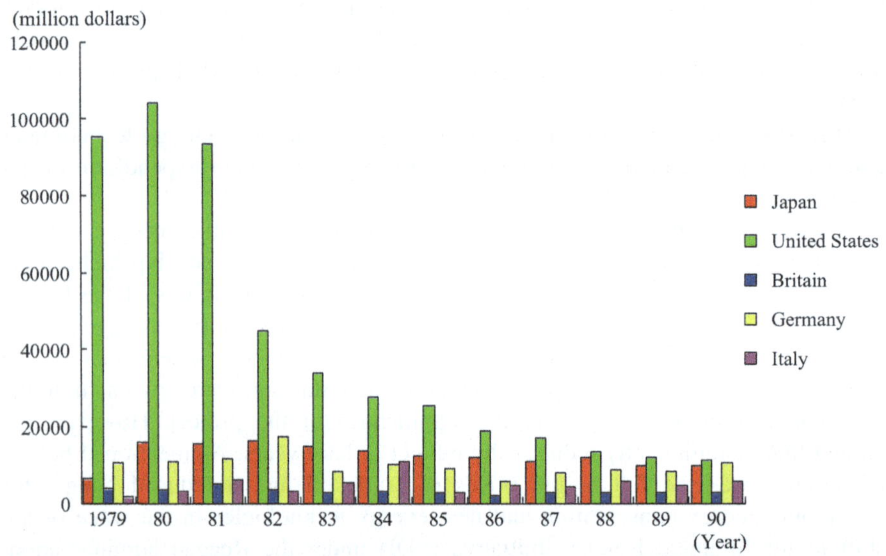

Fig. 2.2 R&D budget for renewable energy. *Note* Renewable energy includes solar heating and cooling, photovoltaic energy, solar thermal energy, wind, ocean, biomass, and geothermal energy. The foreign exchange conversion uses 1990 as the reference year. Coal liquefaction and gasification are not included. *Source NEDO News*, February 1992, p. 43

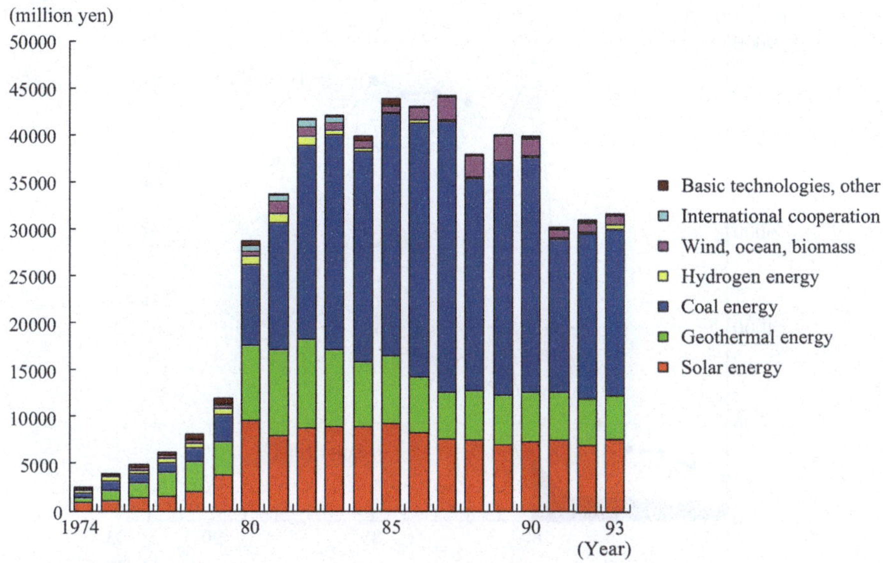

Fig. 2.3 Budget for the sunshine program broken down by technology. *Note* The figures are actual values for each fiscal year. *Source:* Based on MITI Kōgyō gijutsuin, ed., *Kōgyō gijutsuin shōkai* [Introducing the AIST], editions for the respective years

developing new energy (renewable energy) technology in Japan was definitely not small. It was also maintained over a long period of time.

Next, let us look at the budget allocation by research theme for the Sunshine Project (Fig. 2.3).

As indicated by Fig. 2.3, a surprisingly large share of the budget for the Sunshine Project was actually spent on coal liquefaction and gasification. Of course, a simple comparison is not possible because the required budget per plant differs depending on the technology, but considering that renewable energy was at the heart of the project, it is baffling that such a large share of the pot for the Sunshine Project was spent on coal energy.

The budget for the development of solar energy technologies ranged from six to seven billion yen per year. Below, the focus is on the Sunshine Project budget earmarked for developing solar energy technologies. The following is a detailed breakdown of the budget (Fig. 2.4).

As indicated by Fig. 2.4, when the project was first launched, nearly all the budget was allocated to solar thermal energy (including solar systems for industrial use), but in the early 1980s the target for technology development was quickly shifted to photovoltaic energy (solar cells and photovoltaic energy systems). By this time, it was understood that solar thermal energy would not be profitable in Japan even though some results had been obtained after completing the construction and testing of experimental plants. Some efforts to explore overseas potential where the type of solar radiation was better suited to thermal energy coincided with the return

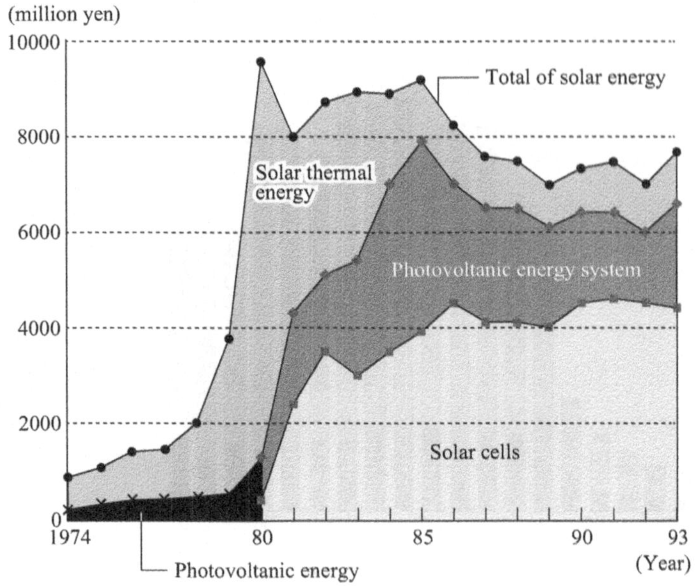

Fig. 2.4 Breakdown of expenditure on developing solar energy technologies under the sunshine project. *Note* In the 1970s, the budget for solar energy was mainly allocated to solar thermal energy. *Sources* From 1974 to 1984: MITI Kōgyō gijutsuin Sanshain keikaku suishin honbu [MITI AIST, Sunshine Project Promotion Office] (1984). From 1985 to 1993: Shigen enerugīchō (1985, 1986, 1987, 1988, 1989, 1990, 1991, 1992, 1993). The budget breakdown is based on MITI (1993)

to the stage of basic research. By the early 1980s, the development of solar energy technologies, mainly focused on solar cells and photovoltaic energy, accounted for the largest share of the budget, including the kind of system demonstrations that the average company could not afford. Even though the technologies are different, both thermal energy and photovoltaic energy were slotted under the solar energy framework, so the development of photovoltaic energy benefited from being on the receiving end of the sizeable thermal energy budget (Fig. 2.5).

In the United States, the government budget for photovoltaic energy was substantially decreased in the 1980s, but Japan continued to maintain a certain standard without reacting to external circumstances. In contrast to the budget for photovoltaic energy in the United States, which was greatly influenced by changes in the prices of crude oil and government policy, the Japanese government safeguarded the budget over a long period of time.

In the 1980s, there were no major variations in the research budget for solar energy under the Sunshine Project. In particular, the budget for commissioning corporations to research solar cells and photovoltaic energy systems averaged six to seven billion yen throughout the decade. During the period from 1974 to 1992 when the Sunshine Project was implemented, the total budget for commissioning research on photovoltaic energy systems channeled via NEDO reached 85.4 billion

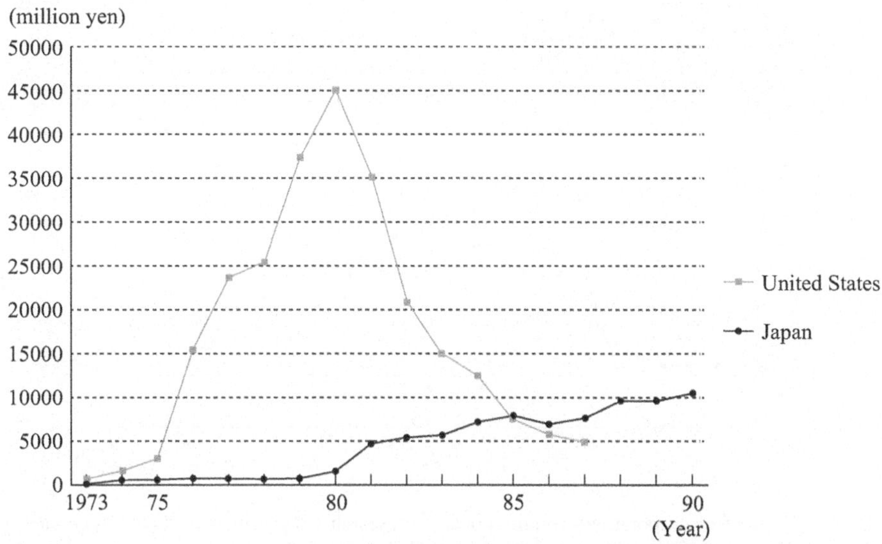

Fig. 2.5 Comparison of the U.S. and Japanese government budgets for developing photovoltaic energy technologies. *Note* The figures for the United States refer to the department of energy budget. The figures for Japan refer to NEDO expenditure on research and development to implement photovoltaic energy systems. Since the NEDO budget for solar energy includes more than the share allocated to the agency of industrial science and technology, the amounts are higher than the budget for the sunshine project. The amounts have been adjusted with GDP deflator (1985 = 100). *Source NEDO News*, May 1987, p. 21 and p. 25, December 1987, p. 39. NEDO (1990)

yen.[5] Incidentally, the VLSI Research Association, which is noted for its research output, was given subsidies of approximately 30 billion yen between 1976 and 1980.[6] Compared to the VLSI Research Association, the budget invested in photovoltaic energy was nearly three times as large.

2.3 Corporate Investment in Photovoltaic Energy

Other than the government budget spent on the Sunshine Project, the corporations participating in the project contributed funds to the budget assigned to them to carry out research into and development of specific technologies. Consequently, we also need to add the corporate investment amounts as input to the Sunshine Project. As a preliminary step, let us first confirm the number of corporations participating in the Sunshine Project.

[5]MITI (1993, p. 64).

[6]Wakasugi (1986, p. 157). Also, Sakakibara (1986, p. 294).

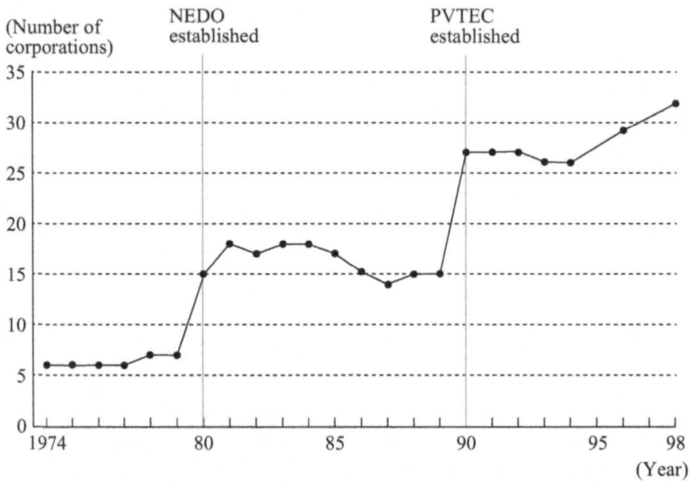

Fig. 2.6 Number of corporations commissioned to research solar cells and photovoltaic energy systems. *Source* PVTEC (1996); and the annual *Seika Hōkokusho* summaries compiled by the Kogyō gijutsuin Sanshain keikaku suishin honbu

A total of 166 corporations contributed commissioned research to the Sunshine Project as a whole (all technology themes) from 1974 to 1995. Concerning the number of corporations actually participating in any given year, the figure of 61 participating corporations in 1992 will be used as reference.[7] In 1995 when the New Sunshine Project was launched, 79 corporations (including corporations that had already completed their assignments at the time) were contributing to ongoing projects, and 32 of the 79 were contributing to photovoltaic energy. These figures tell us that approximately 40% of all corporations participating in the Sunshine Project focused on this theme, and that the photovoltaic energy theme had the highest number of corporations participating in the project. Figure 2.6 shows the fluctuations in the number of corporations contributing to photovoltaic energy.

When the project was launched, six corporations were contributing to solar cells and photovoltaic energy systems, but by the end of the 1990s the number had risen to above 30. Figure 2.6 tells us that there were dramatic increases in the number of participating corporations in 1980 when the New Energy Development Organization, currently the New Energy and Industrial Technology Development Organization (NEDO) , was established, and in 1990 when the Photovoltaic Power Generation Technology Research Association (PVTEC) was established.

As of 1980, corporate investment in researching and developing photovoltaic energy exceeded the budget provided by the Sunshine Project. This means that

[7]Tekunorisāchi Kenkyūjo [Technoresearch Co., Ltd.] (1997, pp. 7–8). The number of research associations have been used where the precise number of corporations could not be determined due to the nature of the technology theme.

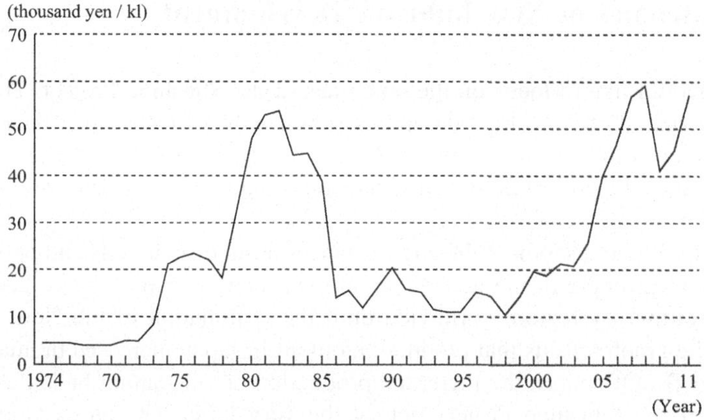

Fig. 2.7 Fluctuations in CIF prices (on arrival in Japan) for imports of crude oil. *Source* Shigen enerugīchō [Agency for Natural Resources and Energy] (2013)

corporations voluntarily augmented funds in excess of the national budget for research commissioned by the government.[8] The existence of the government project also encouraged companies that had not been commissioned to do research to invest in photovoltaic energy.

However, compared to the government budget, the corporate investment amounts varied greatly. In the late 1980s corporate investment decreased due to the drop in crude oil prices. The dramatic drop in prices in the 1980s is made clear by Fig. 2.7, which shows the price of crude oil imports in this period.

However, by the 1990s there was a marked rise in corporate investment in research and development despite the substantial decline in the prices of crude oil. The reason for the change was that, although new energy development was a resource energy issue, the impact of the environmental problems was also starting to make itself felt. As of this period, the government also introduced measures such as licenses for grid-connected systems and monitoring systems, so it can be surmised that this had an impact on corporate investment in research and development. What then were the outcomes of the research commissioned under the Sunshine Project?

[8]The following are remarks by Hamakawa Yoshihiro (Professor, Osaka University), a university researcher involved in photovoltaic energy research commissioned by the Sunshine Project. His remarks provide anecdotal support for Fig. 2.7: "Japanese corporations did not take this [the American] approach. The budget provided by the government was primarily used to cover research costs while the companies covered personnel and other costs. Considering that companies also matched the government budget and added in personnel costs on nearly the same scale, generally speaking, where the United States spent ten billion yen, Japan invested close to 20 billion yen for every ten billion spent if you looked at the actual content of the research" (Hamakawa Yoshihiro, interview by Tomae Hisao, December 7, 1995, transcript, Tomae 1996, p. 85).

2.4 Outcomes of New Industry Development

Below, I would like to focus on the outcomes of the Sunshine Project. The project had two goals: (1) to develop new industry by means of technology development, and (2) to substitute oil with new energy. Consequently, I would like to confirm what outcomes the project delivered as a result of budget investment from these two perspectives.

To start with, let us look at the circumstances around solar cells and photovoltaic energy as an example of the goal of developing new industry. Figure 2.8 outlines solar cell output by region worldwide until the termination of the New Sunshine Project. The graph tells us that production output in Japan increased dramatically in the late 1990s. Of course, the increased production output cannot be laid entirely at the door of the Sunshine Project but, at the very least, we can assume that the government's systematic investment in technology development had some form of positive impact on the development of the technologies in question.

By the time the New Sunshine Project was terminated in the late 1990s, solar cell production output for the whole world had reached 287.65 MW. Japan accounted for 128.6 MW (44.7%) followed by the United States at 75.1 MW (26.1%) and the European countries with a total of 61 MW (21.2%).

As Fig. 2.9 indicates, the leading companies were Sharp, Kyocera, and Sanyo Electric (currently, Panasonic).

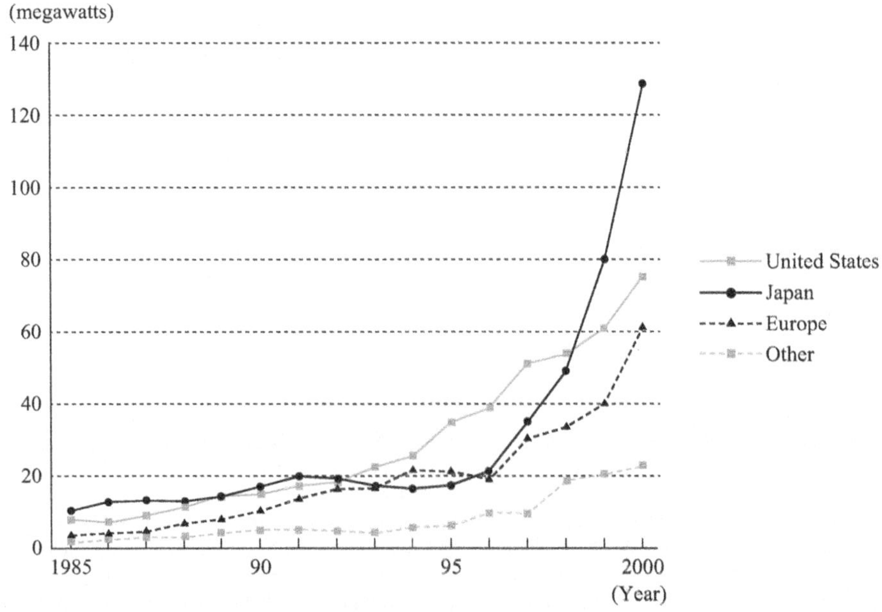

Fig. 2.8 Solar cell production by region. *Source* NEDO (1996), and *PV News*

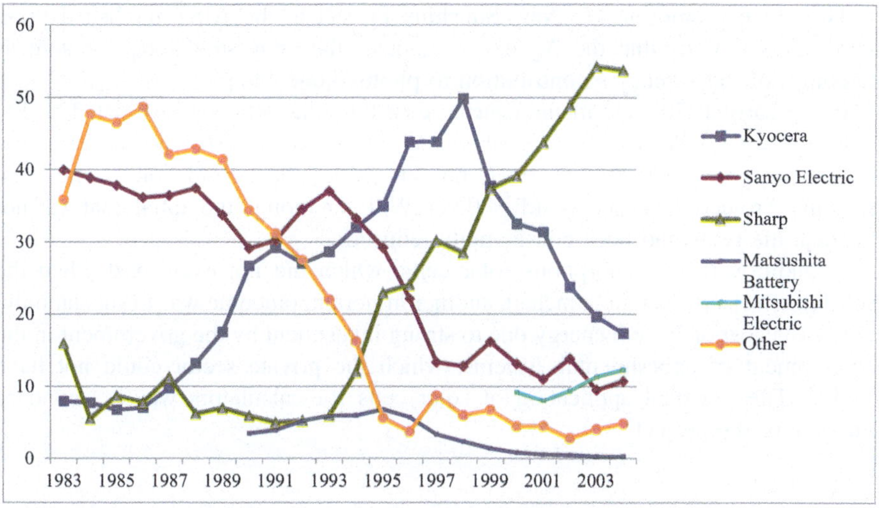

Fig. 2.9 Production output for major Japanese solar cell corporations. *Source PV News*

In the 1980s, Sanyo Electric maintained a high share of the market for amorphous solar cells used in electric calculators and watches. Later, Kyocera expanded its market share with polycrystalline solar cells, but when the production of photovoltaic energy systems gained momentum in the late 1990s, Sharp expanded its market share and, by the time the New Sunshine Project was terminated, Sharp was the market leader among the Japanese corporations.

How did these three companies approach the commercial development of solar cells and photovoltaic energy systems? What role did these corporations play in the Sunshine Project? The case studies in this book will shed light on why these three corporations among so many others succeeded to create commercial products and to corner most of the market.

Regarding the growth of photovoltaic energy, the AIST cites the following as outcomes of the Sunshine Project (New Sunshine Project) (Table 2.1):

Table 2.1 Twenty years of the new sunshine project: outcomes

1	Succeeded in cutting manufacturing costs for solar cells to approx. 1/30th (20–30,000 yen→600 yen/W
2	Achieved best conversion rate (12% for 10 cm square cells) and large-area cells (10.5% at 30 × 40 cm) for amorphous solar cells
3	Succeeded in lowering the cost of photovoltaic system power generation to just over 1/15th (approx, 2000 yen/kW)
4	Partial implementation of special purposes (calculators etc.) (The data for FY1994 in Japan was approx. 17,000 kW)

Source MITI Kōgyō gijutsuin Sanshain keikaku suishin honbu (1997, p. 36)

The office promoting the New Sunshine Project at the AIST published these items. This tells us that the Agency recognizes the above-mentioned content as outcomes of the Agency's contribution to photovoltaic energy.

In the early 1970s, the manufacturing cost for solar cells was estimated at 20–30,000 yen per 1 W. The initial goal was to cut the cost to 1/100th (200–300 yen/W) by the year 2000 but, in 1997, not long after the termination of the New Sunshine Project, the cost was still 600 yen/W. Even though the initial goal was not reached, the costs had been considerably reduced.

In addition to the amorphous solar cells, which did not even exist when the Sunshine Project was first launched, another important outcome was to substantially reduce the cost of system energy due to strong investment by the government in the development of experimental systems, which the private sector could not have funded. The practical application of solar cells for calculators was a secondary outcome of the project.

2.5 Substituting Oil with New Energy

Next, let us look at Japan's energy supply and fluctuation in the ratio of new energy, as an outcome of the second goal of the Sunshine Project. In 1973, oil accounted for as much as 77% of the whole supply of primary energy. By 1992, the year before the Sunshine Project was reorganized, oil had dropped to 58.2%. In this sense, we can perhaps say that the long-term goal of lowering the degree of dependence on oil for Japan as a whole produced results. However, this goal was mainly achieved with contributions from nuclear power (up from 2% in 1977 to 10% in 1992) and natural gas (up from 0.9% to 10.6% in the same period) while new energy, for all its expectations, varied hardly at all from the 1% level throughout the period (Fig. 2.10). In this sense, the initial goal of the project to substitute oil with new energy was not achieved.

When the Sunshine Project was first launched, it was declared that a substantial amount (approx. 20%) of all energy would be covered by new energy by the year 2000. So, what is the reason for the plateau at 1% that continues today? Every few years, the MITI Advisory Committee for Energy reviews the outlook for long-term energy supply and demand. Since the goals for new energy supplies are presented at the time, it is possible, by looking at how the estimates have changed, to shed some light on the changes in recognition of how urgent and necessary it is to develop new energy technologies. Figure 2.11 shows the fluctuations in the actual supply of new energy as well as targets for the supply of new energy in the interim outlook for long-term energy supply and demand.

In 1979, the targets for new energy supply were drawn up very quickly in the wake of the two oil crises. However, by the 1980s, the targets were steadily brought down as indicated by the increasingly gentle inclination of the line segments connecting actual values with target values. For example, immediately after the start of the New Sunshine Project in 1994, the target for the ratio of new energy supply

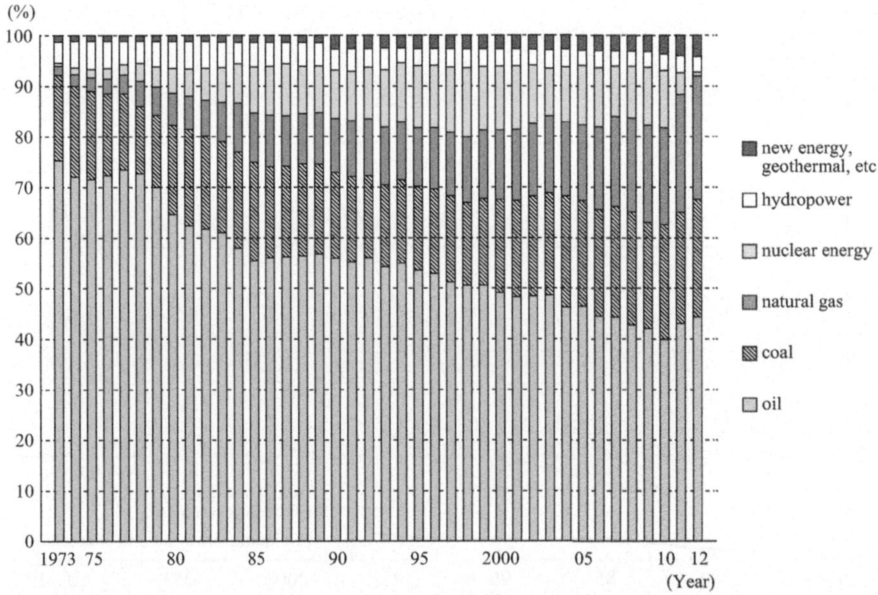

Fig. 2.10 Primary energy supply ratios in Japan as of 1990. *Notes* (1) Comprehensive energy statistics has changed the calculation method since 1990. (2) "New energy, geothermal, etc." includes solar, wind, biomass, and geothermal, etc. *Source* Shigen enerugīchō [Agency for Natural Resources and Energy] (2014). The data were originally from comprehensive energy statistics compiled by agency for natural resource and energy

in 2000 was 2% for new capacity (12.1 million kl oil equivalent), and 3% by 2010 (19.1 million kl oil equivalent) (Table 2.2).[9] For example, the goals for photovoltaic energy were 400,000 kW (400 MW) in 2000 and 4,600,000 kW (4600 MW) in 2010.[10]

However, as of 1994, installations accelerated quickly due to the residential photovoltaic system monitor program.[11] The context was the launch of government subsidies and companies at the receiving end of subsidies setting up systems to increase production.

---

[9]The targets for 2010 were unchanged in interim report by the Supply and Demand Subcommittee of the Advisory Committee for Energy in June 1998.

[10]Shigen enerugīchō [Agency for Natural Resources and Energy] (1995, p. 17).

[11]The monitor program subsidized approximately half the cost of installing residential photovoltaic power systems. The intent of the program was to improve equipment performance to conform to consumer needs by requiring the system owner to act as monitor. The program spent 2.03 billion yen in FY1994, 3.31 billion yen in FY1995, 4.06 billion yen in FY1996. Where photovoltaic systems were installed in schools, community centers, museums or other public buildings, government subsidies covering two thirds of the cost were available through NEDO (Shin'enerugī zaidan 1997).

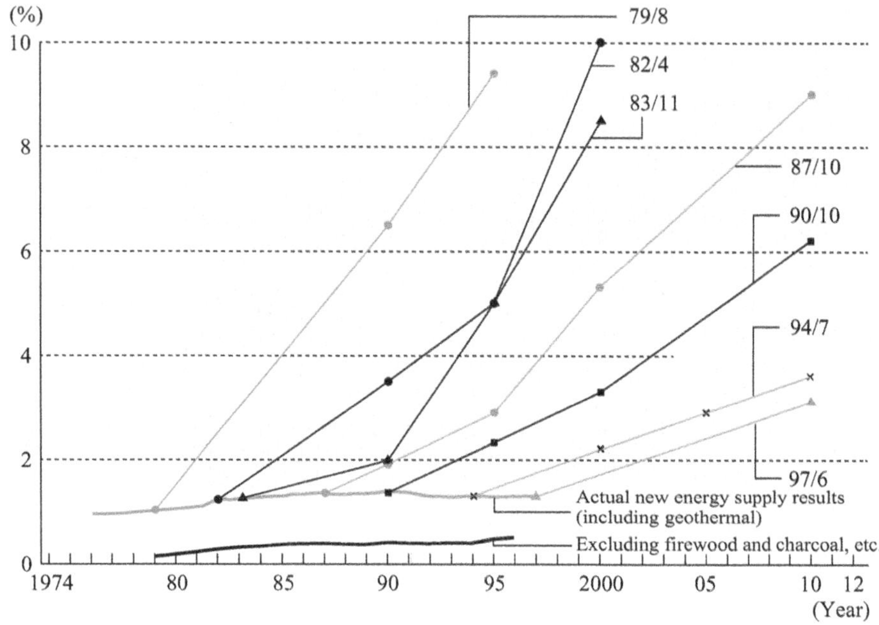

Fig. 2.11 New energy supply targets and performance in the interim outlook for long-term energy supply and demand. *Note* The forecasts are the aggregate total for petroleum alternatives, new energy, other (including solar energy, oil shales and tar sands, alcohol fuel, coal liquefaction, wood and charcoal etc.) and geothermal. The lower line showing new energy supply performance is the aggregate total for supplies from solar thermal energy, refuse-derived fuel, and geothermal, but firewood is excluded. If firewood were added, the supply ratio would be around 1%. *Sources* MITI (1979, pp. 62–63) (October 1978 forecast). MITI Kōgyō gijutsuin Sanshain keikaku suishin honbu (1980, p. 67) (August 1979 forecast). MITI (1979, pp. 42–43) (April 1982, November 1983 forecasts). Shigen enerugīchō (1987, p. 21) (October 1987 forecast). Shigen enerugīchō (1992 and 1995 editions) (October 1990, June 1994 forecasts). Shigen enerugīchō, ed., *Shigen enerugī dētāshū* (Tokyo: Denryokushinpōsha, annual editions) (new energy supply performance). Kikkawa Takeo and Tsūshō sangyō seisakushi hensan iinkai [Editorial Committee on the History of Japan's Trade and Industry Policy] (2011)

The aim of this book is to throw some light on how the investments described above are connected to the outcomes of the project, and on the internal workings of the black box that is the national project. The following chapters will look at the issue in the light of case studies.

Table 2.2 The outlook for long-term energy supply and demand after the end of the new sunshine project(unit: %)

	Fiscal year (FY)							
	1990	1999	2000	2005	2010	2020	2030	
July 2001	1.3			1.6				Standard goal
				3				Higher goal
March 2005	2		2	3				Reference data
				4				With current measures
				4				With additional measures
May 2008			3		3	4		With current effort
					4	4		With additional effort
					5	7		With maximum effort
August 2009					3	5		With current effort
					4	5		With additional effort
					5	7		With maximum effort

Note The figures until 2005 are actual values for each fiscal year. The figures after the year are target values
Source Kikkawa Takeo and Tsūshō sangyō seisakushi hensan iinkai [Editorial Committee on the History of Japan's Trade and Industry Policy] (2011, p. 99, p. 101, pp. 104–107)

References

Kikkawa, T., & Tsūshō sangyō seisakushi hensan iinkai [Editorial Committee on the History of Japan's Trade and Industry Policy]. (Eds.), (2011). *Tsūshō sangyō seisakushi 1980–2000 [History of Japan's Trade and Industry Policy 1980–2000]* (Vol. 10) Natural Resources and Energy Policy. Tokyo: Keizai Sangyō Chōsakai.

MITI. (Ed.). (1979). *Nijūisseiki eno energī senryaku*. Tokyo: Tsūshō Sangyō Chōsakai.

MITI. (Ed.). (1993). *Nyū sanshain keikaku handobukku [The handbook of new sunshine project]*. Tokyo: Tsūshō Sangyō Chōsakai.

MITI Kōgyō gijutsuin Sanshain keikaku suishin honbu. (Ed.). (1980). *Sanshain keikaku no kasokuteki suishin senryaku: Sangyō gijutsu shingikai shin energī gijutsu kaihatsu bukai chūkan hōkoku o chūshin toshite [Acceleratory promotion strategies for the sunshine project: Based primarily on an interim report by the industrial technology advisory committee's new energy technology development subcommittee]*. Tokyo: Tsūsan Seisaku Kōhōsha.

MITI Kōgyō gijutsuin Sanshain keikaku suishin honbu. (1997). *Nyū Sanshain Keikaku*. Tokyo: Nihon sangyō shinkō kyōkai [Japan Industrial Technology Association].

MITI Kōgyō gijutsuin Sanshain keikaku suishin honbu [MITI AIST, Sunshine Project Promotion Office]. (Ed.). (1984). *Sanshain keikaku jūnen no ayumi [Ten year history of the sunshine project]*. Tokyo: Sanshain keikaku jusshūnen kinen jigyō suisin konwakai [Sunshine Project 10th Anniversary Commemorative Projects Promotion Committee].

NEDO. (Ed.) (1990). *NEDO jūnen no ayumi [Ten year history of NEDO]*. Tokyo: NEDO.
NEDO. (1996). *Shin enerugī gijutsu kaihatsu kankei dētashū sakusei chōsa: Taiyōkō hatsuden [Research for preparing collections of data related to the development of new energy technologies: Solar power generation]*. Self-pub.
PVTEC. (1996). *PVTEC gonen no ayumi [Five year history of PVTEC]*. Tokyo: PVTEC.
Sakakibara, K. (1986). Kyōdō kenkyū kaihatsu project no soshiki to management: Chō LSI gijutsu kenkyū kumiai no kēsu. In K. Imai (Ed.), *Inobēshon to soshiki*. Tokyo: Tōyō Keizai Inc.
Shigen enerugīchō. (Ed.). (1992 and 1995). *Sōgō enerugī tōkei*. Tokyo: Tsūshō Sangyō Chōsakai.
Shigen ererugīchō. (Ed.). (1985). *Shigen enerugī dētā shū*. Tokyo: Denryokushinpōsha.
Shigen ererugīchō. (Ed.). (1986). *Shigen enerugī dētā shū*. Tokyo: Denryokushinpōsha.
Shigen ererugīchō. (Ed.). (1987). *Shigen enerugī dētā shū*. Tokyo: Denryokushinpōsha.
Shigen ererugīchō. (Ed.). (1988). *Shigen enerugī dētā shū*. Tokyo: Denryokushinpōsha.
Shigen ererugīchō. (Ed.). (1989). *Shigen enerugī dētā shū*. Tokyo: Denryokushinpōsha.
Shigen ererugīchō. (Ed.). (1990). *Shigen enerugī dētā shū*. Tokyo: Denryokushinpōsha.
Shigen ererugīchō. (Ed.). (1991). *Shigen enerugī dētā shū*. Tokyo: Denryokushinpōsha.
Shigen ererugīchō. (Ed.). (1992). *Shigen enerugī dētā shū*. Tokyo: Denryokushinpōsha.
Shigen ererugīchō. (Ed.). (1993). *Shigen enerugī dētā shū*. Tokyo: Denryokushinpōsha.
Shigen enerugīchō [Agency for Natural Resources and Energy]. (Ed.). (1995). *Shin enerugī binran heisei shichi-nendo ban [New energy handbook, FY1995 edition]*. Tokyo: Tsūshō Sangyō Chōsakai.
Shigen enerugīchō [Agency for Natural Resources and Energy]. (2013). *FY2012 annual report on energy (Energy White Paper 2013)*. Tokyo: Nikkei Printing.
Shigen ererugīchō [Agency for Natural Resources and Energy]. (2014). *FY2013 annual report on energy (Energy White Paper 2014)*. Tokyo: Nikkei Printing.
Shin'enerugī zaidan. (Ed.). (1997). *Shin enerugī kaihatsu riyō jittai chōsa hōkokusho sōgōhen*. Tokyo: Shin'enerugī Zaidan Keikaku Honbu.
Taiyōkō Hatsuden Gijutsu Kenkyū Kumiai [Photovoltaic Power Generation Technology Research Association]. (Ed.). (1998). *Taiyōkō hatsuden: Sono hatten to tenbō [Solar power generation: Its development and prospects]*. Tokyo: Art Studio 76.
Tekunorisāchi Kenkyūjo [Technoresearch Co., Ltd.]. (1997). *Nyū sanshain keikaku purojekuto taishō gijutsu no dōnyū jōkyō chōsa hōkokusho*. Shizuoka: Tekunorisāchi Kenkyūjo.
Tomae, H. (1996). Kyōdō kenkyū ni okeru ritateki shinrai (*Altruistic trust in joint research and development*) (PhD dissertation). Hitotsubashi University.
Wakasugi, R. (1986). Gijutsu kakushin to kenkyū kaihatsu no keizai bunseki. Tokyo: Tōyō Keizai Inc.

Chapter 3
Case Study: Managing Technology Development

Case Study 1: The Sunshine Project from the Perspective of Project Management

3.1 The Origins of the Sunshine Project

3.1.1 Draft Proposal Preceding the Oil Crisis

In 1973, the Ministry of International Trade and Industry (MITI) selected solar energy research as the development theme for the Large-Scale Project System. The decision by MITI to adopt this theme was based on the importance of new energy technologies. This decision was the first step in setting up the Sunshine Project.

Work on the Sunshine Project did not start after the oil crisis, though the draft plans were already in place before the oil crisis ever happened.[1] Why was the Japanese government able to draft the project plans before the oil crisis occurred? What was the historical background and what was the judgment of policymakers that led to this decision?

MITI was the government agency that planned and implemented the Sunshine Project. As the name suggests, MITI was responsible for all trade and industry policy for Japan. The ministry was not only involved with direct industrial policy for the stable development of the economy and industry, but policies to provide lateral support were also vital. For example, important policy issues included the stable supply of essential energy sources for industrial production, or support for corporations researching and developing technologies that would become the cornerstone for new business. For this purpose, MITI established the Agency for Natural Resources and Energy to develop resource energy policy, and Kōgyō gijutsuin (the Agency of Industrial Science and Technology (AIST)) to develop policy for industry technologies.[2]

---

[1]Sawai Minoru and Tsūshō Sangyō Seisakushi Hensan Iinkai [Editorial Committee on the History of Japan's Trade and Industry Policy] (2011, p. 246).

[2]As explained in Chap. 1, industrial technology policy in this context refers to technology policy at MITI, which aimed for industry application. The term is specifically used to set it apart from science and technology policy at the Science and Technology Agency.

M. Shimamoto, *National Project Management*,
Advances in Japanese Business and Economics 25,
https://doi.org/10.1007/978-981-15-3180-4_3

Seeing the context, we might imagine that the Sunshine Project was launched as a policy for resource energy because the aim of the project was to introduce and disseminate new energy. However, when you trace its history, it is clear that the project began as an industrial technology policy. In other words, the aim of project implementation was to successfully develop technologies through innovation in new energy. The introduction and spread of new energy was an expected ripple effect of successful development. For MITI, new energy was a promising technology theme that would support the development of industrial technology.

In the 1970s, the Large-Scale Project System constituted the core of industrial technology policy at MITI. The system supported an extremely broad range of large-scale industrial technology. As mentioned above, new energy technology was originally one of the development themes proposed for the Large-Scale Project System. The Sunshine Project was spun off from the Large-Scale Project System as an independent development project dedicated to new energy.

As already mentioned, the work of drafting the proposal for the Sunshine Project was undertaken in the spring and summer of 1973 prior to the oil crisis. Why was the project drafted ahead of the oil crisis? To clarify the reasons, we need to turn our attention to the energy issues confronting Japan at the time, as well as government perceptions and measures taken to counter the situation.

3.1.2 *Symptoms of an Energy Crisis*

The following is a description of the energy situation in Japan on the eve of the oil crisis. In the mid-1950s, Japan had entered the period of rapid economic growth, achieving remarkable growth over a period of a dozen years. In 1968, the gross national product of Japan surpassed West Germany to become the second largest free market economy in the world. Due to such rapid economic growth, the energy configuration in Japan had also undergone major change since the 1960s. Briefly, it was a case of a structural change from coal to oil.

By 1968, the Japanese government's policy of switching the energy supply from coal to oil, which dated back to the early 1960s, had positioned oil at the heart of industry due to the ease of transport. The proportion of oil as a share of all primary energy in Japan was 20.2% in 1955, 37.7% in 1960, 58.4% in 1965 and, having increased rapidly through the 1960s, 70.8% in 1970. By the early 1960s, the consumption of coal and oil had already been reversed and the gap had steadily expanded since then (Fig. 3.1). For a country lacking in resources, the overemphasis on oil was a situation that would have to be remedied as soon as possible to guarantee energy security.

Energy self-sufficiency is strongly linked to issues of security guarantees and economic growth for the nation. The fact that Japan was poor in oil resources and depended on other countries for the majority of its domestic oil requirements was, of course, a major unresolved question for the government.

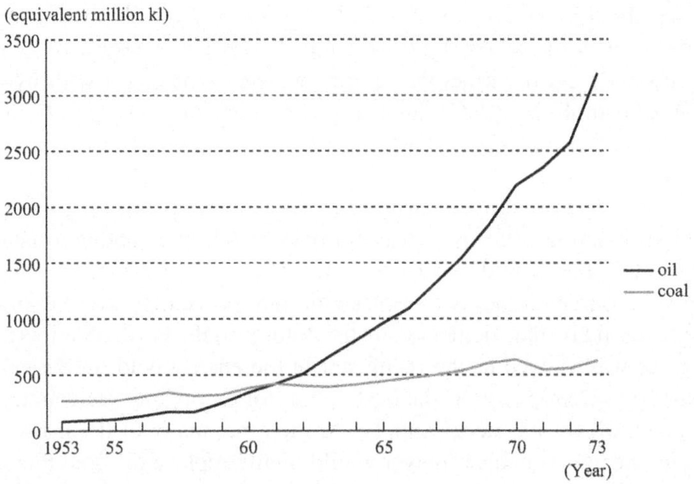

Fig. 3.1 Comparison of coal and oil supplies in Japan. *Source* MITI (1973, pp. 83–89)

To start with, the stable supply of oil was the cornerstone of the nation's security guarantees. If, for argument's sake, Japan were thrust into a situation where a country in possession of oil resources stopped exports, Japan would be placed in an extremely weak position in any diplomatic negotiations. In fact, during the first oil crisis, Japan found itself squeezed hard in the diplomatic discussions when the producing countries used oil as a weapon to come away with favorable terms for themselves. To avoid such a crisis, the government had to maintain a constant sense of crisis when responding to issues of energy security.

Oil was also the wellspring of economic growth for the nation. Any delays in the stable supply of oil would have devastating consequences for economic activities in Japan. Oil had become increasingly important for postwar Japan since the period of rapid economic growth. Postwar Japan had achieved economic development by importing raw materials and energy, and trading in processed goods. Coinciding with this process, coal, previously the main source of energy, had rapidly been replaced by oil, which had grown increasingly important.

In this context, the government was considering what could be done to rectify the vulnerabilities of such a system of energy supply. For example, one solution might have been to propose a reduction in the amount of oil consumed in Japan as a whole. However, if we consider the situation in Japan in the 1970s, a period of remarkable economic growth, this was hardly a feasible approach to the practical problems. Of course, further energy conservation policies were a key issue, in particular researching and developing technologies. In fact, the government vigorously promoted energy conservation policies and the requisite technology research and development as of this period.

So, what should the government do? The idea of developing some kind of energy produced in Japan emerged as a more assertive measure to avoid excess dependence on oil. At the time, the options were to replace oil with hydroelectric power, liquid natural gas (LNG), nuclear power, or natural energy. At first glance, the options may seem plentiful, but each inherent advantage also had an inherent disadvantage.

For example, the installment of hydroelectric power plants is governed by the topography of potential sites, so it was not possible all of a sudden to increase new installation sites that would have cost advantages. Liquid natural gas was also promising as a source of energy to replace oil, but gas is after all a fossil fuel and it was thought unlikely that depletion in the future could be avoided. At the time, nuclear power was viewed as the rising star in the energy field. In the early 1970s, MITI regarded nuclear power as the best option for extricating the country from the dependency on oil for its energy supply. At the time, many of the parties involved assumed that the age of nuclear power would finally replace oil-fired thermal power and, to some degree, the shift to nuclear power was already envisaged.[3] Japan had been constructing nuclear power stations since the late 1960s and six of them were in operation by fiscal year 1973. However, as is still the case today, there was already a mountain of unresolved issues relating to nuclear power including the issue of the final processing of radioactive waste, and measures to counter serious accidents. Thus, each method of generating power had its advantages, but since there were also many disadvantages, none of them held the trump card that could single-handedly resolve the energy problem. In this context, energy derived from natural resources, in particular solar energy, had many advantages when compared to other sources of energy: the site restrictions were not as great as that of hydroelectric power; unlike nuclear power, it was a clean energy that did not generate waste; and unlike natural gas there were no concerns about future depletion. Considered from this perspective, it was only natural that a few insightful researchers would focus on solar energy.

However, solar energy may seem wonderful at first glance, but that does not mean it is all plain sailing. The technical issues posed greatest difficulty. Even though the total amount of solar energy is great, it is difficult to make efficient use of the heat and light because it is dispersed wide and far across the planet. The technical standards of the day being what they were meant that solar energy was unable to compete with other methods of generating power in terms of cost. To turn solar energy into a source of energy with practical applications required technical innovation to efficiently convert the heat and light of the sun into electricity. However, it was difficult to make a profit, and the government anticipated that the technology would never reach the implementation stage if it were left to the efforts of the private sector. To reach implementation, it was necessary to continue to provide government funds and to coordinate policies to promote innovation. Since

[3]Kikkawa (2008, p. 22).

this was the situation at the time, MITI came up with the idea of developing solar energy under the auspices of a national project.

3.1.3 Energy Conservation Policies and Setting Up the Agency for Natural Resources and Energy

Alongside the long-term problem of rising oil consumption described above, another major short-term problem threatening the energy system in Japan in the early 1970s was shortages in the electricity supply. At the time, the supply of electricity came repeatedly under pressure during the summer periods. Having looked into the problem, the government introduced and enacted legislation to restrict the use of electricity as a specific countermeasure in April 1971. Developing new energy sources was not so simple due to the problems of pollution and site availability. Therefore, the government determined that taking legal steps to restrict the nation's electricity usage might be effective in the immediate term to avert the electricity shortages. Based on this line of thinking, the government prepared regulations to introduce limits in crisis situations. Fortunately, the summers of 1971 and 1972 were relatively cool, so Japan avoided actually putting these restrictions into action. However, the summer of 1973 was extremely hot and in August 1973 the government was finally forced to establish specific standards for electricity usage during the midday peak hours.[4] The governments' appeals for energy conservation covered the full gamut, from switching off lighting in window offices in government buildings, to opening and shutting household refrigerators. As a result, promoting nationwide efforts to conserve energy alleviated the energy shortage crisis.

Now, let us look at the changes in the organization of MITI at the time. The measures to counter the indications of energy problems in the early 1970s also manifested themselves in the form of restructuring at MITI. As it happened, this was also the time when MITI set up the Agency for Natural Resources and Energy, which was singlehandedly responsible for energy policy (July 1973). As a result, energy policies that had previously been the responsibility of other agencies or the Minister's Secretariat were now centralized within the MITI organization (Fig. 3.2). The preparations had started as early 1970 in anticipation of organizational restructuring at MITI. At the time, there was no denying that energy issues would grow increasingly important and MITI clearly recognized the need to create an organization that would become the control tower for energy policy.

The facts described above suggest that the Japanese government was already aware of the potential for a range of problems and had taken some countermeasures prior to the first oil crisis. The New Energy Development Organization (NEDO),

---

[4]MITI Editorial Committee on the History of Japan's Trade and Industry Policy (1991, pp. 126, 184).

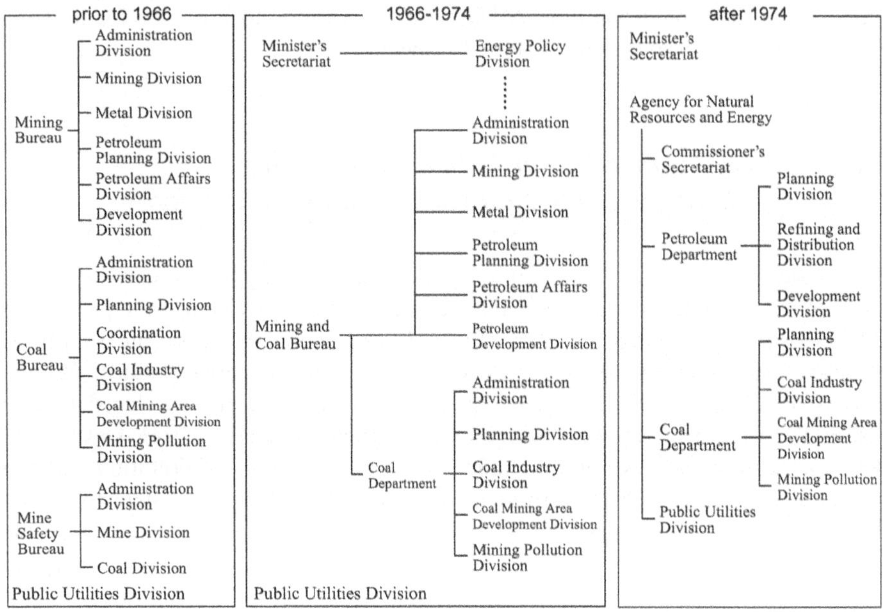

Fig. 3.2 Changes in the energy-related organization at MITI. *Note* Organizational structure prior to 1966 at left, 1966–1974 at center, and after 1974 at right. *Source* MITI shigenererugīchō (1993, pp. 225–226)

the agency that took charge of project management at the Sunshine Project at a later stage, explained the responses to the energy issue by the Japanese government at the time in a publication about the history of solar power generation at the organization:

> The Agency for Natural Resources and Energy, which was singlehandedly responsible for energy policy, was established in July 1973, shortly before the full-scale budget negotiations for the Sunshine Project got underway in the autumn. There were already opportunities within MITI to investigate the nature of future energy as indicated by a consultation with the Industrial Technology Council (currently, the Industrial Technology Committee at the Industrial Structure Council) initiated by Nakasone Yasuhiro, Minister of International Trade and Industry, on the question of how to promote the development of energy technology. Even earlier in the year, in May, the AIST deliberated the need to set up a new energy development section to manage long-term research and development, and the significant sums invested in energy development.[5]

A comprehensive evaluation of the situation described above suggests that by this time the Japanese government had prepared policy responses and organizational restructuring in order to respond to any energy problems that might occur in the future.

[5]NEDO BOOKS Editorial Committee (2007, p. 82).

In the spring of 1973, when researchers at national research institutes adopted energy-related themes and applied for research and development under the auspices of the Large-Scale Project System, the AIST decided to provide policy support for the research subject. The agency was in charge of technology policy at MITI at a time when tackling new energy research and development clearly agreed with perceptions at the ministry itself as described above. For MITI, the theme matched the government's policy of future energy security for Japan and, at the same time, they judged that a long-term project would increase the technical feasibility of successful development.

3.1.4 *The Emergence of Long-Term, Large-Scale Projects*

In the spring of the year of the oil crisis, several national research institutes submitted development proposals based on new energy topics to the Large-Scale Project System. MITI acknowledged the importance of these energy topics from both technical and social perspectives. Therefore, MITI decided not to leave the development of new energy to the efforts of private sector enterprise, but to use the Large-Scale Project System to support development and to give the state a central role in the research and development of the technologies.

However, the following issue came under scrutiny at MITI at the time. The conventional development period under the Large-Scale Project System was five years, but new energy research is technically demanding, and research and development is time-consuming. There were concerns that the project term would come to an end before any fundamental results were achieved in the five-year period. The decision-making process within MITI at the time is explained in the *History of Japan's Trade and Industry Policy*:

> Even though the Office for Research and Development at the AIST decided to adopt solar energy as a theme for the Large-Scale Project System based on the rising dependency on oil and the power shortages during the summers, it proved difficult to advance the development of solar energy within the budget framework and time restrictions of the Large-Scale Project System, and there were no agencies or organizations that attempted constructive work on this theme.[6]

To overcome this problem, MITI started to investigate a new system that would take a longer-term perspective on facilitating the implementation of technology development projects. Taking account of the technical difficulties of new energy research and development, the investigation led to a decision by MITI to pursue energy research through a scheme that was independent of the Large-Scale Project System. All new energy-related themes pursued at the affiliated national research institutes were centralized and promoted under the new plan.[7]

---

[6]Sawai Minoru and Tsūshō sangyō seisakushi hensan iinkai (2011, p. 246).

[7]Sawai Minoru and Tsūshō sangyō seisakushi hensan iinkai (2011, p. 246).

Research on solar energy and hydrogen energy had already been submitted to the Large-Scale Project System. At this point, it was decided to establish a four-pronged approach by adding the research on geothermal energy, and coal gasification and liquefaction, which was already underway at the national research institutes, to the development of new energy technologies, and to spin off these research themes from the Large-Scale Project System. At the same time, the implementation period for research and development was established with an ultra-long time frame extending to shortly before the start of the twenty-first century, far exceeding the five-year development period of the conventional Large-Scale Project System. NEDO described the circumstances as follows:

> Even today most projects in Japan are expected to show results in five years. The long-term project planning extending from 1974 to 2000 was also a first for MITI. This suggests that the issue was considered so important that it had to be realized even if it took a quarter century to do so. At the same time, the goal could not be achieved unless this much time was spent on it.[8]

Based on the national importance of developing new energy technologies and the high degree of technical difficulty, MITI advocated a new project that was separate from the Large-Scale Project System. This moment marked the emergence of the national project that came to be known as the Sunshine Project.

3.1.5 The Club of Rome Warning

When deciding on the overall image for the new system and the specific development themes, MITI zeroed in on something that would feel familiar to the nation. This is why the project was nicknamed the Sunshine Project.[9] It was the first ever case of assigning a nickname to the government's technology development projects in Japan. As a result, the project became associated with the bright image of the sun, conveying a strong impression to the nation of a project for developing new and clean energy technologies. Giving the project a nickname helped to increase its visibility through the subsequent media exposure.

As will be revealed later on, the fact that this project was already in existence at the time of the first oil crisis in the autumn of 1973 meant that government was able to tell the nation that it had worked out long-term measures for countering the oil crisis. Together with the Sunshine Project name, the MITI pamphlet featured an image of a large-scale solar power plant as the symbol of the project, illustrating in

[8]NEDO BOOKS Editorial Committee (2007, p. 80).

[9]MITI Kōgyō gijutsuin Kenkyū kaihatsukan shitsu (1987, p. 51).

an accessible and visual way the project vision of solar energy as an oil substitute that would support Japan in the future.[10]

For the theoretical foundation of the new energy development project, MITI also made effective use of the Club of Rome's *The Limits to Growth*, which had attracted a great deal of attention when it was published in Japan in May 1972.

Established in 1970, the Club of Rome is a private organization with the aim of implementing research and recommendations concerning the depletion of natural resources, environmental pollution due to contamination, the rising population in the developing world, and other predicaments facing humanity. *The Limits to Growth* was a report on the results of research conducted by the organization in collaboration with the Massachusetts Institute of Technology from 1970 to 1971. Published in May 1972, the book claimed that without measures such as resource recycling, and developing new products to reduce the use of resources, the world was heading for the "limits to growth" because of resource depletion due to the exponential growth of population and industry.[11] Seen from the present viewpoint, it was definitely a forerunner to the sustainability theory that argues for sustainable economic activity that does not harm the environment. The warnings about oil depletion by the Club of Rome provided the scientific arguments for new energy development, which would demand huge sums over the long term.

The following description is taken from the draft proposal for the Sunshine Project, first published in the name of the Office for Research and Development at the AIST at MITI:

> The Sunshine Project is a national project comparable to the Apollo Project. It is an ambitious national technology development project that will attempt to overcome the energy crisis caused by the depletion of oil resources, and to restore the blue planet with its bright sunshine by replacing the current oil-based energy system with a permanent green energy system by implementing solar energy, hydrogen energy, geothermal energy, and other non-polluting and inexhaustible energy supplies before the year 2000.[12]

In the project proposal, MITI cites the example of how the United States pulled together as a nation to realize the dream of traveling to the moon through the Apollo project, while declaring that the Sunshine Project should be a comparable national project. By means of this project, MITI flagged up the national goals and showed its resolve to get the nation committed to finding a solution to the energy problems, to return the blue planet to a state free of any environmental pollution by the twenty-first century, and to realize a stable energy situation where Japan was not excessively dependent on oil resources.

In August 1973, the Sunshine Project was finally made public. Nakasone Yasuhiro, Minister of International Trade and Industry, published a summary of the Sunshine Project as promoted by the AIST, and there were consultations at the

[10]MITI Kōgyō gijutsuin Sanshain keikaku suishin honbu [MITI AIST Sunshine Project Promotion Office] (1980). Frontispiece, construction site photographs, and conceptual drawings.

[11]Meadows et al. (1972).

[12]MITI Kōgyō gijutsuin Kenkyū kaihatsukan shitsu (1973).

Special Committee for Energy Technology at the Industrial Technology Council led by Dokō Toshio (Keidanren vice-president) about the nature of the new energy technology development, and how to advance the technology development.

On August 23, 1973, the *Asahi Shimbun* published a favorable introduction to the Sunshine Project, referring to "a national project for overcoming the energy crisis caused by the depletion of oil resources, a huge project that would require massive investment well over one trillion yen by the year 2000 to secure Japan's energy sources in the twenty-first century".[13] This is how the Sunshine Project was first presented to the nation.

3.1.6 The Energy Crisis Transformed into Reality

Announced in August 1973, the Sunshine Project suddenly began to attract attention around the world by the autumn of the same year. The reason was the first oil crisis in October 1973, which meant that the feared energy crisis had finally become reality. For oil-dependent Japan, the interruption in oil imports was a serious political and economic crisis for the state.

The clashes between Egypt, Syria, and Israel that had begun on October 1, 1973 developed into the fourth Arab–Israeli war when Iraq and Kuwait became involved, followed by Saudi Arabia in the middle of October. In the name of the "oil war", Arab nations seized the opportunity to raise the price of crude and to suspend oil supplies to countries supporting the United States and Israel. As a country allied with the United States, Japan was, of course, also subject to the sanctions. In January 1974, the United States invited the European countries and Japan to a meeting of the oil-consuming countries to press for a response. The price of crude had quadrupled between October 1973 and January 1974. In Japan, the crisis caused people to panic and to start hoarding toilet paper and detergent as early as November 1973.[14]

In the context, the existence of the government's draft proposal for the Sunshine Project certainly threw out a lifeline in the eyes of a nation caught up in the oil crisis.

This is also apparent from the media coverage. For example, on October 16, 1973, immediately after the outbreak of the fourth Arab–Israeli war, the *Nihon Keizai Shimbun* ran an article introducing all the particulars of the project with the headline "Rising Expectations for the Sunshine Project". The following illustrates the tone of expectation in the article:

---

[13] *Asahi Shimbun*, August 23, 1973.

[14] Kishida Fumitake, "Kiki o norikoete [Overcoming the Crisis]," in *Shōgen daiichiji sekiyu kiki: Kiki wa sairaisuruka?* [*Testimony from the First Oil Crisis: Will There be Another Crisis?*], ed. Denki Shimbun (Tokyo: Nihon Denki Kyōkai Shimbunbu [The Newspaper Division of the Japan Electric Association], 1991), p. 138.

If the new energy is implemented, we will no longer depend on overseas resources. In short, the project is not limited to achieving technical success; it is about breaking away from the status of a country without resources and taking our position among the countries that possess resources. For Japan, in particular, it means that we can rid ourselves of the problems associated with the "have-not nation" dating back to the Meiji period.[15]

The Sunshine Project was highly acclaimed in the media for having the foresight to set goals that would contribute to overcoming the energy problem, not just developing a single technology as had been the case in the past.

As described above, the whole nation was soon gazing fervently at the Sunshine Project due to the outbreak of the Arab–Israeli war in the autumn of 1973 and the associated first oil crisis. After the program launch in the following year (1974), staff at the Office for Developing the Sunshine Project at the AIST reflected on the start-up period for the project, describing the passionate responses around the world as follows:

Everyone involved with the Sunshine Project was deeply moved by the international interest in the project. Since last summer [1973] when the concept for launching the Sunshine Project in fiscal 1974 was put together, we have had a succession of inquiries about the Sunshine Project from elementary school students and university professors, and from every circle of society including corporate entrepreneurs. Quite a number of enthusiasts have even taken the trouble to come to MITI to listen to the explanations of the project leaders.[16]

We can infer from the comments by project staff that citizens from every reach of society were interested in the Sunshine Project at the time, and were looking forward to its success.

On December 18, 1973, the Industrial Technology Council published the conclusions of the investigations, underway since August in response to a consultation by Minister Nakasone, in a report about how to promote the development of new energy technology.[17]

The findings made specific recommendations for several important issues including (1) the urgency of new energy technology development; (2) the launch of new energy technology development programs; (3) the basic concepts of formulating and implementing projects; (4) specific ways of promoting the project; (5) setting up a semi-governmental corporation as the core institution; and (6) establishing and maintaining the implementation structure. In the report, experts in each field verified each theme in the draft project based on their own expertise, adding detailed investigations of the project feasibility.

---

[15]*The Nikkei/Nihon Keizai Shimbun*, October 16, 1973.

[16]MITI Kōgyō gijutsuin Sanshain keikaku suishin honbu (1974, p. 3).

[17]The report starts with a grand statement: "Energy is the source of activities in human society. Energy is indispensable for improving the quality of life and developing industrial activity. The ceaseless development of human society has been built together with a constant increase in energy consumption since Prometheus brought fire." (Sangyō gijutsu shingikai, "Shin'enerugī gijutsu kaihatsu no susumekata ni tsuite," in *Shin'enerugī gijutsu kenkyū kaihatsu keikaku (Sanshain keikaku)*, ed. MITI Kōgyō gijutsuin (Tokyo: Nihon Sangyō Gijutsu Shinkō Kyōkai, 1974), p. 372.

The negotiations to secure the budget for the Sunshine Project were also held in the autumn and winter of 1973. This was, of course, at the height of the oil crisis. Reflecting the crisis situation, the budget request for the MITI project proposal was granted nearly in full. The Ministry of Finance, which controlled the allocation of the national budget, recognized the claims made by MITI and approved the launch of the project.

In the end, the entire project budget of 2.27 billion yen was approved in the general account budget for fiscal year 1974. The breakdown was 873 million yen for solar power, 560 million yen for geothermal energy, 263 million yen for coal (synthetic natural gas), and 332 million yen for hydrogen power. (The balance was allocated to general research etc.) Another 172 million yen was added to coal from the Special Accounts for Coal and Petroleum to reach a total of 435 million yen. Consequently, when the general accounts and the special accounts are combined, the budget for the Sunshine Project in its first year was 2.442 billion yen.[18] This is how the Sunshine Project obtained backing for its budget.

MITI set about designing the project organization for implementing the program. In keeping with the findings of the December 1973 report, the Sunshine Project Preliminary Office was set up at the AIST in February 1974. In March, the Guidelines for the Sunshine Project to advance specific plans were agreed by ministerial decision. As of this time, the AIST cooperated with the national research institutes, research and development commissions were launched, and research and development themes were assigned to corporations. In April, the Preliminary Office was renamed the Promotion Office and the work to formulate basic polices and implementation plans in more detail started in July. By August 1974, the large-scale technology research and development project kicked off even as expectations were rising across the board.

So, how did corporations in the private sector respond to the launch of this national project?

Corporations showed a strong interest in the Sunshine Project from the start and many corporations put themselves forward for commissions. For the corporations, the national project was attractive from the viewpoint of improving technological strength.

In its policy for the Sunshine Project, MITI had stated that energy was an issue for the state and that they did not intend to burden the private sector with risk.[19] Therefore, corporations were on the whole positive and motivated to take on commissions, thinking that by participating in the Sunshine Project, they would gain major advantages from technology development without taking on any risk. In fiscal year 1978 the total budget for the Sunshine Project was approximately 8.1 billion yen whereas the total development cost related to new energy at major corporations had risen to 15 billion yen at Hitachi, 12 billion yen at Toshiba, and 12

[18]MITI Kōgyō gijutsuin Sanshain keikaku suishin honbu [MITI AIST, Sunshine Project Promotion Office] (1984, p. 12).

[19]*The Nikkei/Nihon Keizai Shimbun*, January 25, 1974.

billion yen at Mitsubishi Electric.[20] The government's attitude and ambition to develop new energy encouraged private sector corporations to voluntarily and proactively invest in new energy development.

3.2 The Start of the Solar Energy Project

3.2.1 Solar Thermal and Photovoltaic Energy: A Two-Pronged Strategy

Below, we will take a look at the actual circumstances of solar energy research in the Sunshine Project from the government's perspective.

As already mentioned, when the project was launched, the main target technologies were solar, geothermal, coal, and hydrogen. As reflected by the Sunshine Project name, solar energy became the symbol of the project.

There are two approaches to solar energy research—heat utilization and light utilization. Today we are so familiar with photovoltaic power generation that it may seem surprising that when development of the technologies started in the 1970s, the overriding emphasis was on generating power from the heat of the sun. Featured in pamphlets and other materials, illustrations of giant solar power plants in the cities of the future became the face of the Sunshine Project.[21]

Naturally, as part of solar energy research, solar cells were also targeted at the start of the project based on their applications in space power generation and other future technologies, but the budget scale was quite small in the 1970s, and the amounts were insignificant compared to those for solar thermal power. The project targeted solar thermal power generation because of the low degree of technical difficulty while large-scale photovoltaic plants, like the recent Mega-Solar Plant, were goals for the distant future. Hitachi and Toshiba bid to develop the technologies for solar thermal energy at the core of the Sunshine Project. In the end, Toshiba decided to withdraw and the commissions to develop the technologies for solar thermal energy plants were awarded to Mitsubishi Heavy Industries and Hitachi. These two companies constructed large-scale power plants.

At the outset, the research on photovoltaic power maintained a low profile with all the interest focused on solar thermal energy. The initial goal was to improve the performance of solar cells and to bring down the price. There were no high expectations of photovoltaic power generation. A NEDO publication describes the circumstances at the time:

> Due to the oil shock, there was an upsurge in public opinion and with research funds expanding even photovoltaic power was given some funding despite the lack of

[20]Imai (1982).

[21]MITI Kōgyō gijutsuin Sanshain keikaku suishin honbu (1980). Frontispiece, construction site photographs, and conceptual drawings.

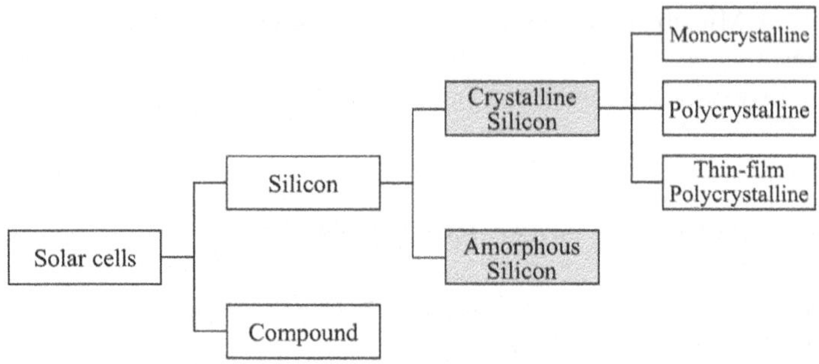

Fig. 3.3 Types of solar cells. *Source* Sharp (1996, p. 64)

expectations. Nobody anticipated that a quarter century later Japanese photovoltaic technology would be unrivalled in the world.[22]

This description suggests that there were no high expectations of photovoltaic power when the Sunshine Project was launched.

The following is a brief outline of the solar cell structures at the heart of photovoltaic power generation. Solar cells are classified by manufacturing technology and divided into the cells that use silicon and the ones that use chemical compounds. The former are further divided into solar cells that use crystalline silicon and solar cells that use amorphous (non-crystalline) silicon. In 1974, when the Sunshine Project started, the assumption was that crystalline silicon would be the winning technology for solar cells. By the 1980s, the Sunshine Project also incorporated amorphous silicon and the two camps started to compete with each other on development. Each manufacturer of solar cells was either developing projects based on crystalline silicon technologies, or amorphous technologies, and they competed with each other within the Project to demonstrate the superiority of their own manufacturing technology. Figure 3.3 outlines the main categories of manufacturing technologies for solar cells. However, when the Sunshine Project was launched, amorphous solar cells were not targeted.

Next, we will briefly review the history of the development of solar cell technologies. It is possible to trace the source of the technology for solar cells back to 1876 when the selenium solar cell was unveiled, but the prototype for the solar cells that are generally the subject of research and development these days was announced by Gerald Pearson, Daryl Chapin, and other researchers at Bell Laboratories in the United States in 1954. These solar cells were based on the principle of generating electricity from light by mixing boron and phosphorus with silicon to create P-type and N-type semiconductors (PN control) and joining them together (PN junction). When light strikes the semiconductor, the bond between the nucleus and electrons is weakened creating pairs of electrons and holes. Since the P-type semiconductor attracts holes and the N-type semiconductor attracts electrons, a force that shapes the

[22]NEDO BOOKS Editorial Committee (2007, p. 81).

movement of the electron is generated where the surfaces join, making it possible to produce electricity if conductors are connected to both poles. That is, the principle of the crystalline silicon solar cell was essentially an extension of semiconductor technology, which meant that corporations in possession of semiconductor technologies potentially had the leeway to enter the market.

For a time, Japanese corporations also flocked to solar cell research because of Pearson's announcement. As of 1959, Sharp started to develop single crystalline silicon solar cells using this principle. In the same year, NEC supplied solar cells for unmanned lighthouses to the Ikadase lighthouse in Suōnada at the request of the Maritime Safety Agency.

As described above, the technical principles of crystalline silicon solar cells were clarified at an early stage, and it was entirely feasible for corporations to manufacture the solar cells. However, silicon was an expensive raw material and it was not easy to produce something that was economically viable. Therefore, thoroughly reducing the cost of the manufacturing process was considered essential before it would be possible to disseminate practical solar cells on a large scale. Since crystalline silicon was very expensive at the time, the cost of generating electricity from solar cells was estimated at several tens of thousands of yen per watt. Practical applications were developed for manmade satellites, lighthouses, wireless relay stations in remote areas, or other unique usages where the high cost was of no consequence, but even high-end estimates put the market scale in the tens of millions of yen at the most. Despite expectations of future growth in the market for solar cells, it was generally considered that it would be a long time coming. In fact, before the Sunshine Project, the market was still so small that Sharp had it to themselves with no other serious contenders entering the market.

3.2.2 *Taking Advantage of Specialized Technologies at Corporations*

To decide what research to commission from which solar cell manufacturer, the AIST and the Electrotechnical Laboratory (ETL), which was central to the process, listened to the proposals and opinions of engineers at each company. The goal of solar cell development was to be able to compete with existing power sources by reducing the cost to one hundredth of the level at the time. Since the special characteristics of the market for solar cells meant that it was still unclear what the price per watt would be if solar cells were mass produced, ELT requested an estimate of production costs from NEC when setting the value. After accepting the request, NEC calculated a figure of 20–30,000 yen per watt, resulting in a specific reduction goal of 1/100 to bring the cost down to the level of 300 yen per watt. It was not difficult to imagine that the cost could be reduced by improving the process, as the figure of 20–30,000 yen was due to the limited market scale and the single-batch production process that simply diverted transistor and IC technologies.

However, none of the company engineers thought it would be easy to achieve the 1/100 goal. It was necessary to once again carry out a fundamental review of the solar cell manufacturing technologies to achieve the goal.[23] In order to deliver the goal of reducing the cost of solar cells, the AIST cooperated with ELT and listened to the company engineers to classify the processes for solar cell development as outlined below, and selected companies to take charge of each process.

The biggest issue for the development of solar cells under the Sunshine Project was to establish technologies to manufacture the silicon raw material at an affordable price. The price of solar cells would fall if some new innovative manufacturing technologies were developed to improve the cost of manufacturing crystalline silicon. This is why the leading large-scale electronics manufacturers in Japan mobilized to improve the techniques for manufacturing silicon and the modules using silicon. The six participating corporations were Hitachi, NEC, Toshiba, Matsushita Electric Industrial, Sharp, and Toyo Silicon (which changed its trade name to Japan Silicon in 1978, and is now SUMCO). Hitachi, NEC, and Toshiba, in particular, were the leading semiconductor manufacturers, and it was expected they would leverage their outstanding technical skills to produce results in solar cell research and development.

To be able to implement integrated technology research and development across the board, the AIST allocated each area of research and development among these electronics manufacturers. The agency targeted the technologies that caused cost-related bottlenecks in solar cell production, and created a structure where each company was put in charge of its own area of expertise. First of all, the agency identified the solar cell manufacturing technologies where breakthroughs were necessary, and then it assigned the tasks to the electronics manufacturers, taking account of their specialization and technology expertise.

As already mentioned, the high cost of manufacturing solar cells was mainly due to the steep price of the high-purity silicon raw material. The main manufacturing method for semiconductors was the Czochralski method (CZ), which was used to form expensive mono-crystalline silicon substrates of an excessively high quality for solar cells. Consequently, for the Sunshine Project, the key issue was to reduce the amount of silicon used to cut overall manufacturing costs.

The ribbon crystal technique and thin film polycrystalline, a more advanced technology, were developed in parallel to reduce the amount of silicon. Ribbon crystal is a technique for pulling thin ribbons from crystalline silicon, while thin-film polycrystalline is a technique for forming thin films of silicon. If either technology were successfully developed, it would be possible to greatly reduce the amount of silicon.

NEDO explained the research and development of silicon manufacturing techniques at the time:

> The silicon used for solar cells had a high degree of purity and was expensive. To lower cost, it was important to somehow reduce the use of such expensive materials. Therefore,

[23] NEDO BOOKS Editorial Committee (2007, p. 82).

> technologies to splice the silicon as thinly as possible were developed. Another idea was to melt the silicon and to spread it on thin sheets to make sure nothing was wasted. Previously, blocks of silicon had been cut thinly with a saw, but the scraps went to waste. If the silicon was melted down, nothing would be wasted.[24]

In addition to silicon manufacturing techniques, other research looked at process simplification and automation for solar cells, or developing basic technologies for module manufacturing in order to reduce the costs incurred in the assembly process. Whether the raw material was silicon or the chemical compounds discussed below, the idea was that they could be used with any basic technology.

At the time, silicon was the most promising technique for manufacturing solar cells, but there was also another method of producing solar cells with compound semiconductors without using silicon. Since the manufacturing process was both simple and cheap, it was developed concurrently with the project. The decision was taken to commission research into basic technologies for manufacturing group II–VI compound semiconductor solar cells using chemical compounds, and other new styles of solar cells, and to develop evaluation technologies.

When selecting corporations, the Agency of Industrial Science and Development brought a global perspective to the project, assigning each company responsibility for a particular technology. Finally, after the deliberations in the spring of 1974, the Agency assigned (1) ribbon crystals to Toshiba and Toyo Silicon; (2) thin-film polycrystalline to Hitachi and NEC; (3) module manufacturing technology to Sharp; and (4) compound semiconductor solar cells to Matsushita. This was the sequence of events up to the start of commissioned research aiming to cut manufacturing costs to one hundredth of what they were at the time.

3.2.3 The Outcome of the Photovoltaic Energy Project of the 1970s

In the late 1970s, the scale of solar cell development was very small compared to what it would become. Compared to the 1980s, the budget scale was also extremely small, and development did not go beyond basic research carried out at ELT and corporate research institutes. However, despite starting from an extremely low technical standard, the research enjoyed continuous financial support from the government and, by the end of the 1970s, one important outcome after another emerged from the basic technologies. In this period, six companies were assigned themes that were hardly changed at all with each company continuing to research and develop the technology themes until the first review period from 1979 to fiscal year 1980.

Toshiba took on the challenge of manufacturing silicon using the ribbon crystal technique. There are two approaches to the ribbon crystal technique, the vertical

[24]NEDO BOOKS Editorial Committee (2007, p. 83).

pull and the horizontal pull. The two techniques were developed concurrently for the Sunshine Project with Toshiba taking charge of the vertical pull technique and Toyo Silicon handling the horizontal pull.

As a result of the research on pulling vertical strips of silicon ribbon crystal, Toshiba successfully pulled 2-cm wide strips in fiscal year 1974 and 3-cm wide strips in fiscal 1975 to achieve a conversion efficiency rate of 11%.[25] The horizontal pulling technique developed by Toyo Silicon had the advantage of high speed, achieving 20 cm strips in fiscal year 1974 and 40 cm strips in fiscal year 1975. But, whatever the technique, the crystal substrates were extremely rough and the problems of achieving practical applications remained unresolved.

Hitachi and NEC tackled thin-film polycrystalline, which is another method of reducing the amount of silicon used. Here as well, there were two approaches: Hitachi handled non-accelerated particle growth while NEC was in charge of accelerated particle growth. In fiscal year 1976, Hitachi achieved a conversion efficiency rate of 7.3% for polycrystalline silicon substrates. In the same fiscal year, NEC obtained a rate of 4.4% for polycrystalline silicon substrates. The two companies used different approaches to compete with each other.

As of fiscal year 1975, Matsushita Electric Industrial selected cadmium sulfide (CdS) and cadmium telluride (CdTe) for its compound semiconductor solar cells, and posted success with a method of screen printing and annealing in fiscal year 1976. Sharp developed new types of solar cells, producing good results by developing technologies for spin coating, and electrode formation by printing and annealing.

The efforts of each company were evaluated at fixed periods. At the end of the 1970s and based on the outcome of the basic research, the Technical Committee on Solar Power System Technologies established at NEC decided on the research and development approaches for the next stage starting from 1980. Based on the outcomes of technology development, the committee selected which themes to transition to the implementation stages and determined the directions of development.[26]

[25]The description below references MITI Kōgyō gijutsuin Sanshain keikaku suishin honbu (1984).

[26]The specifics of the evaluation are summarized below. As a result of comparing the outcomes of vertical pull and horizontal pull for ribbon crystal, the vertical method was transitioned to implementation research because of its exceptional performance in terms of width, thickness, and control, while the horizontal method was discontinued. The verdict was that the Toshiba method was superior. At first, NEC worked on research and development of thin-film polycrystalline, but the method did not deliver the expected results. However, the researchers realized that the method of casting silicon substrates that they had developed for the research theme could be used for low-cost solar cells. Therefore, together with ribbon crystal, the method of using slices of polycrystalline silicon ingots for solar cells was advanced to the implementation research stage instead of thin-film polycrystalline. Meanwhile, junction formation using the ion implantation method developed at Hitachi was advanced to the implementation research stages as dry PN junction formation method due to its high-speed and regulatory properties. Since the junction and anti-reflection co-firing technology based on the coating method developed by Sharp combined with the electrode penetration firing technology made it possible to shorten the number of processes, it was advanced to implementation research in the guise of the wet PN junction formation method. MITI also continued the applied research of modules at Sharp and the Group II–VI

Of course, not all the outcomes expected of the development themes materialized in the end, but the expectation was for the new technologies and processes developed en route to link to breakthroughs in some other form.

Here it is important to note that the AIST had commissioned different companies to develop multiple promising methods relating to important technologies for the solar cell project. By doing so, the agency created a forum where several corporations competed with each other on developing improved technologies. In a government-led national project, it is important that corporations both cooperate and compete with each other. For example, several corporations collaborated and coordinated their technology processes in bids to meet the final target of the whole project: to bring down manufacturing costs to one hundredth of what they had been. On the other hand, there was also a system of competing on outcomes where several companies tackled research and development of different technologies when exploring optimum technologies for achieving the overall target. To generate innovation in a national project, it is essential to implement a style of management that delineates cooperation and competition in an appropriate manner.

Based on the outcomes achieved by the late 1970s under the Sunshine Project, research and development of crystalline solar cells entered the second stage, aiming for full-fledged implementation under NEDO, which was set up in 1980. The individual technologies were combined in pursuit of new technologies for mass production of low-cost solar cells under the direction of the new organization. NEDO was, as it were, set up to implement the policies formulated by the AIST. What led to the decision to set up NEDO? The second oil shock in 1979 and the ensuing strategy of accelerating the Sunshine Project provided a big boost. So, how did it all come about?

3.3 Establishing NEDO and Accelerating Plans Due to the Second Oil Shock

3.3.1 The Need for a Project Implementation Unit

The second oil shock was triggered when strike action by oil workers at the end of 1978 stopped oil exports from Iran, and the production of crude oil was interrupted in the wake of the Iranian Revolution. Since Japan imported large quantities of oil from Iran, the general expectation was that Japan would be heavily impacted. The energy issue was even the main topic of discussion on the agenda at the Tokyo Summit in June 1979. The development of new energy technologies was suddenly back in the limelight as an abnormal situation repeated itself.

---

compound semiconductors at Matsushita (MITI Kōgyō gijutsuin Sanshain keikaku suishin honbu (1984, pp. 70–72).

Table 3.1 Development targets in the acceleration strategy

	Project	Acceleration targets
Priority project	Coal liquefaction	Aim for early establishment of implementation technologies, strive to accelerate development steps, develop large-scale pilot plants and demo plants with the goal of providing the oil equivalent of at least 1500 kL/day (approx. 290,000 barrels/day) by 1990
	Geothermal	Aim for early establishment of probing, extraction technologies to facilitate large-scale, wide-area deep geothermal development with the goal of providing the oil equivalent of at least seven million kiloliters by 1990
	Solar	Aim to expand measures to promote solar housing, early development of amorphous elements and materials, and the early establishment of production technologies to rapidly expand solar cell implementation with the goal of providing the oil equivalent of at least seven million kiloliters by 1990
Overall targets		Expect to play a large part in strengthening Japan's energy supply structure, contribute to developing alternative energies, and provide at least 5% of total energy (1.6% in current development timetable) from the Sunshine Project in 1990 based on the above-mentioned acceleration targets

Source MITI Kōgyō gijutsuin Sanshain keikaku suishin honbu, ed., *Sanshain keikaku no kasokuteki suishin senryaku: Sangyō gijutsu shingikai shin enerugī gijutsu kaihatsu bukai chūkan hōkoku o chūshin toshite* [Acceleratory Promotion Strategies for the Sunshine Project: Based Primarily on an Interim Report by the Industrial Technology Advisory Committee's New Energy Technology Development Subcommittee] (Tokyo: Tsūsan Seisaku Kōhōsha, 1980, p. 16)

In August 1979, the Supply and Demand Subcommittee under the Advisory Committee for Energy once again stressed the importance of new energy in its interim report, *Long-Term Energy Demand and Supply Outlook*, hurriedly and substantially raising the supply targets.

Shortly after, in November 1979, the Industrial Technology Council recommended a strategy of accelerating the Sunshine Project. Referring to the above-mentioned *Long-Term Energy Demand and Supply Outlook*, the recommendations redefined the targets, aiming for early implementation of the Sunshine Project (Table 3.1). The recommendations advocated the need for stronger government support to increase the scale of the project targets and to achieve them at an early stage.

In accordance with these government policies, the organizational and budget preparations related to the development of new energy technologies were hastily put in place. The Act on the Promotion of Development and Introduction of Alternative Energy (the Alternative Energy Act), based on the acceleration policy and enacted in May 1980, was especially important in this phase. The Act stipulated the establishment of a new organization to promote the Sunshine Project and, in October 1980, NEDO was set up on this basis. As of this time, NEDO handled all project management functions for the Sunshine Project. Almost simultaneously, at the end of September, the New Energy Foundation was set up with the aim of studying and researching the development and utilization of new energy. Thus, the second oil shock prompted improvements in the organizational structures for new energy development.

When the Sunshine Project was launched, the Sunshine Project Promotion Office at the AIST coordinated directly with the corporations. However, in 1977 the Electric Power Development Company set up a Sunshine Project head office separately from the agency, taking over responsibility for the management of the implementation stages. Based on the 1952 Electric Power Development Promotion Law, the Electric Power Development Company is a special company funded by the government and set up to develop sources of electric power and to improve the facilities for power transmission. The development of power sources supplemented work related to new energy at the company. However, due to the launch of NEDO, the Sunshine Project head office at the Electric Power Development Company was closed down at the end of September 1980 and the services outsourced by the AIST were gradually transferred to NEDO. Later, NEDO entrusted the Electric Power Development Company with the projects that had been continually operated within the company. By this time, the scale of the Sunshine Project had grown too large for a supplementary project at the Electric Power Development Company.

NEDO participated in the implementation stages of the majority of the Sunshine Project themes. These included large-scale and important projects such as technology development for coal liquefaction and photovoltaic power systems, substantive surveys on environmental protection at large-scale deep geothermal power stations, and new battery storage systems under the Moonlight Project (a project to develop technologies for energy conservation). Under the supervision of the AIST, project management at the Sunshine Project was left to NEDO, which enjoyed a degree of autonomy.

Below, we will look at the incipient organization of NEDO. Established as the body implementing the Sunshine Project, NEDO was a semi-governmental corporation under MITI. At its launch, NEDO was capitalized at 135.8 billion yen (including private sector contributions of 450 million yen), had 327 employees, and its offices were located in the eponymous Sunshine 60 building in Ikebukuro.[27] Later, as the project expanded, the scale of the organization also increased significantly.[28]

---

[27]Asahide (later renamed Hinodemachi), and other place names in the vicinity of the Sunshine 60 building were originally associated with the sun.

[28]Current program content has grown substantially from (1) developing and promoting new energy/energy conservation, and (2) coal mining structural adjustment programs at the time when NEDO was established to include (3) alcohol production (from 1987), (4) industrial technology research and development (from 1990), and (5) coal mine damage compensation (from 1996).

The legal basis for the semi-governmental corporation are found in (1) Act on the Promotion of Development and Introduction of Alternative Energy (Act No. 71 of 1980), (2) Act on the Rational Use of Energy (Act No. 49 of 1979), (3) Act on Special Measures for the Promotion of New Energy Use, etc. (Act No. 37 of 1997), (4) Act on the Improvement of Research and Development Systems for Industrial Technology (Act No. 33 of 1988), (5) Act on the Promotion of Research, Development and Dissemination of Social Welfare Equipment (Act No. 38 of 1993), (6) Act on

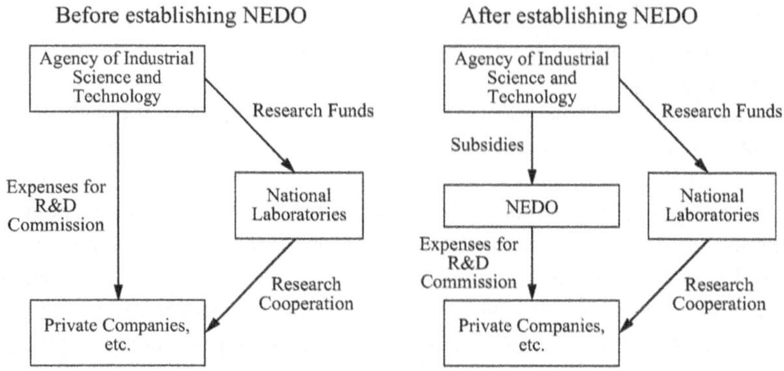

Fig. 3.4 New energy technology development structure. *Note* The agency of industrial technology and development set aside a sum toward the budget for NEDO at the time of the launch. *Source* Compiled by the author based on interviews with the parties involved

Work that had previously been handled by the Office for Developing the Sunshine Project at the AIST was transferred to NEDO, which was now responsible for managing research and development in the project implementation stages. Under the Sunshine Project technology development structure, corporations and the AIST had been in direct contact with each other. After the launch of NEDO, the relationship was mediated by NEDO. The AIST concentrated on formulating the overall project, while NEDO, in its capacity as the body implementing the project, explored themes and commissioned work from universities, national research institutes, and research departments at private sector corporations based on plans formulated by the Agency (Fig. 3.4).

After the launch of NEDO, new energy technology development was transferred from the Sunshine Project Promotion Office to NEDO. Separate departments to develop solar, geothermal, or coal technologies were established within the NEDO organization, bringing together experts in the relevant technologies from industry, government, and academia. For example, at the launch, there were 11 staff members in the department for developing solar technology, including three who had joined from the Electrotechnical Laboratory, one from MITI, three from Electric Power Development Company, two from power companies, and two from other private sector organizations.[29] The organization was set up in such a way that human

Temporary Measures Concerning the Structural Adjustment of the Coal Mining Industry (Act No. 156 of 1955), (7) Act on Temporary Measures concerning the Compensation, etc. for Coal Mine Damage (Act No. 97 of 1963), (8) Temporary Act on Coal Damage Recovery (Act No. 295 of 1952), and (9) Alcohol Sales Act (Act No. 32 of 1937).

[29]Photovoltaic Power Generation Technology Research Association, *PVTEC* (1996, pp. 111–113). Strictly speaking, the ETL researchers were also affiliated with MITI, but unless it presents a problem, we consider researchers at national research institutes to be employees of a separate organization than ministry staff because they were expected to fulfill substantially different functions.

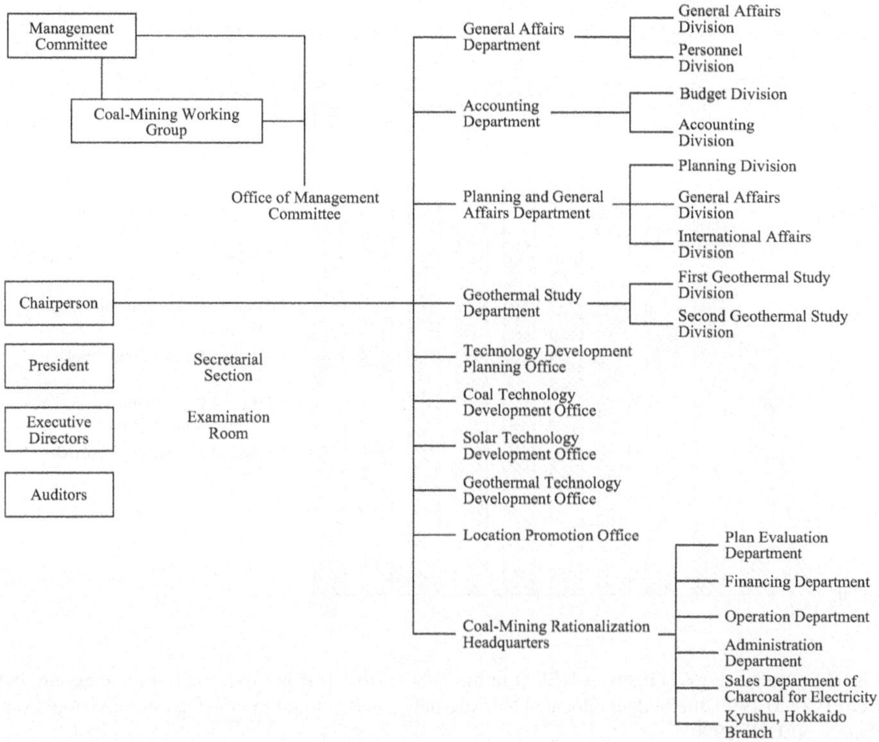

Fig. 3.5 Organizational Diagram of NEDO at the launch (October 1980). *Source NEDO News*, January 1981, p. 5

resources with varied backgrounds consulted with each other to develop the technologies. Figure 3.5 presents a diagram of the NEDO organization at the time it was established.

As already mentioned, the bulk of the budget for NEDO was provided in the form of special accounts based on the Alternative Energy Act. For example, in the draft budget for fiscal year 1983, the sum total of the energy-related special account was 776.4 billion yen with petroleum taxes accounting for 437.2 billion yen (56.3%), the tax on promoting power resources development for 195 billion yen (25.1%), and tariffs on crude and heavy oil for 144.2 billion yen (18.6%) when broken down by source. In short, they were the taxes paid by the nation when using petroleum or electricity, or importing crude oil and heavy oil. In terms of usage, 391.5 billion yen (50.4%) was assigned to petroleum measures, 70.5 billion yen (9.1%) to location provisions, and 314.4 billion yen (40.5%) was spent on

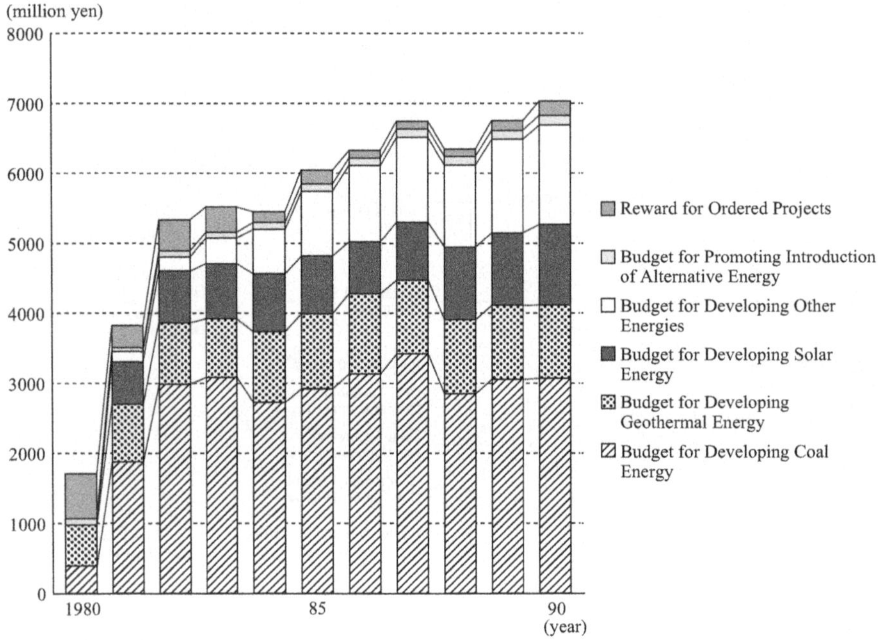

Fig. 3.6 New energy budgets at NEDO in the 1980s. *Note* It is not indicated in the diagram, but from 1988 to 1990, the budget allocated an extremely small amount to information provision costs. *Source* NEDO (1990)

alternative energy measures and coal measures, including 75.2 billion yen (9.7% of the whole) earmarked for NEDO.[30]

In terms of the NEDO budget, the new energy and coal rationalization budgets came out of the special account, while alcohol production (included in the other related business category) was treated as a special case. Consequently, the fiscal year 1983 budget included 51.7 billion yen for new energy, 23.5 billion yen for coal rationalization, and 23.5 billion yen for alcohol production totaling 93.8 billion yen. Figure 3.6 presents the fluctuations in the amounts earmarked for new energy budgets.

The total budget for the Sunshine Project and the Moonlight Project in fiscal year 1983 was 51.6 billion yen, which included 39.1 billion yen (75.8%) allocated to NEDO. That is, three quarters of the budgets for these projects were subsidies paid to NEDO, which meant, in effect, that NEDO was responsible for three quarters of the Sunshine Project and the Moonlight Project. This figure increased every year with the proportion allocated to NEDO reaching 94.6% of the Sunshine Project, and 87.1% of the Moonlight Project in fiscal year 1988.[31]

[30]*NEDO News*, March 1983, p. 9.

[31]*NEDO News*, March 1988, p. 9.

The expectation was that NEDO would act as the core mediator between the Office for Developing the Sunshine Project at the Agency for Industrial Science and Technology, and private corporations, universities, and national research institutes.

The organizational structure of the Sunshine Project was substantially upgraded when the second oil crisis caused the program to be accelerated, leading to the launch of NEDO in 1980. The new energy development project had now entered the next phase.

3.3.2 Testing Solar and Photovoltaic Plants

New energy technology development in the early 1980s was characterized by the construction of demonstration plants for test running solar thermal energy or coal gasification and liquefaction programs. These demonstrations provided the basic data for future construction of over-sized power plants as laid out in the plans for the Sunshine Project. In the early 1980s, the regulatory route favored the construction of such centrally located large-scale power plants.

A large-scale pilot plant to generate power from solar thermal energy was built at this time. In March 1981, a dual-system 1000 kW class solar thermal power plant was completed in Nio Town (currently Mitoyo City), Kagawa Prefecture, with power generation tests commencing in August the same year. By September, the plant had successfully generated power using both the central receiver tower system, and the plane-parabola system. The systems had been under development at Hitachi and Mitsubishi Heavy Industries since they were commissioned to carry out the research in 1974 with Hitachi taking charge of the plane-parabola system and Mitsubishi Heavy Industries handling the system of a central receiver tower. It was often the case under the Sunshine Project that plants fulfilling nearly identical functions were developed with systems using different technologies. These plants continued to run test operations over a period of three years. During this period, plants also carried out demonstrations using heat, light, or heat-light hybrid systems to test the potential for highly efficient power plants.

Expectations for generating solar thermal power at the plant were high but, unfortunately, the plant was subsequently unable to produce the expected results. As early as the summer of 1982, halfway through the test period, coverage of the outcome was pessimistic. The plant was confronted with unfavorable conditions due to fog preventing it from getting the insolation required to generate power, and other adverse weather conditions during the test period. Therefore, the amount of power generated only reached about one third of the original forecast.[32]

As it gradually became clear that the experiment had failed, the question of how to use the solar power generation facilities came under review with early demolition on the cards in February 1983, about one year before the scheduled end of the

[32]Regional Economy pages, *The Nikkei/Nihon Keizai Shimbun*, August 21, 1982.

three-year test period. The town of Nio Town hoped to keep the power generation facilities, but it was impossible for the town to pay the fees associated with the sale, or the annual operating costs of as much as 800 million yen. A plan was put forward to leave at least the tower standing as a monument but, in July, a decision was taken to abandon all efforts to keep the facilities because the cost of maintenance could not be covered by the research budget, and other circumstances.[33] As an alternate plan, Nio Town studied a proposal to establish a regional facility for energy development in the form of the Kagawa Seibu Denen Technopolis on the site of the solar power generation plant. Test operations at the plant finished in March 1984 and, in September, the AIST commended Nio Townfor services rendered by cooperating with the Sunshine Project. As a result of this sequence of events, solar power generation returned to the stage of basic research, and the odds-on favorite for implementing solar energy technologies shifted from solar thermal energy to photovoltaic power generation.

Research and development of photovoltaic power generation began in earnest around the same time. A two-stage strategy was implemented whereby an integrated production plant for solar cells was built to facilitate automated production, and then a test plant for photovoltaic power generation was set up using the solar cells produced in the first stage.

By the end of the 1970s, the major manufacturers in the Kanto area had already played a central role in the advancement of crystalline solar cell development and, using the findings of the basic research, had started to build an automated integrated production plant. The purpose of the 500 kW per year automated and integrated plant, which handled the whole process from the silicon raw material to module assembly, was to build an actual plant and to undertake manufacturing as a preliminary step toward future mass production. Therefore, the aim was to confirm production technologies and, based on the test data, to clarify problem areas in order to bring down costs. Each department at the production plant was allocated to the participating corporation that had been responsible for the associated crystalline solar cell research theme in the late 1970s. The allocations are listed below (Fig. 3.7). Newcomers Osaka Titanium Technologies and Shin-Etsu Chemical, who joined the research in fiscal year 1980, provided inexpensive raw materials, with Toshiba using the ribbon crystal technique, and NEC and Osaka Titanium the cast substrate technique (polycrystalline) to form the crystals. Electrodes were attached to the silicon ribbon crystal or polycrystalline ingot either by dry PN junction (Hitachi), or wet PN junction (Sharp). Finally, Toshiba, NEC, and Hitachi assembled the solar cells produced in this way into panels.

The development of the test plant started in fiscal year 1980 and was completed by the end of fiscal year 1982 (March 1983). Running tests to confirm performance

[33]Regional Economy pages, *The Nikkei/Nihon Keizai Shimbun*, July 23, 1983.

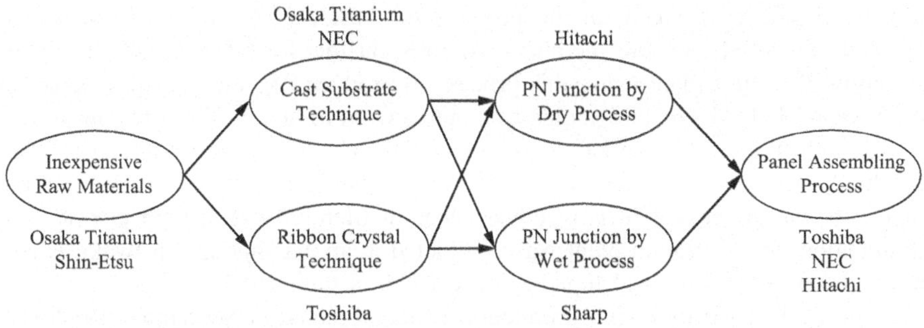

Fig. 3.7 500 kW per year solar cell panel production process and company allocations. *Source* Produced based on MITI Kōgyō gijutsuin Sanshain keikaku suishin honbu (1984, p. 72)

were conducted from the second half of fiscal year 1983 and the experiments continued until March 1985.[34]

Based on the test results, the construction of a 1000 kW class photovoltaic power plant in Saijo, Ehime Prefecture, began in September 1981 with Shikoku Power Company as the main stakeholder. In October, construction began on a test facility for a light-heat hybrid photovoltaic power system in Aki-gun, Hiroshima Prefecture.

The following is the schedule for developing other types of new energy. In January 1982, flow tests were started at large-scale boreholes at a depth of 3000 meters to explore geothermal energy. In June, production tests were successful at an exploratory well in Oita, with further successes with steam jets at Okiura, Aomori in July 1983, at Yuzawa-Ogachi, Akita in October 1983, and at Okuaizu, Fukushima in December 1983. In 1981, the NEDO steering committee promised to provide all necessary support since geothermal development is particularly heavily regulated in Japan:

> Since we believe that NEDO cannot resolve the legal and customary limitations on its own, we would like you to report to the steering committee if any such issues arise. … We believe that in some cases it may not be possible to progress toward solutions to the problems without major political movement.[35]

The steering committee assumed the protective role of a higher instance, backing the seamless expansion of technology development and implementation at NEDO.

There are two approaches to coal liquefaction: the bitumen method or the brown coal method. Parallel development by the private sector had already resulted in the three processes of direct hydro-liquefaction, solvent extraction, and solvolysis for bitumen. In fiscal year 1983, NEDO identified and unified the advantages of each

[34]MITI Kōgyō gijutsuin Sanshain keikaku suishin honbu (1984, p. 73), Ishizaka and Hirono (1985, p. 41).

[35]Remarks by Ashihara Yoshishige, *NEDO News*, October 1981, p. 7.

process in order to speed up the projects for NEDO. The process was named NEDOL, since NEDO staff members were responsible for integration.[36] The construction of a pilot plant using this process to produce 250 tons per day began in fiscal year 1984. Meanwhile, a plan was put forward to liquefy the abundant brown coal found in Victoria in Australia. Construction of a fifty tons per day pilot plant for the Australian brown coal liquefaction project began in November 1981 and the first coal was processed in November 1985. In March 1982, a high caloric coal gasification demonstration plant was completed at an existing site in Iwaki, where research into operating coal liquefaction carried on until 1985.

A pilot plant for hydrogen production was completed in September 1982 and, from August 1983 to March 1984, Showa Denko was at the heart of developing a demonstration plant for water electrolysis techniques. Operational testing of a five mega-calorie long-term storage system made with hydrogen absorbent alloy started in December 1984.

NEDO was also committed to international cooperation where the development of new energy technologies was concerned. Even if a technology was ill suited to Japan, some technologies could be useful to other countries where resources, the climate, and the overall natural environment were different. The needs for new energy systems also varied from country to country.

International cooperation included the coal liquefaction plant in Australia, joint Japan–U.S. development of hot dry rock geothermal power generation, and cooperating with Indonesia on photovoltaic power generation systems. At a meeting about energy research development in November 1980, the governments of Japan and Australia agreed to proceed with the brown coal liquefaction project. As a result, the project was incorporated into the Sunshine Project in April 1981, with NEDO taking on the role of developing and advancing the project. By the late 1980s, NEDO participated in the construction of all types of plants in its capacity as the organization responsible for the implementation stages of new energy technology development. With NEDO at the helm as the organization bringing together the knowledge of industry, government, and academia, the Sunshine Project was taking the first steps on the path to reaching its ambitious goals by the year of 2000.

3.4 The Emergence of Amorphous Materials

3.4.1 The Emergence of Amorphous Solar Cells

Another important solar cell technology emerged in the late 1970s: the amorphous-silicon semiconductor solar cell, a new technology that did not use crystalline silicon. Compared to crystalline silicon solar cells, the conversion efficiency was inferior, but the new technology used only a fraction of the amount of

[36]NEDO (1990, p. 19).

silicon and there were expectations of substantial reductions in manufacturing cost. These outstanding characteristics also raised hopes for the technology within the Sunshine Project. Below, we will consider the technical characteristics and the sequence of events that led to its adoption by the Sunshine Project.

In 1974 when the Sunshine Project started to develop solar cells crystalline solar cells were the main focus because the amorphous approach was not yet an option. The amorphous material was not intended for use in solar cells when it was introduced. Rather, it was originally viewed as a new semiconductor material (chalcogenide glass). In Japan, the material and its promising properties caught the eye of a small number of researchers who started to research it in the late 1960s. By the mid-1970s, it became clear that the material could be used for solar cells.

It is possible to trace the research in amorphous material back to 1948 when Schaffert and Oughton described amorphous selenium (Se) in a paper on xerography.[37] In 1950, Weimer discovered the photoconductivity of amorphous selenium at RCA and, in 1955, Kolomiets blended and fused dithallium selenide with arsenic triselenide to discover a vitreous state. Kolomiets named the substance chalcogenide glass and demonstrated that its properties differed from crystalline semiconductors. In 1960, Joffe and Regel discovered that the short-range order of constituent atoms determines whether a given metal becomes a semiconductor or an insulator. The discovery suggested the potential for using amorphous instead of crystalline semiconductors. This research led to Ovshinsky's paper on electric memory switching in thin-film chalcogenide amorphous semiconductors published in November 1968. This moment marked the creation of the amorphous semiconductor.

Ovshinsky was the director of Energy Conversion Devices (ECD), a U.S. venture business set up in 1960. His paper attracted international attention as it opened up the potential for new amorphous applications. Aiming to develop practical applications in the future, Japanese corporations also started to research the material. However, at the time, the research was still fumbling around for possible product applications and merely looked at the extremely interesting scientific properties of the amorphous substances. Not surprisingly, nobody knew that amorphous semiconductors could be used in solar cells. This is why Japanese research on amorphous substances was limited to a few research institutes and companies in the 1960s.

In the mid-1970s, when scientists discovered that amorphous materials could be used in solar cells, a few corporations started to take solar cell development seriously. Establishing clear aims for researching amorphous material made it easy for companies to promote the research. The fact that amorphous materials could be used in solar cells also meant that the companies could get research contracts sponsored by the Sunshine Project.

Up to this point, the two research themes in the photovoltaic power generation program under the Sunshine Project had been crystalline silicon and compound

[37]The following refers to Kikuchi (1982, pp. 10–15).

semiconductors. However, when the spotlight turned to amorphous materials, the Sunshine Project soon appreciated the scientific possibilities of the new method and launched the research.

The first small-scale projects began in the late 1970s after putting in applications for amorphous materials research to the Sunshine Project. Later, the research advanced when national research institutes centered on the Electrotechnical Laboratory as well as Sanyo Electric, Fuji Electric, Kanegafuchi Chemical Industry (currently, Kaneka), and other companies with ambitions to develop amorphous materials participated in the project. In the early 1980s, when the results of the amorphous materials research emerged and were widely publicized, the budgets for these projects were also greatly increased.

The October 1981 issue of *NEDO News* cites comments by Dokō Toshio and Ashihara Yoshishige on their expectations for amorphous solar cells at a roundtable discussion with members of the NEDO steering committee:

> [Dokō] This is a little while ago, but at the AIST (Electro-Technical Laboratories), I saw a newspaper article about the discovery of amorphous silicon. I understand there are high expectations for good batteries. ·
>
> [Ashihara] Recently, electronic technology has made rapid progress. [...] Is it not time for NEDO to come to grips with amorphous research?[38]

The members of the NEDO steering committee also had high hopes for amorphous solar cells as a new groundbreaking technology. At this time NEDO provided extensive support to advance the amorphous project.

The structures for researching and developing amorphous solar cells had one striking feature: universities were systematically incorporated into the project, which had not been the case with the development of crystalline technologies. University researchers with an interest in amorphous materials also participated in the project. There were hardly any prior examples of universities participating extensively in MITI projects and, in that sense, the amorphous materials research was a collaboration between industry, government, and academia both in name and in reality. Consequently, there was an assumption that universities and corporations would pass the outcomes of basic and applied research back and forth between themselves.

In fiscal year 1980, the first fiscal year of the research program, the Sunshine Project outsourced amorphous materials research to corporations, adding universities in the following fiscal year. In fiscal 1983, NEDO took over the applied research.[39] The research themes for fiscal year 1981 are outlined in Table 3.2.

Why were so many university researchers attracted to amorphous materials? The reason is that, compared to crystalline silicon, many of the characteristics of amorphous materials had yet to be clarified. Amorphous material was untrodden territory, which is precisely why university researchers perceived it as a subject with

[38]*NEDO News*, October 1981, p. 11.

[39]MITI Kōgyō gijutsuin Sanshain keikaku suishin honbu (1984, p. 103).

Table 3.2 Overview of research themes for amorphous solar cell development (FY1981)

	Research theme	Contractor
1	Multilayered structure amorphous solar cell	Sanyo Electric
2	Integrated amorphous solar cell	Sanyo Electric
3	Amorphous solar cell by using ceramic substrate	Kyocera
4	Technology for large-area of Amorphous solar cell	Fuji Electric
5	Amorphous solar cell on flexible film substrate	Teijin
6	Method of monosilane production	Komatsu Electronics
7	Technology of filming high quality amorphous silicon films by plasma separation	Sumitomo Electric Industries
8	Production of amorphous solar cell with mixed phase by CVD method	Hitachi, Mitsui Toatsu Chemicals
9	Local structure of amorphous silicon	The University of Tokyo
10	Optical investigation on electronic states of amorphous silicon	Osaka University
11	Theoretical investigation on electronic states of amorphous silicon	Kyoto University
12	Defect density of amorphous silicon	Hiroshima University
13	Interface investigation of amorphous silicon	Tokyo Institute of Technology
14	New amorphous-silicon materials	Kanazawa University

Source MITI Kōgyō gijutsuin Sanshain keikaku suishin honbu, *Shōwa gōjuroku-nendo seika hōkokusho* [*FY1981 Annual Report*] (self-pub., 1982)

potential to turn up a wealth of results.[40] Normally, corporations aiming for practical applications are risk-averse when there are many unknowns, but researchers perceive it as an opportunity to make their own contributions to the field.

Consequently, the corporations that were participating in the development of amorphous solar cells were supported by researchers at universities and national research institutes, and were able to gain technical information, which is why this is a good example of collaboration between industry, government, and academia.

From this point on, there were two strands to solar cell development under the Sunshine Project—crystalline silicon and amorphous silicon, competing with each other to produce results.

[40]Tanaka Kazunobu, interview by author, October 15, 1998 and Kuwano Yukinori, interview by author, October 29, 1998.

3.5 The Crude Oil Price Slumps and the Project Is Restructured

3.5.1 The Unexpected Slump in Crude Oil Prices

Bearing in mind the links between the first oil shock in 1973 and the formation of the Sunshine Project, and the circumstances of the second oil shock in 1979 that led to the establishment of NEDO, it is clear that oil price fluctuation is a decisive and important factor for the development of new energy technologies. If the price of crude rises, new energy projects benefit from a tailwind. However, if the price of oil declines, questions are asked about the reasons for the existence of such projects. Originally, the major premise for promoting the Sunshine Project was that the price of crude would continue to rise in the future. However, this premise fell apart when the price of crude oil suddenly started to drop in the late 1980s. This immediately resulted in a search for reasons to continue to develop energy technologies.

After peaking in 1981, the price of crude continued to fall throughout the first half of the 1980s.[41] From 1985 to 1986 prices dropped so sharply that the period came to be referred to as the reverse oil shock. As a result, new energy technology development found itself in a difficult situation.

Responding to the drop in the price of crude, the Agency for Natural Resources and Energy also lowered the target values for new energy development in its *Long-Term Energy Demand and Supply Outlook*. In November 1980 the target for alternative energy supply was 4.8% by 1990, a decade into the future. However, the target was lowered to 2.5% (geothermal 1%, total 3.5%) in the April 1982 *Outlook*, and then again in November 1983 when it was 1.7% (geothermal 0.3%, total 2%). As the price of oil fell, government targets for introducing new energy and the public interest in such targets also declined.

What happened with NEDO and the structure for developing new energy technology amid this adversity? Enjōji Jirō (NEDO steering committee member) had already made the following remarks at a roundtable discussion in October 1981 to mark the first anniversary of NEDO: "It's important to note that if NEDO does not prepare a sturdy research and development structure, and if the recent glut of oil continues, there is a risk that people will stop caring because the Japanese have a strong tendency for stopgap thinking."[42]

The fledgling NEDO had not foreseen the unfavorable situation in the mid-1980s. By rights, prices for new energy should have become more competitive due to rising oil prices, which should have opened up a major market when coupled with technical progress and cost cuts due to mass production.

There are limits to the planet's supply of oil and other fossil fuels, or plutonium and other mineral resources. However, there is an inexhaustible supply of new

---

[41]Please see Fig. 2.11 in for fluctuations in crude oil prices.

[42]*NEDO News*, October 1981, pp. 4–15.

energy, and the assumption was that new energy would become relatively advantageous from a cost point of view with the passage of time. In the 1970s, the Club of Rome and many other experts argued that the depletion of oil was a question of time. However, the steep fall in the price of crude in the early 1980s appeared to overturn the assumptions behind this outlook.

The nature of electricity is such that pricing is the only means of competition. One may even say that there is no way of distinguishing one electricity product from another. Whether generated by oil-fired thermal power or nuclear energy, or by solar energy, the electricity is the same and there is no way of telling the difference.[43] Under such circumstances, it is easy to convince people that it makes no sense to adopt new energy or other expensive methods of generating power as long as there is a cheap and stable supply of oil. This is what Mr. Enjōji referred to when he mentioned "a tendency for stopgap thinking".

Although the solar thermal energy and photovoltaic demonstration plants of the early 1980s successfully generated power, they were still utterly unable to compete with existing electricity on cost.

For example, the 1000 kW class solar thermal energy power plant at Nio Town delivered results on several points including the successful generation of a rated output of power and the world's longest continuous operation, but the conclusion was that in terms of cost the limit was 100 yen per kW hour even when multiplying the scale by ten (10,000 kW).[44] This was extremely expensive compared to other methods of power generation. In 1980, it cost nine to ten yen to generate one kW hour using oil-fired thermal power, or five to six yen using nuclear power, so in terms of power generation efficiency, solar power did not stand a chance.[45]

Meanwhile, constructing large-scale plants was the original aim of photovoltaic power generation and, in March 1986, a centralized 1000 kW photovoltaic power generation plant was completed in Saijo, Ehime Prefecture, in order to expand on the results from past demonstration plants. However, the module cost for solar cells still exceeded 1000 yen per watt at the time.[46] Partly also due to the unexpected drop in the price of crude oil in the mid-1980s, it was impossible to construct large-scale solar thermal or photovoltaic power plants that could realistically make a profit. For the Sunshine Project this meant that the initial plan to construct large-scale power plants proved far more difficult than expected. Such technical difficulties were gradually ascertained as a result of accumulating data from experiments in a range of new energy fields.

---

[43] Konno Kunisuke, interview by author, March 11, 1998.

[44] At first, Hitachi had no success with the running tests, but Nebashi Masato, the Deputy Director-General of Technology, lost his temper, summoned the Executive Vice President of Hitachi and ordered him "to get it to work properly". The response from Hitachi was to complain of a budget shortfall, but Nebashi would have none of it. Recalling this episode, Tani Tatsuo, who was in charge of solar thermal power generation, commented "it was extremely clear what the corporation had to do with the budget it received" (Tani Tatsuo, interview by author, July 9, 1998).

[45] MITI Editorial Committee on the History of Japan's Trade and Industry Policy (1991, p. 256).

[46] MITI (1993, p. 48).

Together with the shift from thermal to photovoltaic energy, by the mid-1980s the keyword for the project was distribution, not centralization. Rather than the previous design proposals for centralized solar thermal energy or photovoltaic power facilities on a gigantic scale, suggestive of the "big-ship, big-gun" advocacy of naval warfare, expectations turned to developing distributed energy systems in the vicinity of households and other places of consumption.

In the early 1980s, NEDO had already started to develop technologies for distributed energy systems including photovoltaic systems for individual households or housing complexes, but as of the mid-1980s the development of independent distributed systems tailored to specific applications gained momentum. If system designs are tailored to environments where new energy excels, it is possible to argue against existing energy sources by demonstrating the advantages of new energy. Such were the expectations when systems for mountainous areas, power supply systems for remote islands, seawater desalination systems, offshore systems, methane gas hybrid systems, and wood pulp hybrid systems were developed in fiscal year 1984, followed by photovoltaic power generation systems for tunnel lighting, broadcast satellites, greenhouse farming, and other unique applications in fiscal year 1985.[47] As indicated by the term "hybrid", both NEDO and the AIST were facing up to the unforeseen reality of a slump in oil prices and high costs for large-scale new energy power plants. New energy technology development was no longer a matter of large-scale power plants using a single energy source; the policy was changing to favor optimal mixes of different types of energy and the construction of efficient supply systems. This meant that the mid-1970s vision of a future where alternative energy was supplied by oversized power plants for clean energy was changing.

Although the importance of new energy technologies to Japan, a country lacking in resources, had not decreased in reality, it is an undeniable fact that the sense of urgency to provide energy for Japan by utilizing new energy faded amid the remarkable decline in the price of crude oil.

These were the circumstances when NEDO was restructured and transformed from an organization solely committed to new energy development into an organization responsible for managing projects to research and develop broader high-tech.

The Act concerning the Improvement of Systems for Research and Development in the Field of Industrial Technology was enacted in October 1988. In the same month NEDO changed its name from the New Energy Development Organization to the New Energy and Industrial Technology Development Organization. The name change may seem insignificant but, in fact, the work of the organization grew to include not only new energy, but also other technical development projects recommended by the AIST as a result of the change.

In addition to the Sunshine Project and the Moonlight Project, the AIST also implemented the Large-Scale Project System, the Next Generation Project, and the

---

[47]MITI Kōgyō gijutsuin Sanshain keikaku suishin honbu (1985, 1986).

Medical and Welfare Equipment Technology Project with NEDO given responsibility for schemes promoted under these projects. As a result, the themes handled by NEDO became far more varied as of fiscal year 1988.

NEDO was not only responsible for the Sunshine Project and the Moonlight Project, but also a wide range of other technical development themes including the manganese nodule mining system, and the high-speed computer system for science and technology under the Large-Scale Project; high-efficiency polymer separation membrane materials (new materials), and recombinant DNA technology (biotechnology) under the Next Generation Project; as well as arterial surgery lasers, and reading systems for the visually impaired under the Medical and Welfare Equipment Technology Project.

As a result, in addition to the budget for new energy development, NEDO received an extraordinary allowance for research and development of industrial technology as of fiscal year 1988. The amounts were 4.4 billion yen in 1988 (October 1–December 31), 15.1 billion yen in 1989, and 20.3 billion yen in 1990.[48] Incidentally, in fiscal year 1997, the 74.2 billion yen budget toward the cost of new energy technology development was nearly matched by the budget for costs related to the research and development of industrial technology, which had risen to 62.4 billion yen. Developing new energy technologies was no longer the most important mission at NEDO; rather, NEDO had transformed itself into a comprehensive project management organization for high-tech applications.

3.5.2 *New Energy Development: Outcomes*

The circumstances in the late 1980s hardly favored the Sunshine Project but, even so, there were some successes involving new energy technology development. Despite the difficult circumstances, the development of each theme made steady progress. As of this period, photovoltaic power generation replaced solar thermal energy to take the dominant position. There were advances in the development of distributed systems, and one testing facility after another started operations. A system of photovoltaic power generation for elementary schools was completed and started up in July 1985, followed by a power supply system for remote islands in October. In Indonesia, photovoltaic water pumping systems and photovoltaic power generation systems for medium-sized villages started operating in June 1986 and February 1987 respectively. In August 1986, Oita Prefecture, Shimizu Corporation, and Sharp completed a commission to develop a demonstration plant for Japan's first large-scale offshore photovoltaic power system.

In November 1985, the primary hydrogenation system for the Australian brown coal liquefaction project started operations. In April 1988, the Australian brown coal liquefaction plant started comprehensive test operations. In August 1989, the

[48]See NEDO (1990, pp. 148–157) for the changes in the budget for project cost.

Australian project scored a success with long-term continuous operations. In May 1986, the IEA joint research project on hot dry rock succeeded with a 10 MW circulation extraction heat test and, in October 1986, a project to develop technologies for hot dry rock power generation in Japan carried out hydraulic fracturing. In June 1987, geothermal development surveys had successes with venting in the Minami Kayabe region and at Unzen Nishibe. Hydrogen and fuel cell projects started up a trial system for 1000 kW power storage in October 1986, and successfully generated power using a 10 kW class molten carbonate fuel cell (matrix and paste format) in March 1987. In September, a low-temperature, low-pressure phosphoric acid power plant successfully generated 1000 kW. New energy also diversified as researchers started looking at practical applications for wind power, ocean energy, bio-energy, and other sources.

3.6 Project Outcomes

3.6.1 Managing Technology Development

Here, we will consider the basic policies of managing technology development under the Sunshine Project in this period. The following are three dilemmas that may arise between governments/semi-governmental corporations (consignors) and private sector companies (consignees) where research and development projects are based on a format of government commissions. They are issues involving: (1) development themes (focus on delivering national objectives or focus on business potential); (2) development and technology options (focused development or parallel development); (3) format for sharing outcomes (national ownership or private sector ownership of patents and other intellectual property). In general, governments expect to conduct focused development with little overlap, and to own patents and other outcomes in order to deliver the national goals. Private sector companies, on the other hand, want to pursue parallel development in a format that agrees as much as possible with the strategies of each company and, for them, the ultimate goal is to secure profits through future commercialization. It is also a given that companies expect to own the outcomes of research and development in line with the effort expended. Private sector companies are not interested when terms and condition only favor the government. On the other hand, if contracts are excessively favorable to private sector companies, governments are criticized for assisting only specific companies. Therefore, it is necessary to draw a line between ingenious competition and planning where these three points are concerned. This is the secret of managing technology development under a national project.

The first dilemma relates to technology development themes. Normally, technology or market trends determine the choice of technologies used in the process of corporate product development. However, if the focus is on trends in the market or technologies, the government's technology policy and corporate technology strategy come into conflict if the initial national goals are not achieved. Even if the goal presents a fair number of difficulties from a market or technology viewpoint, the government believes that these are issues that must be overcome to achieve national goals, which is why the government provides subsidies and commissions. On the other hand, corporations will comply with government policy to the extent of receiving commissions and financial guarantees from the state, but they will not invest any of their own resources to develop technology themes with little commercial potential. If the government pressures companies to achieve the goals, the companies will minimize their commitment to any development theme with a poor commercial outlook and focus on amassing technologies and knowledge that can be obtained by participating in the project. On the other hand, if governments are excessively respectful of corporate choices of strategic development themes aimed at future commercial development, the initial goals of the national project may inevitably change. This is the dilemma that governments and corporations are presented with when selecting themes.

The second dilemma concerns the methods of choosing the technology for development. Government avoids investments that overlap with specific corporate technologies, but tends to focus resources on technologies that are in the limelight, and to conduct the technology selection process as early as possible out of recognition that this increases the overall efficiency of the project. If this approach is successful, it is possible to achieve maximum project outcomes with a minimum of financing. However, experience tells us that technology development processes often do not evolve in line with initial projections. Therefore, delaying the process of technology selection while multiple technologies compete against each other may, in the end, lead to better outcomes. This is the advantage of parallel development. Whether to focus resources and gain an advantage by avoiding overlapping investments, or to improve the potential for at least one success through parallel development of tenuous technologies—this is the dilemma around the choice between focused development or parallel development.

The third dilemma concerns how to share the outcomes of developed technologies. The government would prefer to retain possession of the outcomes of technologies developed for public purposes (e.g. patents) because the aim of government policy is not to develop technologies for the benefit of specific corporations, but to disseminate outcomes widely across a whole industry. However, the more certain corporations are that they will be able to monopolize the outcome, the stronger their commitment to developing the technology in question. Consequently, in cases where the government makes a strong claim for national ownership (or ownership by semi-governmental corporation) of the outcomes, corporations will be less willing to develop technologies with limits on how they can use the outcomes. On the other hand, the government would not be able to justify policies launched in the public interest if it allowed corporations to own the outcomes of developing technologies. This is the dilemma between public interest and corporate commitment where the allocation of outcomes is concerned.

If both the state and private sector prioritize their own goals and adopt hardline positions on these points, commissioning research is out of the question. However, both NEDO and the AIST were able to support the basic direction for the crystalline and amorphous materials, and to keep the competing developments going when NEDO announced that the future choice would be between the two technologies, which had emerged as the indicators of the Sunshine Project.

1. At NEDO, all development themes were focused on systems for power generation, and the organization adopted a position focused on delivering national goals. However, solar cells for calculators and watches were making rapid progress at the time, and NEDO tolerated the corporate focus on their potential for commercial development on the premise that these were projects that would link to solar cells for future power generation systems. As a result, NEDO was able to openly support, for example, Sanyo Electric's use of the advantages of amorphous solar cells to open up the market for solar-powered calculators although such a project did not necessarily benefit the development of photovoltaic power generation systems. By producing amorphous compound semiconductor solar cells for electronic calculators in the early 1980s, Kyocera, Sharp, and Matsushita Electric Industrial were able to develop commercially feasible solar cells. As a result, the commissioned research generated specific outcomes, which, by rights, should not have been linked to corporate earnings.
2. Concerning the choice of technologies and development, NEDO did not change the focused development route, but allowed corporations to compete in the period leading up to project evaluation. The unexpected outcome was competitive parallel development driven by the rivalry between engineers working to develop different underlying technologies, and corporations making additional investments in commissioned research. Whether companies were developing crystalline or amorphous materials, they had to demonstrate the validity of their own technologies to NEDO to obtain dominance. The market for photovoltaic power systems did not exist at the time; rather, the solar cells used for consumer goods were proxy indicators.
3. Setting aside the issue of integrating crystalline and amorphous materials, the format for sharing outcomes posed a problem for NEDO from an early stage. Later, NEDO no longer retained the monopoly on patents, as was the case initially. Rather, a structure was set up whereby NEDO split ownership with the company that accepted the commission. As a result, companies no longer handed over the entire outcome of research supported by NEDO payments, but could now use the outcomes in company technologies. MITI constructed a logic that facilitated the sharing of patents with the private sector by regarding the funds paid to NEDO as subsidies to the operator of an autonomous business, rather than funds provided to NEDO to commission research.[49] As a result, participation in commissioned research took on increased significance for corporations.

[49]Ishikawa Fujio, interview by author, June 4, 1998.

3.6.2 Competition in a Tri-Polar Structure

The following is a summary of the technology development directions adopted by manufacturers participating in solar cell development in 1983 when NEDO started to commission research: (1) Hitachi, Toshiba, and NEC developed crystalline materials, whereas (2) Sanyo Electric, Fuji Electric, and Mitsubishi Electric developed amorphous materials. Occupying the neutral ground between these two opposing groups, (3) Sharp developed amorphous materials with the focus on mono-crystalline silicon, while (4) Kyocera (Japan Solar Energy) and its subsidiaries developed the ribbon crystal technique and amorphous materials. In addition, two companies within the Matsushita Group (Matsushita Electric Industrial, Matsushita Battery Industrial) worked simultaneously on developing compound-semiconductor solar cells and systems.[50] With the rapid growth in electronic calculators, companies working on amorphous materials had access to a practical market for their solar cells and, according to the research, the potential for amorphous materials was endless. Corporations working on crystalline materials were still unable to develop commercially viable solar cells, but the accumulation of semiconductor technology, suitable for crystalline-semiconductor solar power generation systems, was regarded as a future strength. After the failure of the Japan Solar Energy concept, Sharp, Kyocera, and Matsushita developed and produced several technologies in parallel to secure alternatives for the future. Such was the tri-polar structure of technology development approaches at companies involved in developing solar cells in 1983.

In this environment, NEDO eventually narrowed down the options to either crystalline or amorphous materials, but considered competition between the two options preferable. Therefore, in 1983, NEDO officially announced plans to conduct an evaluation of crystalline and amorphous materials in fiscal year 1985 before narrowing the choice down to one or the other.

This decision increased tensions in the tri-polar structure in the early 1980s. Up to that time improving conversion efficiency had been an important issue, but now there was the added possibility that NEDO might discontinue research commissions depending on the outcomes.

In the 1980s, solar cell development progressed rapidly amid a climate of confrontation between the companies involved in the Sunshine Project. Figure 3.8 traces the rise in new records for conversion efficiency achieved due to NEDO research commissions.

The price of a completed solar cell module was also reduced from 5000 to 6000 yen per kW in 1980, to 2000 yen in 1983, and further to 1200 yen in 1985 (Fig. 3.9).

---

[50]Commissioned research into compound-semiconductor solar cells ended as of fiscal year 1981.

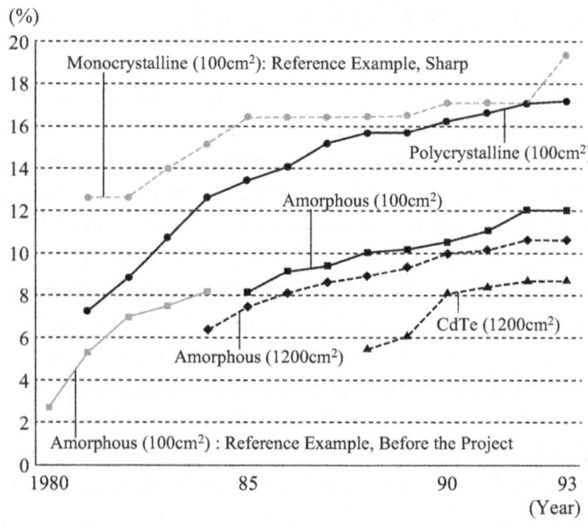

Fig. 3.8 Conversion efficiency trends by solar cell technology. *Note* Cell-based. Polycrystalline for fiscal 1981 refers to the thin-film polycrystalline technique. Amorphous refers to initial efficiency. *Sources* Based on Kyocera Corporation Solar Energy Division (1994, pp. 151 and 158), Sharp (1996, p. 90), and Taiyōkō Hatsuden Gijutsu Kenkyū Kumiai [Photovoltaic Power Generation Technology Research Association] (1998, p. 45)

Fig. 3.9 Price per kW for solar cell modules. *Note* The figures refer to the NEDO purchase price. The value for fiscal 2000 is an estimate. *Source* Taiyōkō Hatsuden Gijutsu Kenkyū Kumiai [Photovoltaic Power Generation Technology Research Association] (1998, p. 40)

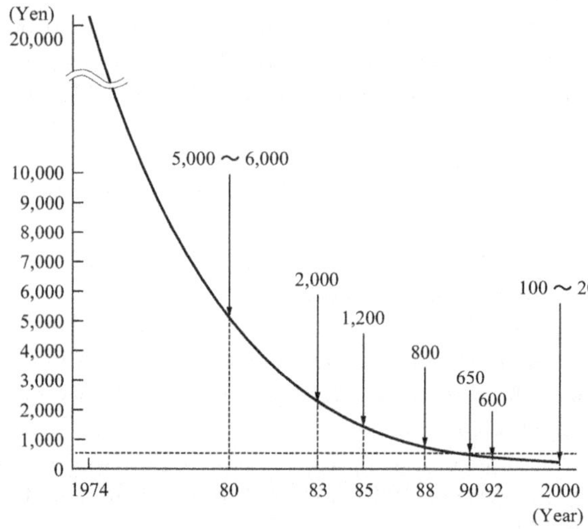

What were the reasons for these performance and cost improvements?

It would appear that the situation emerged due to NEDO and the AIST intentionally masterminding a development race. In fact, the agency had supported crystalline and compound-semiconductor silicon solar cell development since the

1970s, and had invited amorphous solar cell developers to participate in the Sunshine Project from the early 1980s. As a result, the companies had cooperated to achieve the overall project goals while competing against each other on technical excellence.

This is how project management under the Sunshine Project was designed to protect the public interest, on the one hand, while offering companies incentives for development and commercialization. This method was certainly not successful with every new energy theme, but in fields such as photovoltaic power generation where earnings might be expected from commercialization, the conditions were right for drawing out independent development efforts and commercialization on the part of corporations. Each company proceeded to work on practical applications for solar cells under the government's guidance.

3.7 Environmental Issues and the New Sunshine Project

Throughout the 1980s, new energy technology development kept a low profile, but by the late 1980s and early 1990s new perspectives on solutions to environmental problems emerged, taking the project in new directions. There was closer scrutiny of the adverse impact on the global environment of carbon dioxide emissions generated when burning fossil fuel, and new energy technology was recognized as a useful means of finding solutions to the problem. The rising concern about global issues revitalized the research and development of new energy.

From the early 1990s onward, there was much anxiety about global warming, acid rain, destruction of the ozone layer, desertification, and a range of other global environmental issues. It was also the start of a period of consultations to devise solutions through international cooperation. Concerning global warming, in particular, there were ongoing efforts to establish international agreements on reducing the carbon dioxide emissions generated when burning coal, oil, and other fossil fuel since such emissions are the main cause of global warming. To deal with the issue, Japan formulated an action plan to prevent global warming in October 1990. Premised on concerted efforts by the principal advanced countries, the plan set a goal for the year 2000 and beyond of stabilizing carbon dioxide emissions per capita at the 1990 level. New energy technology development was one important method of delivering this goal. To contribute to solutions to global environmental problems had been one of the tenets of the Sunshine Project since its launch. When international cooperation and efforts at the level of individual countries became reality, more was also expected of new energy technologies.

The Industrial Technology Council took note of the importance of environmental issues as early as July 1990. The council requested a radical review of the framework for the Sunshine Project, and the addition of new goals in response to the global environmental problems.

Based on these considerations, the council indicated a medium to long-term plan looking toward the twenty-first century and stressed that the outcomes of new energy research and development should benefit energy technology innovations worldwide. The report pointed out the need to proceed with research and development in a systematic manner, sorting future research themes into three categories by timescale: (1) short-term issues by the year 2000, (2) medium-term issues by 2020, and (3) long-term issues. Specific research themes in the short term were photovoltaic power generation, large-scale wind power generation, hot dry rock geothermal power generation, and others. The themes for the medium term were super-efficient solar cells, hydrogen combustion technologies, and bio-technologies to generate power from industrial and household waste, while the long-term themes included exploitation of space or magma for power generation.[51] In this way, the report emphasized the importance of establishing systematic project goals.

In addition to the rise of these global environmental problems, another event occurred that worked to the advantage of the Sunshine Project. The Gulf War broke out in January 1991 following the Iraqi invasion of Kuwait in August 1990. Japan had become accustomed to cheap oil, but this event reconfirmed the instability of the Middle East. Fortunately, the Gulf War did not cause the same kind of panic as the oil shock thanks to scrupulous advance preparations including emergency stores of oil. However, there was no denying the fear that as long as Japan's oil dependency on the Middle East remained high, international conflicts would immediately cause energy problems. In fact, in 1992, Japan's dependency on imports of crude from the Middle East was 75.1%, which is extremely high when compared to 27.4% for the United States, 17.7% for Germany, 21.9% for Britain, and 45.2% for France.[52] These two factors—the attention on global environmental problems and the instability of the Middle East—suggested to the nation that new energy development in Japan was still a significant issue.

Based on such changes in circumstances, MITI set out to restructure the policy program by clarifying the close relationship between the global environmental problems and the development of new energy and energy-saving technologies.

In January 1992, MITI settled plans to create the Energy and Global Environmental Technology System (provisional name) by integrating the Sunshine Project, the Moonlight Project, and the Research and Development Project on Environmental Technology. The increase in international demand for all countries to make a contribution to the global environment provided the context for the new project proposal. For example, the United Nations Framework Convention on Climate Change to reduce emissions of greenhouse gases was adopted in May 1992, signed by all countries at the UN Conference on Environment and Development (the Earth Summit) held in Brazil in June, while efforts to ratify the framework were announced at the Munich Summit in July. By December 1993, 50 countries had ratified the framework, which came into effect in March 1994. The world had taken a major step toward protecting the environment.

---

[51] *The Nikkei/Nihon Keizai Shimbun*, July 4, 1990.

[52] Shigen enerugīchō (1995, p. 11).

The new direction of integrating the energy-related projects as suggested by the Industrial Technology Council entered the final round of concrete policy discussions in May 1992. The discussions moved toward the launch of the new program based on the main point of triggering synergy by consolidating the development of new energy, energy-saving, and global environmental technologies. Finally, in September 1992, it was decided that the AIST would launch the New Sunshine Project, a comprehensive program to develop energy and environmental technologies, as of fiscal year 1993.

In April 1993, the AIST launched the New Sunshine Project as a comprehensive technology development project in the energy and environment field. The goals of the project were to provide one third of Japan's energy consumption by 2030 and to reduce carbon dioxide emissions by half.[53] Under the systematically developed New Sunshine Project, a range of new energy development programs would create synergy by incorporating measures to counter environmental problem, and going beyond the framework for resource energy.

3.8 Project Contributions

3.8.1 Infrastructure Development

The environment around new energy also underwent major changes in the early 1990s. To respond to these changes, MITI took note of the shared nature of energy issues and environmental issues and decided to integrate several projects as of fiscal year 1993, as described above. The former new energy technology research and development project (the Sunshine Project), the energy conservation technology research and development (the Moonlight Project), and global environmental technology development were integrated and restructured as the New Sunshine Project. Resource energy issues and environmental issues were now one and the same and, if the development of energy and environmental technologies were comprehensively and systematically promoted, rather than considered separately, there was more potential to implement effective technology development. This is why several programs were restructured under the new project. Research and development of photovoltaic power generation, which had been included in the Sunshine Project up to fiscal year 1992, was now integrated into the New Sunshine Project and moved in new directions.

In the late 1980s when crystalline silicon and amorphous silicon were developed in parallel, the central theme for the development of photovoltaic power generation under the Sunshine Project was to improve solar cell performance. However, the technology evolved from a situation where crystalline and amorphous materials opposed each other to a structure that aimed for technical breakthrough by fusing

[53]Shigen enerugīchō (1995, p. 43).

both technologies to focus on a new type of thin-film solar cell. Increasing the conversion efficiency of solar cells and reducing manufacturing costs remained important issues, but this was also the period when the development of solar cell technologies achieved some success, and it was considered increasingly important to improve the overall performance of the system and to focus on developing practical applications, such as panels for generating power in the future. In particular, regardless of whether a manufacturer was using crystalline or amorphous silicon, the issue of how to introduce photovoltaic power generation systems to general households was common to all.

Therefore, MITI promoted a series of policies including updating the power supply infrastructure to encourage the spread of photovoltaic power generation systems in general households, and subsidizing those who wished to install the systems. As a result, photovoltaic power generation systems for households spread rapidly, which boosted Japan's solar cell industry to become number one in the world in terms of installations and production.

Developing technologies for photovoltaic power generation systems was not only a matter of solar cells; it was also necessary to research the support systems. The Sunshine Project had started to research photovoltaic power generation systems as of the end of the 1970s, reserving a considerable amount of the budget for this purpose.

However, there were three major obstructions to the spread of photovoltaic power generation systems at the time: (1) unfinished grid connections, (2) restrictions under the Electricity Business Act, and (3) the high cost of photovoltaic power generation systems.

To start with, there were major problems with the means of disposing of excess electricity generated by solar cells for households. There are two approaches to power generation systems using solar cells. The independent system saves excess power generated in the daytime in storage batteries for use during the night. The advantage of this system is that it can be used in locations away from the grid but, on the other hand, the expensive storage batteries add to the cost of generating power. The other approach is the grid-connected system, which is interconnected with the normal power grid and sends excess power generated by photovoltaic systems via the power company's power lines. With the latter approach, power companies buy excess power generated by household systems during the daytime and it is possible to use normal power during the night when solar cells cannot be used. Compared to the independent system for household use, the grid-connected system offers easy maintenance management and a reliable power supply, and the expectation was that the system would rapidly disseminate if it could be implemented. However, in the late 1980s none of the power companies supported this approach.

The reason was that until March 1990 Japan imposed legal restrictions on systems for photovoltaic power generation, obstructing the introduction of the systems. At the time, anyone who wanted to install a photovoltaic system with a voltage in excess of 30 volts was required to complete several formalities including an application for business planning permission.[54] The law applied this rule to both electric facilities for household use and electric facilities for electric utilities. When operating such systems, a licensed electrician was required to carry out maintenance even in case of an independent power source system separated from the grid. Such strict regulations hindered the introduction of photovoltaic systems in general households.

When the Electricity Business Act was revised in April 1990, photovoltaic power generation facilities were also given legal standing and the procedures for installing photovoltaic systems were greatly simplified. Building permission and pre-service inspections were no longer required for facilities generating less than 100 kW per month; rather, permissions were granted on submission of a safety regulation report and a report by a licensed electrician. It was also possible to use the services of an officially recognized body such as the Electrical Safety Inspection Association in place of a licensed electrician. These revisions to the Electricity Business Act prepared the environment for commercial applications in Japan.

The updates to the guidelines for reverse power flow began in June 1990. All guidelines were completed in March 1993 when the guidelines for reverse power flow in high-voltage and low-voltage power lines were formulated. In April 1992, the methods of power company buyback were also established. Products incorporating solar cells had been regarded as electric facilities for private homes, but in October 1991 solar air conditioners were the first to be recognized as general electrical appliances. Such measures to relax the conditions for use were necessary to be able to distribute products equipped with solar cells without requirements for special controls.

In the 1980s, power companies were fully engaged with photovoltaic power generation. Premised on the introduction of grid-connected systems, all companies —including the Central Research Institute of Electric Power Industry—implemented many types of projects. The Shikoku Electric Power Company (1000 kW power station), the Kansai Electric Power Company (Rokkō Island), the Okinawa Electric Power Company (the Enetopia plan), and others implemented large-scale projects under the Sunshine Project.

3.8.2 The Significance of the Sunshine Project

Gijutsu kakushin no keiryō bunseki [Quantitative Analysis of Technology Innovation], edited by Watanabe Chihiro, points out that under the Sunshine Project, government investment in research and development not only contributed to developing the technologies in question, but also provided the private sector with investment incentives.[55] For example, since the 1980s the private sector has

[54]Taiyōkō Hatsuden Gijutsu Kenkyū Kumiai [Photovoltaic Power Generation Technology Research Association] (1998, p. 47), Kuwano (1992, p. 210).

[55]Watanabe (2001).

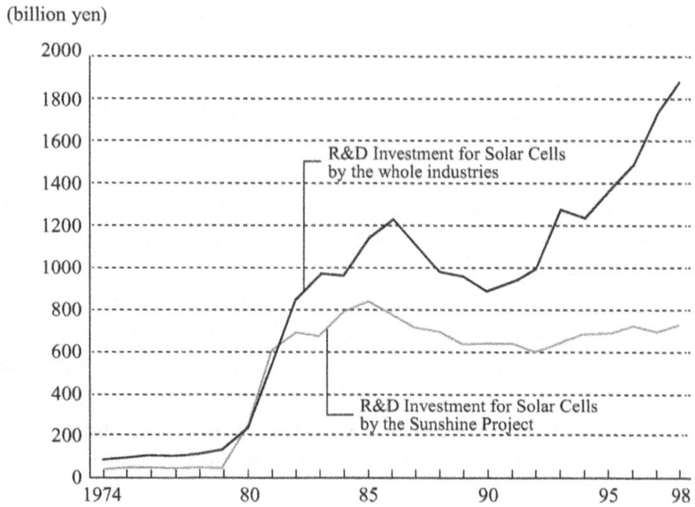

Fig. 3.10 Fluctuations in Investment in solar cell research and development in Japan (FY1974-98). *Note* 1985 real prices. The amounts refer to solar cells only, not photovoltaic power generation as a whole. *Source* Watanabe (2001, p. 140)

increased research and development expenditure on solar cells proportionally to the budget increases for national projects. In the late 1980s, corporate investment decreased for a time due to the unexpected reduction in oil prices, but the government stayed the course and continued to allocate a budget to the Sunshine Project. When the global environmental problems gained prominence in the 1990s, private sector investment once again increased rapidly as indicated by the upward sloping line in Fig. 3.10.

Figure 3.10 shows how government investment in solar cell research and development under the Sunshine Project also stimulated private sector investment by companies that were not participating in the project. *Gijutsu kakushin no keiryō bunseki* describes it as a virtuous cycle. Due to the solar cell research and development under MITI's Sunshine Project, industry players increased their research and development expenditure. As a result, the rise in the stock of research and development technologies led to an increase in solar cell production. This lowered the cost of producing solar cells, creating more demand, and increasing the production of solar cells.

Since the rise in production led to industry increasing its expenditure on research and development, the stock of technologies for solar cell research and development also expanded. This is how the initial budget allocated under the Sunshine Project encouraged corporations to invest independently. By providing the initial budget that launched solar cell research, MITI created a virtuous cycle by attracting private sector investment, thus increasing the accumulation of solar cell technologies. This led to an increase in the production output, which attracted more investment.

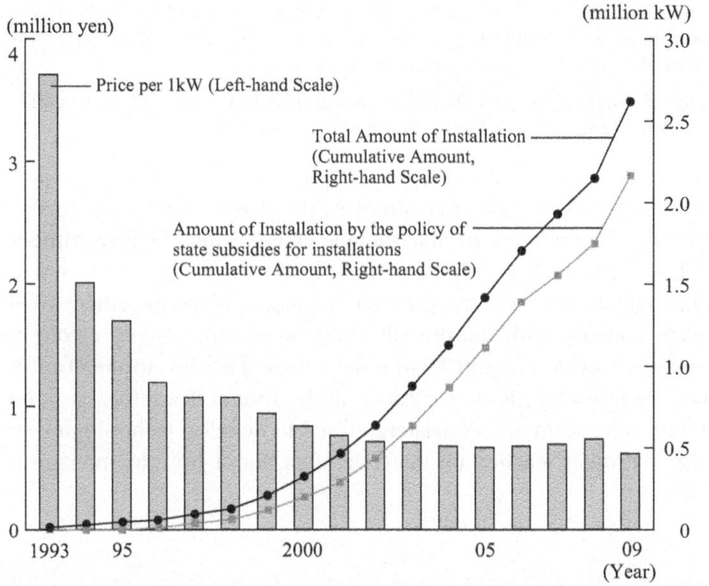

Fig. 3.11 Photovoltaic installations in Japan and fluctuations in system prices. *Note* The figures are based on surveys by the agency for natural resources and energy. *Source* Shigen ererugīchō [Agency for Natural Resources and Energy] (2011)

In the late 1990s, the government updated the transmission lines connected to photovoltaic power systems and other social infrastructure. At the time, there was a rapid increase in the quantities of domestic transmissions of power generated by photovoltaic means. Coupled with a policy of state subsidies for installations, photovoltaic power generation also made rapid headway due to a drop in the system price per kW as mass production developed (Fig. 3.11).

When the Sunshine Project ended as of fiscal year 2000, the projects were transferred to their successors.

The following are remarks made by parties involved with contributions to the Sunshine Project:

> Generally speaking, I doubt that solar cell research would have started without the Sunshine Project. (Electrotechnical Laboratory Energy Department, NEDO Solar Technology Development Office)[56]

[56]Kurokawa Kōsuke, interview by author, April 20, 1998.

> These wide-ranging initiatives would not have been possible without the Sunshine Project. Had there been no Sunshine Project, I don't think photovoltaic power generation would have come this far. (Matsushita Battery Industrial)[57]

> This research would not exist without Sunshine and NEDO. It's not likely that our company would have done it. (Hitachi Central Research Laboratory)[58]

The government managed the programs under the Sunshine Project based on rational evaluation. However, the initial goals of the Sunshine Project were fully implemented amid a number of unavoidable changes that the government could not possibly have predicted. Even so, photovoltaic power generation was one of the remarkable outcomes of the government's successful management of technology development coupled with the private sector with ambitions to create commercial applications, as well as support from universities. The development of new energy will remain an important issue for Japan in the future. Therefore, it is important to continue with long-term and consistent efforts to develop technologies. The case of the Sunshine Project teaches us many lessons about how to manage technology development.

References

Ishizaka S., & Hirono, T. (Eds.). (1985). *Nijūisseiki eno enerugī: gijutsu ga hiraku shin enerugī [Energy for the 21st century: New energy developed by technologies]* (p. 69). Tokyo: Tsūshō Sangyō Chōsakai.

Imai, K. (1982). Shin'enerugī Kaihatsu no senryaku to soshikiron. Institute of Business Research, Hitotsubashi University, Discussion Paper no. 103.

Kikkawa, T. (2008). Nihon no genshiryoku hatsuden: Sono rekishi to kadai. *Hitotsubashi Review of Commerce and Management, 3*(1).

Kikuchi, M. (Ed.). (1982). *Amorufasu handōtai no kiso [The basics of amorphous semiconductor]*. Tokyo: Ohmsha.

Kuwano, Y. (1992). *Taiyō denchi o tsukaikonasu: Taiyō denchi ga hiraku shinjidai [Using solar cells efficiently: A new era of solar cells will unfold]*. Tokyo: Kodansha.

Kyocera Corporation Solar Energy Division. (Ed.). (1994). *Taiyō enerugī eno chōsen: Taiyō denchi no jidai ga yattekita [A challenge to solar energy: The era of solar cells began]*. Tokyo: Seibunsha.

Meadows, D. H., Meadows, D. L., Randers, J., & Behrens, W. W., III. (1972). *The limits to growth: A report for the club of Rome's project on the predicament of mankind.* New York: Universe Books.

MITI. (Ed.). (1993). *Nyū sanshain keikaku handobukku [The handbook of new sunshine project]*. Tokyo: Tsūshō Sangyō Chōsakai.

MITI. (Eds.). (1973). *Nihon no enerugī mondai.* Tokyo: Tsūshō Sangyō Chōsakai.

MITI Editorial Committee on the History of Japan's Trade and Industry Policy. (Ed.). (1991). *History of Japan's trade and industry policy, vol. 13, (IV) the age of diversification (1971–1979) Part 2.* Tokyo: Tsūshō Sangyō Chōsakai.

---

[57]Murozono Mikio, interview by author, September 11, 1998.

[58]Saito Tadashi, interview by author, September 7, 1998.

MITI Kōgyō gijutsuin Kenkyū kaihatsukan shitsu. (1973, July). *Atarashii kurīn enerugī gijutsu no kaihatsu keikaku: Sanshain keikaku*. Self-pub.

MITI Kōgyō gijutsuin Sanshain keikaku suishin honbu. (Ed.). (1974). *Sanshain keikaku: Shin'enerugī gijutsu eno chōsen* [*The sunshine project: A challenge for new energy technologies*]. Tokyo: Daiichi Hōki Shuppan.

MITI Kōgyō gijutsuin Sanshain keikaku suishin honbu [MITI AIST Sunshine Project Promotion Office] (Ed.). (1980). *Kagayakeru taiyō enerugī: Sanshain Keikaku – taiyō enerugī kenkyū kaihatsu no genjō* (rev. edn.). Tokyo: Ōkurashō Insatsukyoku.

MITI Kōgyō gijutsuin Sanshain keikaku suishin honbu. (Ed.). (1984). *Sanshain keikaku jūnen no ayumi* [*Ten year history of the sunshine project*]. Tokyo: Sanshain keikaku jusshūnen kinen jigyō suisin konwakai [Sunshine Project 10th Anniversary Commemorative Projects Promotion Committee].

MITI Kōgyō gijutsuin Sanshain keikaku suishin honbu. (1985). *Showa gojūkyū-nendo seika hōkokusho taiyōhatsuden no kachi hyōka ni kansuru chōsa kenkyū* [*FY1984 Annual Report*]. Self-pub.

MITI Kōgyō gijutsuin Sanshain keikaku suishin honbu. (1986). *Shōwa rokujū nendo seika hōkokusho taiyōhatsuden no kachi hyōka ni kansuru chōsa kenkyū (2)* [*FY1985 Annual Report*]. Self-pub.

MITI Kōgyō gijutsuin Kenkyū kaihatsukan shitsu (Ed.). (1987). *Ōgata Purojekuto nijūnen no ayumi: Wagakuni sangyō gijutsu no ishizue o kizuku* [*The course taken by large projects in the last 20 years: Laying foundations for industrial technologies in Japan*. Tokyo: Tsūshō Sangyō Chōsakai.

MITI Shigen ererugīchō. (Eds.). (1993). *Enerugī seisaku no ayumi to shinten*. Tokyo: Tsūshō Sangyō Chōsakai.

NEDO BOOKS Editorial Committee. (Ed.). (2007). *Naze nihon ga taiyōkō hatsuden de sekaiichi ni naretanoka* [*Why Japan was able to become the world number one at photovoltaic power generation*]. Kawasaki: Shin'enerugī Sangyō Gijutsu Sōgō Kaihatsu Kikō.

NEDO. (Ed.) (1990). *NEDO jūnen no ayumi* [*Ten year history of NEDO*]. Tokyo: NEDO.

PVTEC. (1996). *PVTEC gonen no ayumi* [*Five year history of PVTEC*]. Tokyo: PVTEC.

Sawai, M., & Tsūshō sangyō seisakushi hensan iinkai [Editorial Committee on the History of Japan's Trade and Industry Policy] (Eds.). (2011). *Tsūshōsangyō seisakushi 1980–2000* [*History of Japan's trade and industry policy 1980–2000*] (vol. 9) Industrial Technology Policy. Tokyo: Keizai Sangyō Chōsakai.

Sharp. (1996). *Taiyō denchi no sekai kaiteiban* [*The world of solar cells revised edition*]. Self-pub.

Shigen enerugīchō. (Ed.). (1995). *Shin enerugī binran heisei shichi-nendo ban*. Tokyo: Tsūshō Sangyō Chōsakai.

Shigen ererugīchō. (2011). *FY2010 annual report on energy (Energy White Paper 2011)*. Tokyo: Nikkei Printing.

Taiyōkō Hatsuden Gijutsu Kenkyū Kumiai [Photovoltaic Power Generation Technology Research Association]. (Ed.). (1998). *Taiyōkō hatsuden: Sono hatten to tenbō* [*Solar power generation: Its development and prospects*]. Tokyo: Art Studio 76.

Watanabe, C. (Ed.). (2001). *Gijutsu kakushin no keiryō bunseki: Kenkyū kaihatsu no seisansei shūekisei no bunseki to hyōka* [*Numerical analysis of technological innovation*]. Tokyo: Nikka Giren Shuppansha.

Chapter 4
From the Rational Model to the Natural System Model: Changing Perspectives I

4.1 Case Study Summary

The previous chapter traced the history of the Sunshine Project from its creation to completion through a study of facts. It is clear that the project did not reach its original goals relating to the introduction and take-up of new energy technologies despite achieving some positive results in terms of research and development. The following summarizes this conclusion.

Wherever possible, the planners who were responsible for drafting policy for the Sunshine Project adopted a rational approach to promoting the technology development project. Unfortunately, it proved impossible to achieve the initial goals for introducing new energy due to events such as the unforeseen slump in oil prices and miscalculations relating to the level of technical difficulty involved. However, industry, government, and academia moved forward with the development of photovoltaic energy technology as a result of rational planning on the part of the government, resulting in the successful introduction and spread of photovoltaic energy due to a well-organized social infrastructure. A phenomenon whereby the initial goals are not attained, but specific development themes meet with success, emerges from this sequence of events. The factors here include the government's management of rational technology development and its limitations.

Let me review the main points of the case. The Sunshine Project was formulated before the oil crisis. Having recognized the signs of an energy crisis, the government had already promoted energy conservation policies and set up the Agency for Natural Resources and Energy. Anticipating the possibility of a future energy crisis, long-term and large-scale plans aimed at developing new energy were also formulated under the framework of industrial technology policy in 1973. Since the first energy crisis occurred in the autumn of the very same year, causing panic among the general population, the budget for these plans was approved amid a favorable reception by grassroots opinion and the project started the following year.

M. Shimamoto, *National Project Management*,
Advances in Japanese Business and Economics 25,
https://doi.org/10.1007/978-981-15-3180-4_4

Solar energy, geothermal energy, the use of coal, and the use of hydrogen were the four development objectives for the new energy project. Development of the technology for solar energy started promptly in 1974. At first, the targets were to build large-scale thermal power stations to generate solar energy and to develop photovoltaic technologies to produce inexpensive, high-performance solar cells. Several techniques for solar cells were developed in parallel by dividing the fields of expertise among all corporations involved in the project with each technique achieving a level of success.

When the second oil crisis struck at the end of the 1970s, the decision was taken to accelerate the Sunshine Project and to establish NEDO as the task force for project management. As of this period, proof of concept of large-scale plants proceeded at full tilt under the direction of NEDO. Unfortunately, it proved impossible for the solar thermal energy plants to reach the initial targets for energy generation due to a number of setbacks, including adverse weather conditions. On the other hand, the appearance of amorphous solar cells to compete with conventional crystalline silicon solar cells improved the performance of photovoltaic power generation. To increase the incentives for all corporations, the government set up a system of passing on the benefits to the private sector.

However, by the late 1980s new energy development was struggling in a headwind due to the slump in the price of crude oil. Still, the government acted to continue with the plans. Since there were no prospects for results that would balance out the low price of crude in the short term, NEDO continued to develop new energy while taking on the role of project manager for the development of other advanced technologies.

By the start of the 1990s, rising awareness of environmental problems led to the realization that new energy development contributed not only to the issue of resources, but also to the issue of the environment and, once again, interest in development increased. This provided the context for the government to support the introduction and spread of photovoltaic energy by revising legislation, introducing subsidies, and making efforts to build the infrastructure. By the second half of the 1990s, the systems developed by manufacturers for generating photovoltaic energy began to increase their production output and, subsequently, Japan became the world number one in terms of the production output and installation rates for photovoltaic energy.

As described above, the Sunshine Project was launched early to develop technologies through cooperation at the level of industry, government, and academia because the government sensed an impending energy crisis. Unfortunately, it was not possible to reach the goals for new energy introduction due to the unforeseen slump in the price of crude, but Japanese corporations built up technological strength in photovoltaic energy and other fields, developing products that came into widespread use. As suggested above, it was the government's skillful handling of national project planning that influenced the project outcome.

By studying this case we learn that the main turning points that separate planning success from planning failure are the accurate understanding of the external environment and appropriate responses on the part of planners. How swiftly and accurately planners recognize the external environment and take rational countermeasures is extremely important for project management.

Nevertheless, this chapter sheds light on implicit theoretical assumptions through a reexamination of the cases described in the previous chapter. To anticipate the conclusion, this is referred to as the "rational model" in management studies and organizational theory. It is one of the fundamental approaches dating back to the dawn of management studies and it is linked to approaches that understand organizations as mechanisms for achieving specific objectives. W. R. Scott defines rationality as the degree to which a series of actions is systematized to bring about a specific goal with maximum efficiency, arguing that the identification and formulation of goals is a characteristic of a rational system.[1]

Below, I will refer to the prior research to demonstrate that this approach carries a certain degree of validity with regard to the goal of managing technology development. However, while fully recognizing the advantages of this approach, this chapter also highlights some of the social phenomena that are difficult to explain or difficult to put into perspective using this approach and suggests directions for analysis in the next and subsequent chapters.

4.2 The Rational Explanation for Case Studies

In the descriptions of most of the cases in the previous chapter, the subject is the government or, more specifically, the Ministry of International Trade and Industry. The manner of description suggests that planners recognized the external environment, selected the appropriate methods, and delivered the development outcomes. However, it should be noted that adopting this point of view means that we regard implicit decision-making in a project organization as the action of a single rational entity. If we take this approach, it follows that the reasons for the success or failure of a project depend on whether or not the project is executed in a rational manner in light of the overall planning purpose. Therefore, the understanding and responses of the planners explain project outcomes. For this reason, the methods for guiding national projects toward success depend on the validity of the selection of policy methods and the understanding of the external environment on the part of the planners.

Decision-making by a single rational protagonist tends to simplify reality, but the techniques for making rough estimates are also simplified. Since a project organization is the sum of the actions of many individuals, it is advisable to consider the combined outcome based on an understanding of the actions of all participants. However, this would require considerable time and manpower. A simpler method is to look at the actions of the whole organization as if they were the

[1]Scott (1998, p. 33).

rational actions of an individual, a conjecture with major advantages in terms of the economic efficiency of surveys.

Based on this approach, it is possible to regard the actions of the government or other large-scale organizations as the actions of a single protagonist. In fact, when we use expressions such as, for example, "This is what Japan thinks" in everyday conversation, there is already an implicit understanding of the actions of the nation as the equivalent of the actions of a rational individual. In such cases, it is possible to draw analogies and to judge whether the actions of an organization in a particular situation are rational based on whether or not a rational individual would act in such a manner.

In *Essence of Decision*, Allison introduces political science research that employs the rational model (also referred to as the rational actor model or the classic model).[2] Below, I will clarify the characteristics of the rational model and shed some light on its advantages for case studies.

How do we go about finding clarification when confronted with events where the underlying reasons are not understood? Whether a question of war, conflict, or other examples from the realm of politics, or diversification, insolvency, or other examples from the realm of management, the analyst normally starts by trying to clarify the goals that the members—in particular, the leaders—of the administrative bodies, private sector enterprises, or other relevant organizations envisioned during the course of the incident. To be able to eliminate the goals that seem unlikely, the analyst starts by investigating the problems confronting the organization and the actions taken by the organization. If the analyst is able to demonstrate the calculations that form the basis for choosing a particular action in light of a specific situation and a specific goal, he or she is in a position to explain the intentions behind the actions in a way that is universally understood.

Let us look at some examples of the use of such explanations cited in Allison. If we use the rational model, it is possible to easily explain even the complicated actions of nation states during world wars or other major incidents in history. This is where the attraction of this model lies. For example, this is how Morgenthau employs the rational model to explain the reasons for the outbreak of the First World War.[3] Why did Austria declare war on Serbia? At the time, the European nations feared a disturbance in the balance of power and a decisive advantage for the enemy alliance in the Balkans. Therefore, Austria needed to strengthen its alliance with Germany by adjusting its relationship with Serbia. On the other hand, Russia supported Serbia and France supported Russia. As a result, the European powers were in balance and the danger of Austria being isolated was diminished. The fulfillment of such a rational goal is offered as the explanation for the actions by Austria.

So, is the rational model capable of explaining all phenomena? In fact, there are phenomena in the world where the rational model provides no answers. Hoffmann

[2] Allison (1971).

[3] Morgenthau (1972, pp. 191–193).

attempts to explain the choices of nations that are attributed to disengagement from rational decision-making or its norms.[4] Citing the example of American policy toward underdeveloped countries, he describes cases where actions by the government contain clear and obviously irrational contradictions as "schizophrenic" (a mental disorder or, in current usage, an integration disorder). This means that they cannot be explained using the rational model. However, in many cases, such phenomena are the result of exceptional or pathological situations. Unless we abandon the hypothesis that organizations act in a rational manner if they have an accurate understanding of the external environment, we must assume that there is some cognitive error in how they perceive the external environment, or some miscalculation in the decision-making process. Schwenk pays particular attention to such cognitive bias and the measures to prevent it when he applies the rational model to management phenomena.[5]

On the other hand, rational measures are more evident in cases where there is comparatively little cognitive bias. For example, when countries confront an obviously dangerous situation, such as a crisis of all-out nuclear war. Schelling explains that the likelihood of nuclear war is lowered by a stable equilibrium where neither side is capable of destroying the second-strike capability of the other side by striking first.[6] If the equilibrium were simply a matter of maintaining the stability, a rational country would perhaps consider ways of destroying the second-strike capability of the other party and then strike first. However, in a situation where the equilibrium is stable, i.e. when a second-strike attack is considered a guaranteed response to a first strike, no rational protagonist is likely to make a choice that would be the equivalent of the suicide of their own country. This is the obvious conclusion based on the idea that organizations act rationally.

Allison argues that even though Morgenthau, Hoffmann, and Schelling look at different subjects in their research, each study is based on the rational model. Organizations rationalize the reasons for choosing a particular course of action when analyzing social phenomena that are difficult to explain in the first place. Behavior is action based on intent and all that is required in terms of explanation is to indicate what goals that the government pursued at the time of taking action, or how this action was rationalized in the context of national goals. If the action is rooted in the rational model and as long as the organization is not schizophrenic, it should be possible to objectively explain the conduct of the organization.

One of the simplest models among all the possible variations is, for example, a situation where an organization is forced to respond to something completely external, such as a natural phenomenon. Whatever measures are implemented, natural phenomena do not change the way they respond. In such cases, the logical approach for the decision-making entity is to properly recognize the external environment and to take the most appropriate countermeasures. The problem of

[4]Hoffmann (1965).

[5]Schwenk (1988).

[6]Schelling (1960).

explaining an organizational phenomenon can also be reduced to mechanically choosing the most appropriate measures. The appropriateness of an action is determined by whether or not the action is truly suitable as a means of realizing some goal or other.

However, the external environment may also include other people or other organizations and, in such cases, another problem arises. A phenomenon that is often the subject of social studies is the complex situation where the actions of one organization influence another organization, causing the other organization to modify its actions. In such situations, the optimum choice for one organization comes to depend on its predictions of the actions of the other organization.

The rational model is valid in such cases as well. Kahn argues that in cases involving another party, it is possible to explain and predict the actions of the other organization by considering how a protagonist aiming to maximize value across the board would act.[7] When considering their own and the other party's hand, players base their calculations of what the other party will do on the assumption that the other party is a rational actor. Albert Wohlstetter makes the case for the advantages of the minimax analytical method where one side selects the optimum measure and, in response, the other side searches for its optimum response.[8] Based on these arguments, the rational model allows the analyst to effortlessly gain insights that are close to reality by simply imagining their own actions if they were in the position of the other party.

These approaches assume that, no matter what happens, a rational protagonist is unlikely to take measures that run counter to rational calculations and, as a result, decision-makers can convince themselves that their own reasoning is correct. We can predict the actions of other governments or rival corporations without getting off the sofa through a thought experiment where we consider the situation from the viewpoint of the other party.

Citing these cases in a discussion of the advantages of the rational model, Allison makes an extremely interesting point when he introduces Schelling's idea of signals. Allison refers to the idea as an example of the convergence of the rational decision-making of multiple protagonists. However, in terms of the need to explain how people understand signals and to what degree this is, in fact, rational, it can also be perceived as a clue to a deeper questioning of the flimsy grounds for the existence of the "rational" in the rational model.

Let me take Schelling's study one step further. If both parties know that it is impossible to launch a surprise attack in connection with an all-out nuclear war, the probability of a limited war increases. In such a situation, both parties need to recognize the limits they will not go beyond in order to prevent an escalation that would bring destruction to both parties. For example, in the case of the Korean War, the narrow part of the peninsula had pragmatic significance as a short line of defense, but it also signified a geographical point on the map where both parties,

---

[7]Kahn (1965, pp. 12–13).

[8]Wohlstetter (1964, p. 131).

whether occupying or retreating, could demonstrate their resolve. As a result, it functioned as a signal to prevent further escalation of the war. There is a point where it is easy to reach a consensus based on cultural practice or semantics in the actions of multiple rational players. Generally, this becomes the focal point.[9] The rational abilities of multiple protagonists vary depending on how the focal point is created. What then determines the creation of the focal point? This is a problem that must be considered in more depth in terms of an issue that people focus on when it is not possible to state categorically what is rational. In cases where the focal point is already known based on cultural practice, the focus remains on the referential starting point because it already fulfills this function even though it is not necessarily logical for either party. If so, we must interrogate the cultural and customary existence of such focal points in social phenomena. In fact, such matters cover the social gamut from the QWERTY keyboard to wearing a necktie at the height of summer. Is it possible to eliminate and change the *eigenvalue* of such implicitly determined focal points? As of the next chapter, one of the important themes in this book is the analysis of shifting focal points among multiple protagonists.

4.3 National Project Research

According to one line of thinking, the purpose of social science is to deliver things that are desirable to society (for example, better economic productivity, equality, educational opportunities for the disabled, better social solidarity) and to prevent things that are not desirable (for example, war, recession, racism).[10] If we adhere to this line of thinking, it follows that social science is social policy, or what Popper refers to as social engineering.[11] We could also call it social control through policy.

Here, I would first of all like to consider the methods for exploring the optimum measures for projects where goals and time frames are determined in advance, such as the introduction and spread of new energy. More specifically, in order to achieve the goals within the stated time frame, the administration formulates the project and provides the funding, corporations use money held in trust and subsidies to develop technologies, and utility models for new products are developed to spread the use of new energy. What are the requirements when considering such policies? The simplest idea is to collect as many plans as possible from all over the world and to separate the successes from the failures. If you can identify the elements that are shared across all the successful plans, it will be possible to predict the conditions for success. All you need to do is to survey large numbers of policy programs and individual technology development projects within the programs to identify the elements that are shared across the successful ones. If this strategy is successful, it

[9]Schelling (1960, p. 57).
[10]Little (1991).
[11]Popper (1960).

ought to be possible to consistently reproduce project successes. This is a problem for project system design where planners try to find ways to produce deliberate and straightforward results. Such methods are rooted in ideas that shed light on the relationship between the structure and the outcome of logically designed system, and the focus soon shifts to measures to solve engineering problems within the project.

In historical descriptions based on this perspective, the goals for the system are given facts established outside the system. The factual explanation is that, for any given goal, the system should select the most rational method. The idea being that, historically, systems have always behaved in the most rational way. Here, the assumption is that "the given objective" will not be modified by the system in question. The given objective remains consistent from beginning to end and the assumption behind the historical description is that the most logical choices were made. These are the natural assumptions made in past research of engineering project management.

For example, Yashiro Tomonari has summarized the implicit assumptions about conventional engineering issues in the following five points[12]: (1) clear and concise definitions, (2) a single principle can describe the whole system, (3) foreseeable, (4) a limited number of decision-makers, (5) discussions do not extend to standards. The descriptions of the case studies in the previous chapter of this book are largely based on these premises. However, the third point addresses the reasons for the unexpected failures explained by the drop in the price of oil and other events that could not have been foreseen.

According to Yashiro, project management is to "solve" any issues that come under the following premises: (1) lack of clear definitions, (2) inability to describe the whole system with one principle, (3) foresight is difficult, (4) presence of a number of decision-makers, (5) discussions extend to standards in some situations. This book also recognizes that the issue of how to approach decision-making is an important one for project management whenever several of the conditions for finding engineering solutions are undetermined. The discussion of the cases in the previous chapter already took the third condition into account. It was premised on the difficulty of making rational decisions when it is not possible to make predictions, and the inability to make decisions about rational measures for the project as a whole because it is not possible to predict future outcomes. Consequently, as of the next chapter, the attention will turn to the issue of how planners accurately recognize the external environment to the best of their ability and choose the appropriate response based on these premises.[13]

---

[12]Safety and Environment Center (SEC) for Petroleum Development, Engineering Advancement Association of Japan, *SEC Nyūsu*, No. 80, January 2012, pp. 7–8.

[13]The next chapter will consider the fifth condition and some cases where the discussion extends to standards. This is an example of how measures based on rational decision-making are obstructed by the existence of independent standards at subordinate organizations involved in formulating or implementing projects. It is suggested that if such standards influence the direction of a project, there is a need to predicate system design on these standards. In addition, I will look at the fourth

The discussion below considers the prior research on national projects with a focus on the Sunshine Project and similar projects. The rational model appears with some frequency among these examples.

Imai Ken'ichi has done pioneering research relating to the management of photovoltaic power generation under the Sunshine Project. As early as the start of the 1980s, Imai was a severe critic of the government's new energy development, saying that even though the development efforts were widely publicized, the degree of progress was in reality quite modest, pinpointing the lack of connection between the development of new energy and incentives for the promoting entities as the reason.[14] In Imai's opinion, government policy did not increase economic incentives for corporations.

Imai also recommended the following specific coping mechanisms: (1) focus on small-scale, diversified qualities, (2) link to corporate strategy, (3) establish a coordinating organization between the government and the private sector. Specifically, he identified the implementation of parallel development techniques, securing demand by means of tax incentives, and joint development with amorphous solar cells premised on compulsory consent for patent licensing as methods for producing good results for photovoltaic power generation, a technical development theme that had a small-scale and diversified character from the start.

Imai wrote the essay in the early 1980s so there are only references to the early stages of the Sunshine Project. However, his policy proposals are ahead of the times. Although the format differs, securing demand and joint development partnerships were among the policies adopted at a later stage.[15] In the essay, Imai refers to the VLSI Technology Research Association, a focus of attention at the time, and he proposes a joint development partnership for amorphous solar cells. Listing the conditions for success, he mentions setting dissociative development objectives that are detached from basic corporate strategies, selecting suitable members by narrowing down the participating corporations, and subsidies for software development such as public data banks relating to the use of solar power.

In 1985, Imai elaborated on the essay in his research on industrial policy for new energy development, for the information industry including computers, and for advanced technologies, such as biotechnology and new media.[16] Imai identified organizational issues at NEDO and later developments as the reasons for the lack of

---

condition in a later chapter (Chap. 7) where I discuss the difficulties posed for rational decision-making when various decision-makers are involved and politics becomes an issue. When rational measures are obstructed due to the presence of several decision-makers with conflicting interests, some kind of countermeasure is necessary. However, if politics determine the direction of a project, the process of how the members of the organization construct the political discourse to mobilize resources to achieve their own will is an important topic for project management research.

[14]Imai (1982).

[15]In addition to securing demand through tax incentives, subsidies were also offered to users at a later stage. Joint development partnerships included all solar cells, not only the amorphous ones.

[16]Imai (1985).

progress in new energy development. Specifically, he lists four conditions for NEDO, a semi-governmental corporation funded by the government and the private sector, and overseen by the government, to operate efficiently. They are (1) economic rationality in terms of expenses and benefits, (2) appropriate management technologies, (3) bureaucracy and policy-oriented methods (defensive barrier for policy), and (4) a favorable socio-economic environment. Imai claims that none of these conditions were met at NEDO.

According to Imai, the fourth condition was not met due to changing conditions for oil demand and supply up to the mid-1980s, which coincided with a shift among MITI bureaucrats toward a focus on the economic potential of energy selection. Therefore, MITI and NEDO had to abandon the creation of a defensive barrier for new energy development, thus undermining the third condition. As for the second condition relating to management technologies, unlike the VLSI Technology Research Association, NEDO controlled the subsidies throughout the whole process due to the lack of a direct research and development partnership. Imai argues that by weakening the third condition, NEDO also weakened its function as an organizer, turning new energy development into the administration of subsidies while increasing the money spent on technologies such as coal liquefaction technologies where there was little prospect for long-term profitability. In addition, Imai also argued that the relevant policymakers and NEDO would have to demonstrate considerable leadership because any future joint development partnership that involved solar cells had higher potential for conflicts of interest among the participating corporations than in the case of the VLSI Technology Research Association (PVTEC was set up in 1990, five years after the publication of the essay).

Imai's basic approach in these essays is to infer the conditions for the success of new energy development while pointing to the challenges derived from the examples of success at the VLSI Technology Research Association, and to evaluate efficiency and effectiveness by looking at deviations from reality. He places a particular emphasis on how to ensure development incentives for the participating entities. Seen from this perspective, Imai's evaluation of the system for developing new energy was severe. We will find out more about the specific facts that Imai used as the basis for his evaluation when we look at the second case study in the next chapter.

Let us look at another example of research based on the rational model. It is a series of studies by Watanabe Chihiro who was involved in the Sunshine Project in his capacity as a bureaucrat at MITI.[17] As mentioned when introducing some of the results at the end of the previous chapter, Watanabe targets the development of photovoltaic energy under the Sunshine Project and explains the project outcomes from the viewpoint of the technology–economic theory outlined below.

By calculating the budgeted amounts for the Sunshine Project and the amounts of corporate investment in the new energy industry and indicating how they are

[17]Watanabe (1994, 1995, 1998, 2001, 2007), Watanabe et al. (1998).

statistically relevant to each other, Watanabe shows that the Sunshine Project had the effect of inducing corporate investment in new energy. This viewpoint forms the basis for his evaluation of the contributions made by the project. Watanabe shows that the MITI budget investment in the Sunshine Project and the Moonlight Project persuaded the manufacturing industry to invest in the research and development of new energy and energy conservation, which contributed to expanding the stock of technologies and cutting the costs of researching and developing solar cells. Using regression analysis, Watanabe also shows that investment in energy conservation by the manufacturing industry led to a reduction in carbon dioxide, which contributes to resolving the environmental problems. Based on these results, he lists the following functions of the MITI energy technology policy. MITI (1) involved corporations across a broad range of fields in the planning for the Sunshine Project, the Moonlight Project, and other technology policy programs, (2) caused the technologies to have a knock-on effect beyond the industrial field and promoted technology exchanges, (3) attracted dynamic corporate activity to the development of extensive energy technologies, (4) stimulated the accumulation of technologies for energy research and development by means of these incentives, and (5) acted as a catalyst for energy technology substitutes in industry. According to Watanabe, MITI fulfilled these functions through the Sunshine Project and the Moonlight Project.

Watanabe's research is extremely lucid and uses actual numerical data from the Sunshine Project. It envisages corporate action and foreseeable reactions to the design of the government's policy incentives from a rational perspective. The Sunshine Project was a government policy that attracted corporate investment in new energy. It is clear that the results were not limited to the project framework but promoted the development of new energy. The context for the research is the hypothesis that corporations react in a uniform manner to government policy. The assumption is that governments will mechanically hand out incentives and that corporations will always behave rationally with regard to these incentives. As a result, this study should enable us to predict what kinds of incentives induce corporations to make voluntary investment in new energy. Consequently, the implication for policy is that the study makes practical contributions to building predictable models. Such are the practical advantages of the rational model.

4.4 Case Study Shortcomings

Below, I would like to draw attention to some points that have not necessarily been fully explained in the case studies in the previous chapter. Irrespective of the circumstances, the case studies in the previous chapter emphasize the adequacy of government planning for the successful cases while failures are ascribed to unexpected external environments. Such forms of writing are often found in company histories or other official histories of organizations, so it is a very convenient manner of description for the companies concerned as it does not question their own responsibility for failures. However, this means that there are flaws hidden behind this manner of description. The cases described in the previous chapter contain instances where it is doubtful that the government acted rationally.

Why was the timing just right for solar energy research at the time of the launch of the Sunshine Project? What inspired the national research institutes to start researching solar energy? It is unlikely that the government would have been able to launch the project if the research had not already started. There must have been some reason for the national research institutes to start developing new energy before the idea occurred to the government. To confirm these matters, we need to shift our attention to the internal circumstances at the national research institutes.

Why did the project turn out to be such an extremely ambitious one? Certainly, from a technical point of view, it is understandable that that the conventional five-year limit for large-scale projects was too short for developing new energy, but the grand proposal to extend the timeline fivefold to 25 years, setting the year 2000 as the deadline, and to meet the targets before the end of the twentieth century was somewhat lacking in credibility. Did the government actually formulate plans to develop the technology in stages by the end of the century? Was it not naïve to assume that it was possible reliably to foresee technical development into the distant future? To learn more about the government's judgment regarding these matters, we need to refer to the initial project proposal and to confirm how the project was scheduled to moved forward in subsequent fiscal years.

When amorphous solar cells emerged, why was NEDO in favor, incorporating them into the commissioned research under the Sunshine Project, even though the development was not originally anticipated? Of course, the decision may have been based on the judgment of experts thinking it highly likely that the technique would be a success story for technology but, even so, there may well have been other reasons for incorporating the technique into the project in the early stages and before it was properly viable.

In the initial plans, a semi-governmental corporation would be set up to act as the task force for developing new energy. NEDO was expressly created for this purpose, so why restructure the organization a few years later to add responsibilities for other technical development? Of course, the explanation that the slump in the price of crude oil led to a decrease in the importance of new energy development has the ring of authority. However, if NEDO was restructured to such a degree that the reason for its existence was completely changed, was it not also necessary for the development of new energy to turn over a new leaf from a technical perspective? What did the government think about options to discontinue the project?

Why was the Sunshine Project merged with the Moonlight Project without waiting for the original year 2000 deadline? To merge with another project when there were seven years remaining on the timeline only made it difficult to evaluate the final outcome of the project. Certainly, it is a fact that this was a time when people started to pay attention to problems in the environment, but new energy technologies and environmental technologies do not wholly overlap.

It is possible to throw doubt on the case studies in the previous chapter by looking at them from the variety of angles described above. To put it simply, these doubts indicate that the government does not necessarily always act in a rational manner. The government is not a monolith, but it consists of many subordinate organizations and the ways of thinking differ from one subordinate organization to another. The assumption that subordinate organizations always fulfill government directives in the form preferred by the government is also suspect. The case studies in the previous chapter never assumed that a situation might arise where the government appropriately judges the external environment and chooses a particular course of action, but the organizations inside and outside the government do not follow the directives. However, if we attempt to eliminate these assumptions from the discussion of the case studies in the previous chapter, the very rationality of the government becomes suspect and leads us to question whether rationality explains what actually occurred. Consequently, we have to increase the degree of detail and to survey the organizations that participated in the project in order to understand the events connected to the Sunshine Project. In the next chapter, we will look at the self-sustaining trend in organizations who sustain themselves and attempt to expand whenever there is a good opportunity.

References

Allison, G. T. (1971). *Essence of decision: Explaining the Cuban missile crisis*. Boston: Little.

Hoffmann, S. (1965). Restraints and choices in American foreign policy. In S. Hoffmann (Ed.), *Daedalus* 91, no. 4 (1962). Reproduced in *The state of war: Essays on the theory and practice of international politics*. New York: Praeger.

Imai, K. (1982). Shin'enerugī Kaihatsu no senryaku to soshikiron. Institute of Business Research, Hitotsubashi University, Discussion Paper no. 103.

Imai, K. (1985). Sentan gijutsu bun'ya ni okeru sangyōseisaku. In I. Miyazaki & T. Usui (Eds.), *Sentan gijutsu to Nihon Keizai* [*Advanced technology and Japanese economy*]. Tokyo: Nippon Hyōron sha.

Kahn, H. (1965). *On escalation: Metaphors and scenarios*. New York: Praeger.

Little, D. (1991). *Varieties of social explanation: An introduction to the philosophy of social science*. Boulder: Westview Press.

Morgenthau, H. J. (1972). *Politics among nations: The struggle for power and peace* (5th ed.). New York: Knopf.

Popper, K. R. (1960). *The poverty of historicism*. New York: Basic Books.

Schelling, T. C. (1960). *The strategy of conflict*. Cambridge, MA: Harvard University Press.

Schwenk, C. R. (1988). *The essence of strategic decision making*. Lexington, MA: Lexington Books.

Scott, W. R. (1998). *Organizations: Rational, natural, and open systems* (4th ed.). Upper Saddle River, N.J.: Prentice Hall.

Watanabe, C. (1994). Japanese industrial science & technology policy at a turning point. In *International Conference on Understanding Government R&D Investment Decisions*.

Watanabe, C. (1995). Identification of the role of renewable energy: A view from Japan's challenge, the new sunshine program. *Renewable Energy 6*(3).

Watanabe, C. (1998). Systems option for sustainable development. In OECD IEA CERT 18th Meeting.

Watanabe, C. (Ed.). (2001). *Gijutsu kakushin no keiryō bunseki: Kenkyū kaihatsu no seisansei shūekisei no bunseki to hyōka [Numerical analysis of technological innovation]*. Nikka Giren Shuppansha: Tokyo.

Watanabe, C. (2007). *Gijutsu keizai shisutemu [Techno-economic systems]*. Soseisha: Tokyo.

Watanabe, C., Kumiko, M., & Masakazu, K. (1998). *Gijutsu keizairon [Theory of techno-economics]*. Nikka Giren Shuppansha: Tokyo.

Wohlstetter, A. (1964). Analysis and design of conflict systems. In E. S. Quade (Ed.), *Analysis for military decisions: The rand lectures on systems analysis*. Chicago: Rand McNally.

Chapter 5
The Legitimacy of System Survival

Case Study 2: The Sunshine Project as an Activity Trap Mechanism

5.1 The Origins of the Sunshine Project

5.1.1 Long-Term and Large-Scale Plans to Avoid Risk

Solar energy research was submitted to the Large-Scale Project System in 1973, but in terms of the process there were no particular differences compared to earlier years.

The Agency of Industrial Science and Technology (AIST) had up to a dozen affiliated national research institutes including the Electrotechnical Laboratory (ETL) and the National Industrial Research Institute of Nagoya. These national research institutes were tasked with researching advanced technologies that were too difficult for the private sector to undertake.[1]

---

[1]MITI Kōgyō gijutsuin, ed., *Kōgyō gijutsuin shōkai* [Introducing the AIST], *FY1997 Edition* (self-pub., n.d.). The list of AIST-affiliated national research institutes in 1973 was as follows: National Research Laboratory of Metrology, Mechanical Engineering Laboratory, Government Chemical Industrial Research Institute (Tokyo), Fermentation Research Institute, Research Institute for Polymers and Textiles, Geological Survey of Japan, Electrotechnical Laboratory (ETL), Industrial Products Research Institute, National Research Institute for Pollution and Resources, Hokkaidō National Industrial Research Institute, Tōhoku National Industrial Research Institute, National Industrial Research Institute of Nagoya, Osaka National Research Institute, Chūgoku National Industrial Research Institute, and Kyūshū National Industrial Research Institute. The Government Chemical Industrial Research Institute (Tokyo) became the National Chemical Laboratory for Industry in 1981, and was subsequently reorganized again in 1993 together with the Fermentation Research Institute, the Research Institute for Polymers and Textiles, and the Industrial Products Research Institute; creating the National Institute of Materials and Chemical Research and the National Institute of Bioscience and Human Technology. Four research institutions were reorganized into two. Also in 1993, the National Research Institute for Pollution and Resources became the National Institute for Resources and Environment. Also during the same year, the existing Government Industrial Research Institutes were renamed as National Industrial Research Institutes (the emphasis of the original Japanese name "*shikenjo*" was on testing, rather than research). Accordingly, the term "national research institute" as used consistently throughout this chapter refers to the National Industrial Research Institutes in various locations around Japan as of 1993 onwards. Technically, it is more accurate to refer to the older versions of these

M. Shimamoto, *National Project Management*,
Advances in Japanese Business and Economics 25,
https://doi.org/10.1007/978-981-15-3180-4_5

At a specific time each year, the AIST accepted applications from research institutes for new research themes under the Large-Scale Project System. The development officers scored the submissions based on a set of procedures and selected appropriate research themes. Earlier research themes had included electronic calculators and jet engines for aircraft. These technologies became the foundation for industries that were important to the nation as a whole.

Below, I will undertake a more focused survey of the circumstances around the new applications to the Large-Scale Project System on themes related to energy.

Coincidentally, several applications related to new energy were submitted to the Large-Scale Project System in 1973. It was the year when the ETL and the National Industrial Research Institute of Nagoya made a joint proposal for research and development related to solar energy while the Osaka Industrial Technology Laboratory and the Government Chemical Industrial Research Institute (Tokyo) submitted a proposal for hydrogen energy. Until then, each research institute had separately pursued the research and development of these themes.

The AIST evaluated the future prospects and importance of the themes that emerged from the established procedures. As a result of the investigations by development officers, the themes related to new energy were selected for research and development in 1973.

If a development with a high level of difficulty is successful, it makes a contribution to the AIST, which is responsible for technology policy. However, research themes that are relatively difficult also run a high risk of ending in failure. Launching the new energy theme with a project that ended in failure would inevitably have led to public criticism for wasting taxpayers' money. In fact, some themes had fallen by the wayside because of technical barriers or lack of results.[2]

Therefore, a new long-term and large-scale system dedicated to new energy was devised as a method for overcoming such problems. From the technical perspective it is, of course, preferable to produce results as soon as possible, but it is not possible to make perfect predictions about bottlenecks and failures in research and development. In such situations, it is preferable not to evaluate technology development in terms of success or failure at an early stage. If a project is declared a failure in the early stages, this in itself damages the legitimacy of an entire new system and projects that would have succeeded over the long term are axed before they have had a chance to evolve. One effective approach to avoiding such a situation is to inflate the new system and to delay the timing for interrogating the results by placing the deadline for achieving the targets as far into the future as possible.

This new system was obviously linked to an expansion of the budget for industrial technology policy, which, in turn, was linked to the ability of the AIST to develop more varied technology policy on a larger scale than in the past. At the

---

institutions as government testing facilities or "*kokuritsu shikenjo*". However, in order to maintain organizational continuity with present-day organizations (and also given the existence of other institutions that were not using this name), the term "national research institutions" shall be used generically for all such institutions.

[2]Nebashi Masato and Suzuki Ken, interview by author, May 8, 1998.

same time, the new system agreed with the overarching objectives at MITI in this period. Since the Resource and Energy Department linked stable supplies to economic and industrial development, this was an important policy domain at MITI. The new system emphasized the significance of MITI as a whole in the field. In addition, the new system created the potential for setting up a new semi-governmental organization, which was also extremely important for MITI. It was a matter of momentous interest for the long-term survival and expansion of the organization in addition to the success of individual technology development projects.

In order to implement a particular policy continuously and over the long term, it is essential to secure a budget and to set up a structure where the division of labor is well-coordinated. If it were possible to delegate both project management and research and development to a new organization under its own jurisdiction, the AIST would be able to concentrate on drafting new policy and to avoid the complicated work of commissions and regulations. At the same time, an affiliated organization such as a semi-governmental corporation could be utilized as a reemployment destination for those who retired from the ministry. Even though none of the parties concerned made any explicit statement to such effect, it was an extremely important advantage for expanding the influence of ministries and government offices.[3] In fact, MITI later asked for a semi-governmental corporation to be established and continued to put in a budget request every year until the request was granted at the time of the second oil shock.

Here, we see that the survival and expansion of the policy-implementing system is on a different level than simply achieving the goals of a technology development project. As well as achieving the direct project goals, maintaining and expanding projects and their implementing organizations at the level of daily routines is the driving force behind organizational behavior at administrative bodies.

If we look into the matter closely, a situation where the survival of the system and the implementing organization is more important to the technology development project than the actual development of the technology suggests a subverted displacement of the goals. Why do these things happen? Below, we will take a look at how the system keeps itself alive while turning attention to the formalities of complying with in-house rules at organizations.

5.1.2 The Fight for Survival in the Electricity Sector

In the spring of 1973, the ETL submitted the new theme of solar energy to the Large-Scale Project System. A national research institute affiliated with MITI, the ETL has a history dating back more than 100 years to 1891, when the Denki Shikenjo was established as a testing center for insulators. How did the ETL come

[3]Suzuki Norio, interview by author, September 2, 1998.

to start researching solar energy? Below, we will go back several decades in time to trace the sequence of events that led to the ETL proposal for solar energy research as a theme for the Large-Scale Project System.

The late 1960s was a period of major transition at the Denki Shikenjo. In its vision for the 1970s, MITI hammered out the concept of the knowledge-based industry and the arrival of the information society, unveiling policies that pointed Japanese technology in the direction of information machines. By this time, the electron-related departments (the Physics Department, the Electron Department), newly established in the 1950s, had already cemented their status in basic research and development research in the context of a period of evolving electronic technologies and the electronic industries.[4] The electricity-related departments at the Denki Shikenjo (the Electricity Department, the Electrical Goods Department) were part of mainstream awareness and had played a central role for a long time. However, when the Denki Shikenjo as a whole shifted its policy from a focus on electricity to electrons, the old electricity-related departments also needed to work out new research themes in order to demonstrate their own relevance. Unless they could do so, the electricity-related departments would be unable to avoid downsizing or redundancy inside the ETL.

By the end of the 1960s, the national research institutes lost some of their significance as the Japanese economy grew. The reason was that a remarkable number of other types of research institutes were established at this time. To take the example of research related to electricity supplies, the Central Research Institute of Electric Power Industry (CRIEPI), which had been set up in 1951 when the nine power companies were launched, expanded its own facilities and took possession of large-scale generators that dwarfed those at the Denki Shikenjo. The electricity systems departments at the power companies also expanded their research facilities and it was no longer the case that systems and equipment research could only be carried out at the Denki Shikenjo. Amid the relative enhancements at other research institutes and the trend toward computerization at the Denki Shikenjo, the new Energy Department at the ETL had to work out new research themes to replace the conventional electricity research if it was not to lose the reasons for its existence.

On this occasion, one of the themes proposed by the Electricity Department was to research solar energy.[5] The solar energy theme was not originally set out in top-down policy terms with the aim of solving the energy issue. Rather, the proposal for the theme grew out of the organizational circumstances at the research institute. Following the restructuring of the organization in 1970, the Electricity Department and the Electrical Goods Department at ETL had been merged and reorganized as the Energy Department (Table 5.1).

Solar water-heaters using thermal energy, and solar cells for manmade satellites using the energy of light, had already been put to practical use albeit on a small

---

[4]Ōtani (1995), p. 750.

[5]Denki energy problem Energy Issue Study Group at the Electric Power Division of Denki Shikenjo (1970).

Table 5.1 Organization Change at the Denki Shikenjo and the ETL

Denki Shikenjo (as of April 1969)	**Electricity, Electrical Goods**, Electronic Calculators, Electronic Components, Electronic Processing, Physics, Standards, Applied, Materials, Control Departments
Electrotechnical Laboratory (as of July 1970)	**Energy**, Electronic Devices, Pattern Information, Electronic Calculators, Software, Control, Electronic Radio Waves, Quantum Technology, Basic, Materials, Extreme Technologies, Standard Measurements, General Affairs Departments

Note Bold font indicates departments related to electricity and energy
Source Compiled from organizational diagrams in Electrotechnical Laboratory 100th Anniversary Commemorative Projects Executive Committee 100 Year History Committee (1995)

scale and for specialized purposes. However, prior to this, no one had come up with the idea of using these technologies in systems for power generation. Researching solar energy required long-term endeavors on a large scale and, since this would be difficult for the private sector, it was a cause that the national research institutes could claim for the long term. The national research institutes did not have to immediately consider profitability and commercial potential, so it was important for them to indicate the relevance of their organizations by going ahead with research that was impossible for the private sector due to the profitability problem.

Solar energy research at the ETL started with the design of solar thermal energy systems. In 1973 when this research was getting off the ground, the ETL joined the National Industrial Research Institute of Nagoya, which had been researching solar furnaces in the same period, to submit an application for solar energy research to the Large-Scale Project System. This was the first step for the research and development project that later became the Sunshine Project.

5.1.3 Inflating Project Proposals to Obtain a Budget

The AIST accepted the proposal for solar energy and made it the core of a new system for developing new energy technologies, but the agency also decided to add several other themes related to new energy. It was a matter of hedging the risks to ensure the new system as a whole would not be at an impasse in case solar energy ended in failure. The agency promptly started to search out other new energy themes by polling its affiliates among the national research institutes. Since 1973 also happened to be the year when the Geological Survey of Japan (GSJ) launched a baseline survey of geothermal energy nationwide and started to research the hydrothermal systems in the geothermal areas, geothermal energy was hurriedly listed as a candidate for new energy and incorporated in the new system. The GSJ had conducted baseline research on geothermal reservoirs, the distribution of thermal water, and other topics relating to geothermal energy as early as 1958.[6]

[6] Geological Survey of Japan Editorial Committee (1982), pp. 84–85.

Coal gasification was another candidate incorporated into the new system because of a proposal from the National Research Institute for Pollution and Resources to research coal-utilization technologies. As long as some research on the themes had already been done, the risk of failure would be lower than completely new themes where it was difficult to predict success or failure.

Although assigned a low rating at first, hydrogen energy was also incorporated in the new system at a later stage, based on the expectation that it would become an important energy carrier if solar, geothermal, and synthetic natural gas energies were developed. In the end, there were four key themes. This is how the themes for the new program of researching and developing new energy technologies were completed.

Based on the sequence of events described above, it can be observed that in the process of creating the new system, the officers in charge of development did not choose the most suitable new energy themes from the perspective of technology, but sought out and combined related themes from a risk-avoidance perspective.

On May 18, 1973, the AIST organized a conference for research and development officers where the need for an organization to study measures for new energy development was debated.[7] Below, I will study the process of formulating the new system by tracing the process in detail. Observing the process will shed light on how the development officers maneuvered to create the new system that eventually became the Sunshine Project.

To start with, the May proposal positioned new energy development as the ultimate solution to Japan's energy problems and proposed a plan to formulate a "new system for energy technology development" to stimulate technical development.[8] This plan was accompanied by an organizational proposal to establish a New Energy Technology Development Department at the AIST, and a New Energy Technology Development Committee at the Industrial Technology Council (Fig. 5.1). At the time, neither system names nor budget proposals were worked out. At first, the proposal assumed that another system for technology research and development patterned on the frameworks of past large-scale projects, but focused on new energy would be launched.

The scale of the budget for the new system was an important decision at the next stage. At the time of launching the new system, the AIST considered requesting a small budget amount for the time being. The agency thought it best to first budget for a survey of promising new energy development themes and to obtain a budget suited to the technologies after identifying promising targets with a high potential to succeed.

However, the budget required for a survey of research themes is not necessarily large. Therefore, a request for a technically appropriate budget amount in the first

[7]Satō Masumi and Ezaki Kōzō (research and development officials, who both subsequently became Deputy-Director General of Development), who participated in the meeting, testify that the demands of the New Energy Development Department (which later became the Sunshine Project Promotion Office) were determined at the meeting. MITI Kogyo gijutsuin Sanshain keikaku (1984), p. 14.

[8]Suzuki (1973).

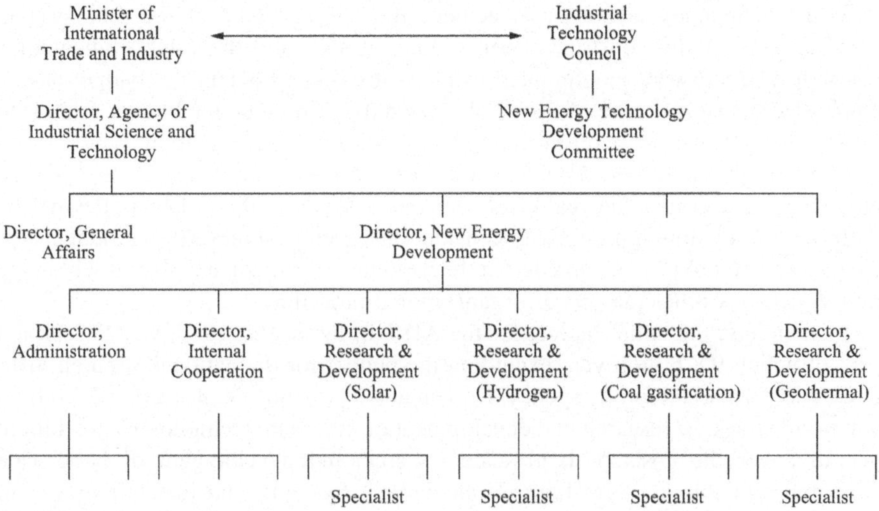

Fig. 5.1 Governing Structure for New Energy Technology Development (May 1973 draft). *Source* Suzuki (1973)

fiscal year might cause the Ministry of Finance, which administers budget allocation, to conclude that the system was of little significance. Based on past experience, the agency knew that asserting significance by boosting policy was advantageous for the budget negotiations. On this basis, the development officers scaled up the new system from the start, estimating a budget of ten billion yen.[9]

Ten billion yen is a substantial sum and it would have been difficult to use up the whole amount in one year by commissioning small-sized and individual research. Therefore, as is evident from the next draft proposal made in June, the proposal was reinforced with the creation of a semi-governmental research institute in the first fiscal year, and a proposal to construct a large-scale solar furnace requiring large amounts of funding.[10] According to the table summarizing research and development costs dated June, approximately 80% (8.8 billion yen) of the draft budget of 11 billion yen for the first fiscal year of the new system was allocated to solar power generation (3.8 billion yen) and to the semi-governmental research institute (five billion yen).[11]

---

[9]"We were told that a 500 million-yen budget is the hardest to secure. If it was 50 million we could secure it as an administrative formality, and if it was 5 billion we could make a political effort to secure it, but around 500 million is the hardest to secure. That's what we were told, so we said, 'OK, understood,' and went away. (After that) we were told to calculate an estimated budget of around 5 billion, but since 5 billion is, you know (i.e. not a nice round figure), we were then told to make it 10 billion. And so we had Okabe calculate an estimated budget of 10 billion" (remarks by Suzuki Ken). Nebashi Masato and Suzuki Ken, interview by author, May 8, 1998.

[10]MITI Kōgyō gijutsuin (1973).

[11]MITI Kōgyō gijutsuin (1973).

While technically necessary to achieve the ultimate goal of the new system, some aspects of the construction of a solar furnace and the establishment of a research institute were incorporated to make it easier to obtain the budget and to emphasize the significance of the whole system by scaling up the budget. That is to say, the budget was inflated to increase the likelihood of setting up the new system, and large-scale plans were established in order to spend the budget. The lesson to take away concerning the reality of national projects is that when policymakers draft plans, they do not necessarily consider only technical reasons, but budgets are also determined to provide an edge in the negotiations, to win agreement within the ministry, or for other reasons of organizational procedure.

Incidentally, the entire budget for the AIST in fiscal year 1973 was 23.2 billion yen, of which 9.8 billion yen, excluding the budget for the national research institutes, was earmarked for the agency. Broken down by item, it consisted of 2.3 billion yen in subsidies for research and development of important technologies, 6.5 billion yen to cover the costs of large-scale research and development of large-scale industrial technologies (the Large-Scale Project System), and one billion yen to cover other costs.[12] The figure of 11 billion yen requested in fiscal year 1974 was nearly twice the total budget for the eight projects under the Large-Scale Project System. Even if the budget did not pass in its entirety, the creation of the new system was certainly a huge step up from the conventional project budgets for the AIST.

5.1.4 Inflating the Project Proposal

The June proposal decided on the name of the project and clarified both the budget scale and specific targets for research and development.[13] This is when the new system came to be referred to for the first time as the Sunshine Project. The name implied that this was a project with due dates and targets, not simply a system (a program). By announcing binding targets and due dates, MITI made a strong commitment to its own project and promised to meet the expectations of public opinion. The expectation was that this approach would prove an effective means of convincing the Ministry of Finance when securing the budget because it emphasized unwavering resolve to achieve the targets.

As for the organizational structure, the plan to establish the semi-governmental New Energy Technology Development Center as the implementing body was noted in the June proposal.[14] The above-mentioned "semi-governmental research institute" had been translated into reality and, according to the proposed plan, the New Energy Technology Development Center was "a semi-governmental organization

[12]MITI Kōgyō gijutsuin, ed., *Kōgyō gijutsuin shōkai* [Introducing the AIST]. *FY1973 Edition* (Self-pub., n.d.).

[13]MITI Kōgyō gijutsuin (1973).

[14]MITI Kōgyō gijutsuin (1973).

that would also conduct joint research by accepting seconded researchers from private corporations and universities" in order to "establish and manage joint research facilities, and to construct and operate prototypes". On an organizational basis, the proposal to establish a semi-governmental corporation is the source of the subsequent New Energy and Industrial Technology Development Organization (NEDO) but, at the time, the proposal envisaged an agency with research functionality unlike the subsequent NEDO (Fig. 5.2).[15]

The June proposal was also the first to record specific targets and cost estimates for research and development. The final target for the project was "to develop by the year 2000 the technologies for new energy sources that would cover the majority of energy supply and demand for several decades to come".[16] As for individual themes, one large-scale power plant after another would be completed, including a 70,000 kW class SNG (synthetic natural gas) power plant by 1985, a 500,000 kW class geothermal power plant by 1990, and a two million kW class solar power plant by the year 2000 (Table 5.2).

The budget for the total cost of research and development was 1.1 trillion yen including 11 billion yen for research and development in fiscal year 1974. At this time, nearly 40% of the total budget by the year 2000 (409 billion yen) was earmarked for solar power generation (Table 5.3).

The next proposal dated July was essentially based on the June proposal, but it emphasized joint research between the Energy Technology Development Center, the national research institutes, and private sector corporations. It was also supplemented with a more detailed project timeline and budget for the long-term research and development plans (Figs. 5.3 and 5.4).

This confirms that in the period from June to July 1973, the Sunshine Project became a key policy at MITI.[17] The project details were defined to give an edge in the budget negotiations with the Ministry of Finance, and to give MITI as a whole the advantage in the event of budget approval. The plan was presented to the public in August the same year.

---

[15]It was as an eventual extension of this proposal that NEDO (New Energy and Development Organization)—a semi-governmental corporation for new energy development—was established. When it was established, NEDO possessed no research function of its own, and was established as an organization tasked with commissioning research and development work.

[16]MITI Kōgyō gijutsuin (1973).

[17]In July of that year, when the task of formulating the project plan had reached a suitable point, Suzuki Ken was scheduled to transfer to a new post as head of the Technical Survey Division, and would originally have had no further involvement with the Sunshine Project. However, the day before his scheduled transfer, Suzuki was called before Ōsono Hideo (who at the time was head of the General Affairs Division) and ordered to take a joint role, with continued responsibility for the Sunshine Project. Evidently, not even a ministry like MITI—with such a fast-paced personnel rotation—could afford to transfer the person who had originally proposed the project partway through its implementation. As a result of this decision, Suzuki continued in his role as a research and development official until July 1974, the following year. (Nebashi Masato and Suzuki Ken, interview by author on May 8, 1998; and "AIST Sunshine Project Promotion Office Genealogy" in *PVTEC gonen no ayumi* [Five Year History of PVTEC], by PVTEC (Tokyo: PVTEC, 1996), p. 110).

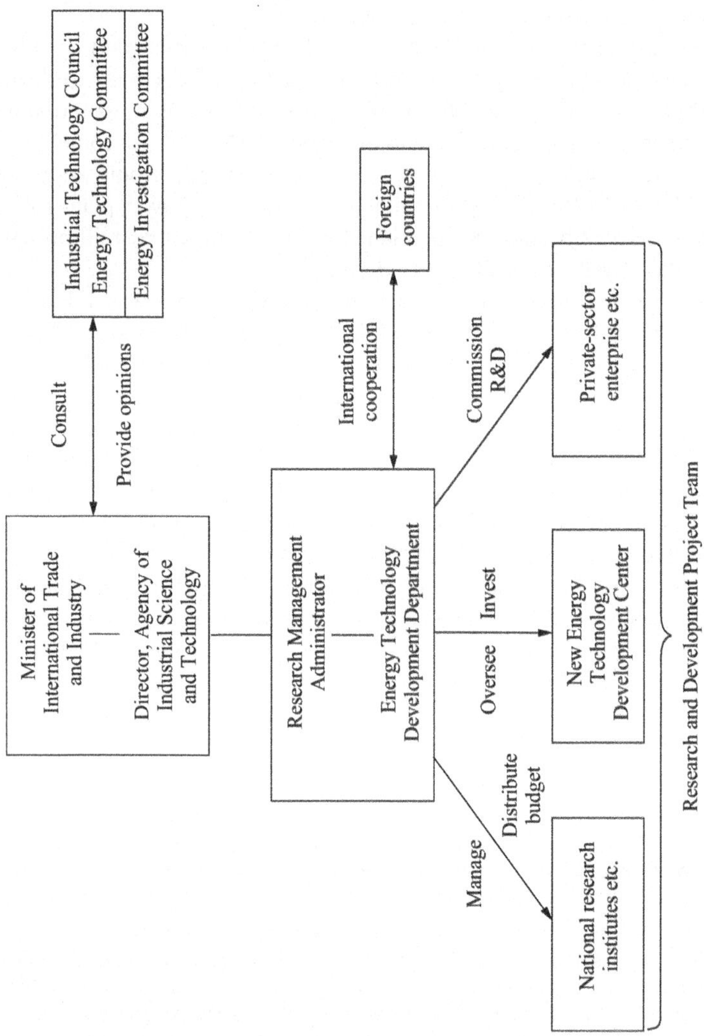

Fig. 5.2 Diagram of the System for Promoting the Sunshine Project (June 1973 draft). *Source* MITI Kōgyō gijutsuin (1973), p. 20

Table 5.2 Provisional Interim Targets by Theme (June 1973 draft)

Energy	Period	Development goal
(1) Solar energy	1980	Determine power generation method, complete basic research
	1999	Complete prototype for 100,000 kW class solar energy power plant
	2000	Complete two million kW class utility-scale power plant
(2) Geothermal	1990	Complete 500,000 kW class geothermal power plant
(3) Synthetic natural gas	1985	Complete 350,000 m^3/day SNG manufacturing plant
		Complete 70,000 kW class SNG power plant
(4) Hydrogen energy	1985	Finish applied research of new manufacturing technology for hydrogen
	1990	Complete commercial version of non-portable fuel cell
	1995	Complete commercial version of fuel cells for automobiles

Source MITI Kōgyō gijutsuin (1973)

Table 5.3 Summary table of research and development Costs (June 1973 draft) (unit: 100 million yen)

Theme	FY1974 R&D cost	Total R&D cost	R&D period
(1) New energy system general research	1	18	FY1974–FY2000
(2) Solar energy power generation	38	4,090	~2000
(3) Solar energy heating and cooling	3	75	~1990
(4) Geothermal steam power generation	2	770	~1990
(5) Volcanic power generation	9	865	~1990
(6) Synthetic natural gas production	3	66	~1985
(7) Synthetic natural gas power generation	1	34	~1985
(8) Hydrogen energy	3	190	~1990
Subtotal			
(9) New Energy Technology Development Center costs	50	1,500	1974–2000
(10) Emerging themes etc.	–	3,400	
Total	110	Approx. 11,000	

Note The research and development period for hydrogen energy differs from Fig. 5.4, but the original source has been left untouched
Source MITI Kōgyō gijutsuin (1973)

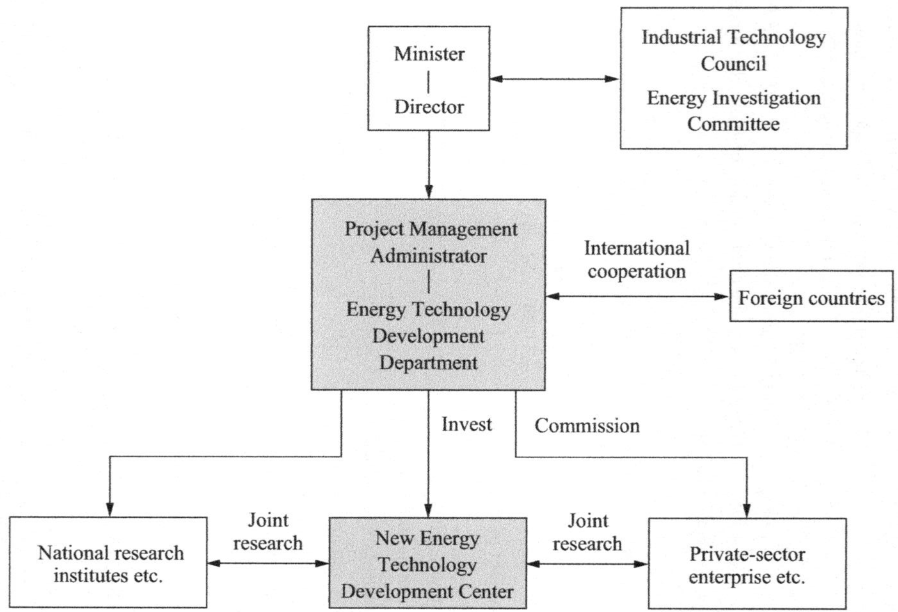

Fig. 5.3 Proposed Framework for Promoting the Sunshine Project (July 1973 draft) *Source* MITI Kōgyō gijutsuin Kenkyū kaihatsukan shitsu (1973, p. 7)

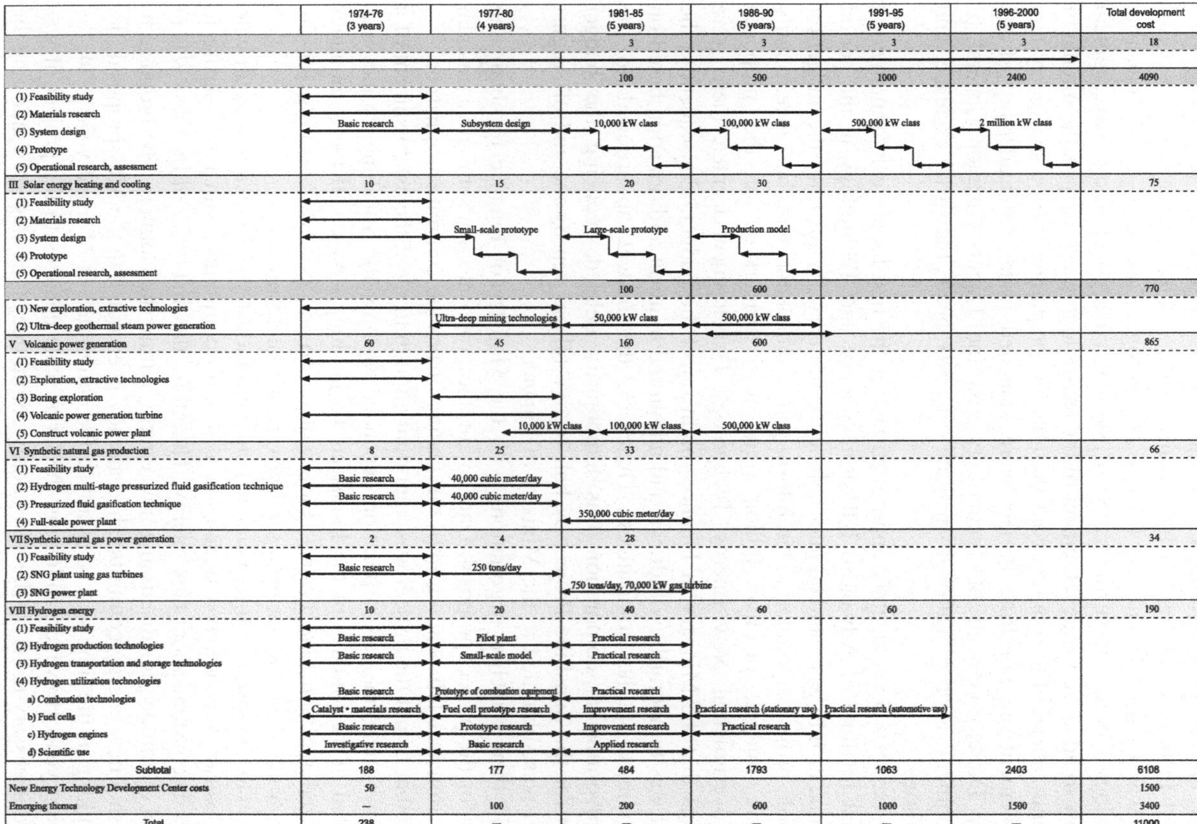

Fig. 5.4 Long-term Research and Development Plan (provisional plan, July 1973 draft). *Source* MITI Kogyo gijutsuin Kenkyu kaihatsukan shitsu (1973, pp. 14–17)

5.1.5 The Process of Consulting with Committees

On August 1, 1973, Nakasone Yasuhiro, Minister of International Trade and Industry, announced the outline of the Sunshine Project under the leadership of the National Institute of Advanced Industrial Science and Technology. The next formality was to consult with the Energy Technology Committee at the Industrial Technology Council about the nature of the new energy technology development and how to advance the development of the technologies. As early as the middle of the month, ahead of the coming consultations, the preparations for a specific system to implement the Sunshine Project were launched under the leadership of the Deputy Director-General of Technology and the Deputy-Director General of Development at the National Institute of Advanced Industrial Science and Technology.

The requests for budgetary appropriations for the first fiscal year formulated at National Institute of Advanced Industrial Science and Technology in the period from June to July were studied internally at the ministry. At this time, the budget for the whole project amounted to ten billion yen, a substantial sum, but since roughly half the budget, or five billion yen, was earmarked for the proposed semi-governmental New Energy Technology Development Center, there were warnings that the Ministry of Finance might reject the budget. Consequently, the request was temporarily withdrawn and adjourned to await authorization based on the findings of the committees. Therefore, the budget apportioned to the semi-governmental corporation was temporarily deleted, reducing the budget request for the Sunshine Project by half to five billion yen by the end of August when it was submitted to the Ministry of Finance.

MITI then launched negotiations with the Ministry of Finance to obtain the budget for the project proposal. Once again, the proposal for a semi-governmental corporation was left pending until the findings from the committees emerged.

Whether or not the project was important enough to merit the establishment of a new semi-governmental corporation depended on how the Sunshine Project was evaluated in the findings of the Industrial Technology Council. If the committees authorized the Sunshine Project, they would provide powerful backing for the request for a budget to set up a semi-governmental corporation.

When the Minister of International Trade and Industry attempted to consult with the Industrial Technology Council, there was no committee with the expertise to take on discussions about new energy. Therefore, the first order of business was to set up a committee for discussing new energy within the Council. The proposal to establish a new energy committee had already been noted in the May proposal, but the thinking at the time was to establish the committee as of fiscal year 1974, the following year. However, following the large-scale changes to the project, it was deemed necessary to set up the new energy committee at the stage of requesting the budget, so the work to set it up was hurriedly launched in August.

This committee was central to formulating the responses to Minister Nakasone's consultations with the Industrial Technology Council. It was not the task of the bureaucrats to freely determine the format of a national project. After all, such projects were based on the will of the nation, and the ultimate responsibility lay with the Minister of International Trade and Industry as the politician representing the people. The structure of the Council also allowed the Minister of International Trade and Industry to reflect the opinions of experts in the plans. Even though the majority of the plans were actually formulated by bureaucrats, this arrangement allowed the National Institute of Advanced Industrial Science and Technology to sidestep any criticism that the plans were arbitrarily created by bureaucrats and to claim that they were based on a general consensus of opinion in the nation.

Considering the size of the Sunshine Project, the agency decided by the end of the discussions that the profile of the head of the new committee must be raised. They came to the conclusion that Doko Toshio, then Vice Chairman of the Japan Business Federation (Keidanren), would be an appropriate appointment to the post of director. Considering the hierarchical relationship with the Chairman of the Industrial Technology Council, the committee was designated the "special" committee for energy technology in order to avoid a situation where Doko was viewed as a subordinate.[18]

During the period from August to December when the Special Committee for Energy Technology at the Industrial Technology Council was debating the answers, the AIST resumed project budget negotiations with the Ministry of Finance.

December was set as the deadline for the Special Committee for Energy Technology because, while waiting for the findings, the AIST had in mind to request that the proposal for a semi-governmental corporation, which had been temporarily postponed, should be lodged in time for fiscal year 1974. Having been unable to submit the request for a semi-governmental corporation to the Ministry of Finance in August, the agency now intended to submit the request to coincide with the end of the discussions in December and the budget proposal decision. The agency had already submitted a memo outlining such a request to the Ministry of Finance in August.[19]

As indicated above, there were no flaws in the formalities for completing the systematic and organizational framework for the Sunshine Project under the leadership of MITI.

5.1.6 The Oil Shock—A Godsend

After the Sunshine Project was announced, the customary budget negotiations with the Ministry of Finance finally began. At the time, the ministry had not yet adopted the use of a ceiling (limit on budgetary requests) and no limits had been set on budgetary requests submitted to the Ministry of Finance by other ministries.

[18]Nebashi Masato and Suzuki Ken, interview by author, May 8, 1998.

[19]Nebashi Masato and Suzuki Ken, interview by author, May 8, 1998.

Therefore, it was customary for each ministry to request budgets based on fairly generous estimates to start with, followed by intense negotiations with the Ministry of Finance that would try to reduce the amounts.

In order to explain the technologies for new energy during the negotiations with the Ministry of Finance, the AIST prepared a lot of resources with lucid drawings and diagrams illustrating the principles in order to explain the budget in detail to the general managers at the level of section head, who normally had no contact with technology. At first, the response from the Ministry of Finance to the large-scale project was indifferent.[20] However, when the first oil shock struck in October 1973, public opinion shifted. The budget negotiations for the Sunshine Project were conducted right in the thick of the crisis.

Originally, the AIST had increased the size of the Sunshine Project to emphasize the importance of the energy issue. The proposal for the first fiscal year confidently requested an unprecedented five billion yen. However, as a result of the energy crisis that followed the first oil shock, the plans gained an unexpected following wind at the time of the negotiations.

The Special Committee for Energy Technology published its findings in December. The report proposed increasing the scale of the project from the July proposal, which had been formulated by the AIST before the consultations. The reason was the first oil shock, which had struck in October, square in the middle between the July proposal and the December publication of the committee's findings.

Concerning the development targets for solar energy, in the process of its investigations, the committee, the authority in the field, had brought the development targets forward and increased the scale of the project. The July proposal by the AIST had planned for a prototype of a power plant in the 100,000 kW class by 1990 but, surprisingly, the committee findings set a 1985 target for developing a prototype of a power station in the 500,000 kW class.[21] The scale of the budget for the national project was not necessarily based on technical judgment alone; rather, it indicates a tendency to get carried away by optimistic targets based on wishful thinking, particularly in the extraordinary circumstances of the first oil shock.

We understand that the budget for the project was not calculated by scrutinizing the technology and building up the costs in minute detail, but the whole project proposal was formulated by counting back from the targets to be achieved by the end of the twentieth century, and the negotiations to obtain the budget were conducted on this basis.

The process of formulating the Sunshine Project was not a simple tale of rational development officers formulating a project with the aim of achieving technical

[20]Suzuki Norio, interview by author, September 2, 1998.

[21]Sangyō Gijutsu Chōsa Iinkai [Industrial Technology Council], "Shin enerugī gijutsu kaihatsu no susumekata ni tsuite [Points Regarding How to Advance the Development of New Energy Technologies]," in *Shin'enerugī gijutsu kenkyū kaihatsu keikaku (Sanshain keikaku)* [New Energy Technology Development Project (The Sunshine Project)] , ed. MITI Kōgyō gijutsuin (Tokyo: Nihon Sangyō Gijutsu Shinkō Kyōkai, 1974), p. 383.

targets. From the start, solar energy was the development theme that would point to the importance to the sector of the Electrotechnical Laboratory and other national research institutes for electricity transmission. Having concluded that it would be difficult to develop new energy within the Large-Scale Project System, the AIST created the new project as a measure to sustain the long development periods of the new energy development themes in a stable and long-term manner. The work was carried out systematically. The routines of an organization attempting to protect the legitimacy of the process, to obtain a budget without problems, and to maintain and expand its own jurisdiction provided the context for formulating the Sunshine Project.

Once the findings were published, what happened to the ins and outs of the budget request to the Ministry of Finance for a semi-governmental corporation, which had postponed by the AIST in August?

The summary of the findings of the committee clearly endorsed the significance of the semi-governmental corporation. According to the perspective presented in the findings "a semi-governmental corporation anchored in the government should be established as the core institution for bringing together the knowledge and vitality held at universities, national research institutes, and private enterprise since the development of these innovative and large-scale technologies will require huge resources, investment in human resources from many disciplines, and systematic management".[22] As a result, the proposal for a semi-governmental corporation was approved by the committee. Consequently, in December, as originally planned, the AIST put in a last-minute request to the Ministry of Finance for a budget for the proposed semi-governmental corporation once it had been endorsed by the committee. However, this was the year when MITI had proposed another semi-governmental corporation, the Japan International Cooperation Agency, and ministry policy had already decided to pass it as the first project. In terms of precedent, it would have been clear to MITI that requesting two semi-governmental corporations at one time was a lost cause, but MITI still took the plunge and submitted the budget request to the Ministry of Finance as planned. However, due to opposition at the Ministry of Finance, the semi-governmental corporation for new energy was excluded from the director level negotiations for the fiscal year 1974 budget.[23]

---

[22]Sangyō Gijutsu Chōsa Iinkai [Industrial Technology Council] , "Shin enerugī gijutsu kaihatsu no susumekata ni tsuite [Points Regarding How to Advance the Development of New Energy Technologies]," in *Shin'enerugī gijutsu kenkyū kaihatsu keikaku (Sanshain keikaku)* [New Energy Technology Development Project (The Sunshine Project)], ed. MITI Kōgyō gijutsuin (Tokyo: Nihon Sangyō Gijutsu Shinkō Kyōkai, 1974), p. 402.

[23]Nebashi Masato and Suzuki Ken, interview by author, May 8, 1998. Incidentally, budget negotiations proceed sequentially through the following stages: division manager negotiations, agency director negotiations, director-general negotiations, vice-ministerial negotiations, ministerial negotiations. The proposal for a new energy-related semi-governmental corporation was scrapped at the director-general level. Afterwards, however, Yamashita Eimei (who was Vice-Minister at the time) discussed the proposal, which had not been amongst the vice-ministerial negotiation themes, with the Vice-Minister for the Treasury. The Vice-Minister showed interest in

In the end, the Sunshine Project was given a budget of 2.442 billion yen for the first fiscal year. Even though this did not reach the 11 billion yen of the first budget proposal or the request for five billion yen submitted to the Ministry of Finance, it gave the AIST a foothold and the potential to expand the framework for the new policy program in the future.

Further, MITI did not abandon the proposal for a semi-governmental corporation when the budget request for fiscal year 1974 was turned down as the ministry submitted the request in every subsequent year including fiscal years 1975, 1976, and 1977.

This is how the Sunshine Project, having obtained a budget, set out on the road to implementation. In February 1974, the AIST set up the Sunshine Project Planning and Preparation Office and started to commission work relating to the technology themes from corporations. The following March, the gist of the Sunshine Project was finalized by ministerial decision. In April the Planning and Preparation Office became the Project Task Force and, in July, the basic policy and implementation plans were formulated. This is how the curtains opened on the project in August 1974.

This is what Kawata Michio, then Director of the AIST, said in 1984, ten years after the start of the project: "Since this year marks the tenth anniversary of the start of the Sunshine Project, I have asked all kinds of questions about the situation ten years ago, but it seems that the people who formulated the Sunshine Project back then did not necessarily launch the project because of the oil shock."[24]

It is evident from this testimony that even the director of the AIST had mistaken the Sunshine Project for a response to the oil crisis. Even though the real leaders of the project were the top managers at the agency, they were bureaucrats who handed over the work to their successors every few years when their posts were rotated. Naturally, with the parties responsible changing every few years, there is no affinity because no one stays in the same post for very long. This is how the memory of why a particular policy was launched fades amid the daily routines of an organization where no one bothers to question the origins of a national project. It turns into a trivial system of simply carrying on the work of one's predecessor within the organization.

---

the proposal, and the creation of a small-scale semi-governmental corporation was discussed. Yamashita then called Mabuchi Naozō—the head of the General Affairs Division—after 24:00 to inform him that a proposal for the semi-governmental corporation had been created. The next morning, however, a strong protest was lodged by the Ministry of the Treasury's Budget Bureau, saying that Yamashita's way of doing things was out of line for a public official. As a result of this, the proposal for the creation of the semi-governmental corporation ended as a one-night apparition.

[24]"Nihon no gijutsu kaihatsu to sanshain keikaku no yakuwari [Japan's Technology Development and the Role of the Sunshine Project]," *NEDO News*, September–October 1984, p. 3.

5.2 The Start of the Solar Power Project

5.2.1 Heat and Light—Association with Sunshine

The following is a consideration of the actual state of solar power research under the Sunshine Project in much more detail and from the perspective of the participating corporations.

At the launch of the Sunshine Project, the focus of solar power research was not on photovoltaic energy, but on solar thermal energy.[25] However, departments working on electronics within the ETL had, from time to time, undertaken research on photovoltaic energy. Heat and light both relate to solar power research, which is why research on photovoltaic energy, being associated with sunlight, was also incorporated as part of the project. Thus, the research plans for thermal and photovoltaic energy were bundled together and proposed in the name of solar power. A NEDO publication testifies to the details around the addition of the photovoltaic power project to the Sunshine Project: "When the application was submitted, solar thermal energy was considered a civic right, which is why we added photovoltaic energy to the application."[26]

At ETL, a group within the energy department was in charge of solar thermal energy, but since there were no experts in solar cells within the energy department, they requested assistance from the electronic devices department.

Since solar power research under the Sunshine Project straddled multiple technologies, the electronic devices department was also involved in the theme, positioned as energy research, and the support of the whole ETL was sought. However, according to statements by employees who worked in the energy department at the time, in comparison to semiconductors, rapidly expanding at the time, the electronic devices department considered the solar cell initiative a minor service they might work on between jobs.[27] VLSI (very-large-scale integration) research is an example of the groundbreaking success of joint research and development mainly carried out at the electronic devices department, which later led to a range of spectacular

---

[25]According to Kurokawa Kōsuke, who was involved in solar thermal research at ETL, "At the time, only a very small portion of people knew that it was possible to generate electricity using sunlight. When you talked about using solar energy, you were talking about the use of heat and solar thermal power generation for residential energy purposes, such as using the sun's heat to boil water in water heaters, to enable heating, and as a further advancement cooling" (NEDO BOOKS Editorial Committee, ed., *Naze nihon ga taiyōkō hatsuden de sekaiichi ni naretanoka* [Why Japan Was Able to Become the World Number One at Photovoltaic Power Generation] (Kawasaki: Shin'enerugī Sangyō Gijutsu Sōgō Kaihatsu Kikō, 2007), pp. 80–81.

[26]Testimony by Kurokawa Kōsuke (NEDO BOOKS Editorial Committee, ed., *Naze nihon ga taiyōkō hatsuden de sekaiichi ni naretanoka* [Why Japan Was Able to Become the World Number One at Photovoltaic Power Generation] (Kawasaki: Shin'enerugī Sangyō Gijutsu Sōgō Kaihatsu Kikō, 2007), p. 81).

[27]Horigome Takashi, interview by author, September 16, 1998, and Tanaka Kazunobu, interview by author, October 15, 1998.

outcomes.[28] This is why the concerns of the electronic devices department lay chiefly in the direction of semiconductors.

In the early days of the Sunshine Project, photovoltaic energy was an afterthought and there was no expectation of the kind of development that followed. At the start of the project, it was by no means obvious that the technology for photovoltaic energy held out any promise.

Given that the principles for solar thermal energy and photovoltaic energy are totally different in technical terms, the decision to include photovoltaic energy research in the Sunshine Project when selecting the themes was based on association—both themes involved sunlight—rather than any clear technical principles. Even though the decision was not necessarily rational from a technology viewpoint, there are many more examples of allowing something to pass in case of a formally legitimized process, or in case people are convinced.

5.2.2 Expectations of Prospective Corporations

In February 1974, the AIST advertised in the government gazette for contract research applicants under the Sunshine Project. At the time, AIST organized briefings at the auditorium of the Ministry of International Trade and Industry (MITI) where they conveyed the proposals, proposal methods, and proposal deadlines to the corporations that wished to participate. However, prior to these briefing sessions, AIST had started to unofficially brief the major manufacturers of the launch of the Sunshine Project and had already sounded out the corporations through their contacts at the MITI.

MITI had structures in place for sharing information with major corporations, so prominent corporations would find out about projects at an early stage. Therefore, the research and development officers at AIST already knew that Hitachi, Toshiba, NEC, Sharp, and Matsushita Electric were likely to bid for the solar cell commission at the start of the financial year. Although it was an open recruitment, in reality it was difficult for new external applicants to enter due to the short duration of the actual application period.[29]

Hitachi, Toshiba, NEC, and Toyo Silicon—with their track record of manufacturing silicon for semiconductor production—were, of course, allocated research

[28]At the time, Tarui Yasuo—who would later become director of the VLSI Technology Research Association's Joint Research Institute—was working in the Electronic Devices Department. It was when Tarui heard about this request from Komamiya Yasuo (director of the Electronic Devices Department) that he agreed to participate in the development of solar cells. (Horigome Takashi, interview by author, September 16, 1998.)

[29]However, there were also companies like Toyo Silicon (now SUMCO Corporation) that applied for the first time, without any prior relationship with AIST, after seeing the details in a government gazette. And so it came to be that the workload for commissioned research into solar cells was shared out between these six companies. (Kurokawa Kōsuke, interview by author, April 29, 1998.)

on the important crystalline silicon manufacturing processes. There is no doubt that technology strength was one of the criteria for selecting a corporation. As proof, Matsushita Electric and Sharp, who had experience of semiconductor production, were contracted to research and develop peripheral or alternative technologies such as module production, compound semiconductor solar cells.

However, if we look at past examples across a longer timeline, we start to perceive a clear pattern. When we look at the kinds of corporations that participated in national projects, we see that specific corporations participated many times. In fact, this trend indicates a conventional method for selecting corporations, already evident under the Large-Scale Project System. Approximately 479 corporations participated in a total of 31 large-scale projects implemented between fiscal years 1966 and 2001. Among them, 17 corporations had participated at least five times or more, and eight companies participated more than ten times—Mitsubishi Electric (18 times), Hitachi (17), NEC (14), Toshiba (13), Mitsubishi Heavy Industries (13), Sumitomo Electric (12), Fujitsu (11), and Ishikawajima-Harima Heavy Industries (11).[30]

In light of this, it is quite natural that Hitachi, NEC, and Toshiba had been sounded out about the photovoltaic energy project prior to the official announcement in the government gazette of the contract research. It is also possible that the fact that Sharp and Matsushita Electric were not counted among the regular project contributors worked against them at the time the selections were made for the research themes. The criteria for deciding which companies to contract, or which important technical development themes to allocate to which company, were based on technical strength in the relevant field at the time, but it also depended on these long-term, continuous, and mutually beneficial relationships of collaboration with MITI.

Ironically, contrary to expectations at MITI, Hitachi, Toshiba, NEC, and other large-scale electric manufacturers in the Kanto area, which were expected to have strong technologies, underestimated the commercial development of solar cells because, at the time, these companies had their eye on the rapidly growing semiconductor business. On the other hand, the commercial development of solar cells presented Sharp and Matsushita with acute operational issues. In the end, MITI entrusted important technical research and development to their regular contributors, which had little appetite for commercial development. We will look at the impact of this below.

[30]Sawai and Editorial Committee on the History of Japan's Trade and Industry Policy (2011), pp. 138–139.

5.2.3 Participation in the Sunshine Project by Corporations with Little Appetite for Commercial Development

Below, I will look at individual initiatives at Hitachi, NEC, and Toshiba—the companies that MITI pinned their hopes on—in terms of research and development of photovoltaic energy. These companies had a high level of technical strength and were expected to produce results under the Sunshine Project but, in all honesty, they had little appetite for commercial development of photovoltaic energy at the time. They were far more interested in semiconductors where substantial growth was expected, and the electronics business in general. At the point of accepting the MITI commission, the companies appeared to cooperate fully, as was the case with other contract research, but it cannot be denied that there was a tendency for half-hearted responses to research and development of products where plans for moving forward with commercial development were not defined. As a result, the engineers at each company made every effort, but the outcomes that were developed did not necessarily link up with voluntary commercial development at the companies. Some engineers even left the companies out of disappointment with the lack of appetite for commercial development.[31]

At the time of the first oil crisis in the fall of 1973, the planning office at the Hitachi Central Research Laboratory had discussed the potential for new energy and the effectiveness of research and development. Therefore, they had extensively investigated solar cells as well as hydrogen storage alloys and other research topics with future potential. Since the oil crisis, Hitachi had discussed responses to new energy research as an issue for the company, and separately from the MITI project. Solar cells had also been a topic for discussion, and the company had started to investigate the present situation and future potential.[32]

Similarly to other manufacturers, Hitachi had embarked on solar cell research in the early 1960s. However, in light of the small scale of the market, they abandoned product development as early as 1965. At the time, the market for solar cells mainly served unmanned lighthouses or space-related official demand, and Hitachi judged the scale to be no more than several tens of millions of yen, even if the company challenged Sharp, the market leader at the time.[33]

However, when the Sunshine Project was inaugurated in 1974, Hitachi took the lead right from the early stages of the project and participated in the contract research on solar power plants and solar cells.[34] Hitachi had already been in charge of research commissioned by the government in a wide range of fields, and had no objection to collaborating with the solar power research in order to maintain a good relationship with the government. As already mentioned, before the project was

[31]Saitō Tadashi, interview by author, September 7, 1998.

[32]Saitō Tadashi, interview by author, September 7, 1998.

[33]Saitō Tadashi, interview by author, September 7, 1998.

[34]MITI Kōgyō gijutsuin Sanshain keikaku suishin honbu (1976).

advertised in the government gazette, Hitachi had already been informed by their contacts at MITI of the contracts under the Sunshine Project.

At the time, Hitachi determined that the commercial development of solar cells would be a long-term commitment and that, for the time being, they had no intention of entering the solar cell market themselves.[35] Even so, MITI valued the technical strength of Hitachi as a leading manufacturer and Hitachi, together with NEC, ended up participating in the development of thin-film polycrystalline silicon, a new technology to replace the Czochralski (CZ) method in order to reduce the amounts of high purity silicon raw material used in production.[36] The thin-film technology is an important method of manufacturing semiconductors, and both AIST and ETL expected these two major Kanto area manufacturers to play a central role in the development of solar cells.[37]

However, as already stated, solar cells were of relatively low importance to Hitachi and other all-round manufacturers with extensive operations in development. In terms of companywide technology strategies, their interests had turned toward the growing area of semiconductors and computers.[38]

The situation was similar at Toshiba. For the Sunshine Project, the company had proposed the vertical ribbon technology for growing silicon. AIST appreciated Toshiba's track record of semiconductor technologies and handed the commission for development to Toshiba.[39] At the time, the ribbon crystal technique had also attracted interest from Sharp and Matsushita as an important method to replace the CZ method and to reduce the cost of the silicon raw material. According to media reports, Toshiba promptly set about development work and succeeded in March 1976 with an experiment to grow ribbon crystals.[40] However, for Toshiba as a whole, such research was peripheral to the preferred semiconductor business. Later, when plant facilities for experimenting with practical applications were required, they rented space at a semiconductor factory.[41]

Meanwhile, NEC was aiming to put solar cells to practical use in its unmanned microwave relay stations, having deemed low-maintenance solar cells convenient for this purpose. In fact, in 1977, NEC used solar cells in its microwave relay systems for the Middle East where sunlight duration is long.[42] In 1980, NEC also adopted solar cells for the microwave relay station linking the cluster of hydroelectric power stations on the Tadamigawa River with the main power station.[43] Although inserting solar cells into their own products had a high degree of utility

---

[35]MITI Kōgyō gijutsuin Sanshain keikaku suishin honbu (1976).

[36]Saitō Tadashi, interview by author, September 7, 1998.

[37]Suzuki Norio, interview by author, September 2, 1998.

[38]Saitō Tadashi, interview by author, September 7, 1998.

[39]Kurokawa Kōsuke, interview by author, April 29, 1998.

[40]*Nikkei Sangyō Shimbun*, March 18, 1976.

[41]Saitō Tadashi, interview by author, September 7, 1998.

[42]*Nikkei Sangyō Shimbun*, January 11, 1977.

[43]*Nikkei Sangyō Shimbun*, October 2, 1980.

value for NEC, the company found it difficult singlehandedly to incorporate solar cells in their flagship business domain of Computers & Communication (C&C).

Recognizing that Hitachi, Toshiba, and NEC possessed outstanding technical skills in the area of semiconductor technologies, AIST expected that these technologies would be fully demonstrated in the development of solar cells. Of course, many engineers at these companies, particularly in the research departments, were enthusiastic about developing crystalline solar cells. Nonetheless, these corporations positioned semiconductors and computers as their strategic targets from the late 1970s through the 1980s and solar cells were only peripheral to this context. Even though the companies had accepted the research and development contracts because of their associations with MITI, creating their own business plans for solar cells remained an uncertain undertaking. In fact, none of these companies developed genuine commercial applications for photovoltaic energy during the course of the Sunshine Project.

5.2.4 AIST Policy of Transition to Domestic Production

The major electrical appliance corporations in Kanto were half-hearted about solar cells, but the Kansai-based corporations had high hopes for solar cells and photovoltaic energy systems. Sharp was one of the companies consistently and enthusiastically pressing ahead with the commercial development of solar cells. The reasons are outlined below.

In the early days of semiconductors in the 1950s when Matsushita Electric, Sanyo Electric, Mitsubishi Electric and others entered the market one after the other, Sharp at first adopted a strategy of wait and see. At the time, Sharp felt it would be futile to enter the semiconductor market and to team up with other companies to compete. The company recognized that semiconductors would become a core technology for the next generation but, at the time, they were forced to abandon plans to enter the market.

In this context, the solar cell was, in terms of technology, a relatively simple application of semiconductors, and a promising product with the potential to give Sharp a way to enter the market. Therefore, Sharp decided to develop solar cells in-house. The company positioned solar cells alongside lasers to be used to convert electricity from light (solar cells), and light from electricity (laser). Perceived as a pair in the domain of optoelectronics, the technologies were targeted for aggressive development as the foundation for prominent products.[44] For these reasons, solar cells became a promising line of technology strategy at Sharp.

As early as 1962, Sharp had launched a transistor radio with a built-in solar cell but, at the time, the cost of the solar cell itself was so high that these products were not sold to the general public. In 1963, Sharp had successfully set up mass

[44]Kimura Kenjirō, interview by author, September 9, 1998.

production of solar cells, installing solar cells in 13 lighthouses nationwide. By 1966, the number of installations of lighthouses, light beacons, and buoys had expanded to around 100 locations across the country and, by 1967, Sharp had also started to develop solar cells for use in space. In 1972, solar cells by Sharp were certified by the National Space Development Agency of Japan. However, from an overall perspective, prior to the Sunshine Project, most demand for solar cells was for special applications such as space, lighthouses, buoys, or unmanned rain gauges. Solar cells were not widely used in consumer products.[45] This is why the solar market never developed on a large scale.

In this context, Sharp was highly encouraged by the government's decision to embark on a policy of comprehensive support for technology training under the Sunshine Project, which is why the company put itself forward for contract research. AIST allocated research into module manufacturing technologies to Sharp. This decision was based on the fact that the company already had experience of manufacturing solar cells.

Matsushita Electric was another enterprising participant in the Sunshine Project. In 1974 when the project was launched, Matsushita Electric immediately stated that it would participate in the contract research and AIST allocated to the company the development of solar cells using Group II–VI compound semiconductors. Originally, the company had researched cadmium sulfide (CdS) and other compounds, but development had been temporarily stopped due to the adverse image of cadmium as a raw material.[46] However, later, when it was demonstrated that it was possible to create a stable substance by adding other chemical elements to cadmium, the idea materialized of using it to produce cheap solar cells. On this basis, Matsushita Electric proposed using compound semiconductors for solar cell development under the Sunshine Project.

Although expectations of compound semiconductor solar cells were high, Matsushita Battery Industrial, another company in the Matsushita Group at the time, was also planning to enter the market for silicon semiconductor solar cells. Since AIST had already commissioned Matsushita Electric, another Matsushita Group member, to research compound semiconductors, Matsushita Battery Industrial was not approved for the Sunshine Project.[47]

This is how Sharp and Matsushita Electric won the contracts to research respectively module manufacturing technologies and compound semiconductor solar cells, but from the overall perspective of solar power research under the Sunshine Project, these themes were peripheral. AIST valued the technical strength in semiconductors at Hitachi, Toshiba, and NEC, and commissioned these corporations to develop technologies for silicon manufacturing techniques with a focus on reducing the costs of solar cells. However, Sharp and Matsushita Electric also

---

[45] Sharp (1996), pp. 112–113.

[46] Tanaka Kazunobu, interview by author, October 15, 1998.

[47] Tanaka Kazunobu, interview by author, October 15, 1998.

needed to develop the technology and to find some way of manufacturing silicon at a cheap price to develop commercial solar cells in-house.

Sharp decided to advance their own method of manufacturing silicon crystals. At this point, Sharp learnt that Mobil Tyco in the United States had developed ribbon crystals using the edge-defined film-fed growth (EFG) technique, and this was the technology Sharp decided to introduce. When Sharp contacted Tyco, they were told that Kyocera had already introduced the technology for a different purpose. As a result, Sharp consulted with Kyocera. When Kyocera found out that the technology could also be used to manufacture solar cells, they resolved to participate in the planning of the business. This is how Kyocera decided to join forces with Sharp to participate in the development of silicon manufacturing methods under the Sunshine Project. They submitted a proposal to AIST, indicating that they would like to research ways to improve the ribbon crystal technique developed in the United States within the framework of the Sunshine Project.

By rights, if the manufacturing method is outstanding from the technology viewpoint, promoting a project to improve the manufacturing method parallel with the Sunshine Project ought to have been a valid proposal, as it would increase the chances of success for the project as a whole. Even though the technique was valid from a technology perspective, AIST turned down the application from Kyocera and Sharp to protect the legitimacy of the system.[48] The reasons are outlined below.

The underlying technology that the two companies were proposing to use was the property of a U.S. corporation. Therefore, if the method proved successful and production of actual solar cells could begin, there were concerns about paying royalties to the U.S. corporation. It was not sustainable for the Japanese government to use Japanese taxpayer money to improve a foreign-owned technology in the name of a national project. After all, it was AIST policy at the time to develop Japanese technologies.

AIST had already commissioned Toshiba to research the ribbon crystal technique in fiscal year 1974 and, from a project policy viewpoint, it was not possible to commission several companies to simultaneously develop the same technology. If things did not go well, there were concerns over censure for duplicate investment when negotiating the budget with the Ministry of Finance. Since AIST was hoping for a stable and long-term continuation of the project, this was something they wanted to avoid. In the end, AIST decided not to accept any proposals that would introduce major changes to the policy for the Sunshine Project, which was already underway, and the Sunshine Project was implemented in line with the processes already determined. As a result, the application from Kyocera and Sharp was rejected.

With some reluctance, Kyocera decided to develop the ribbon crystal technique on their own. Therefore, Kyocera appealed to Sharp, who also wanted to develop the silicon crystal technique, and to Matsushita Electric. Together, the three companies created a new joint venture. In October 1975, the Japan Solar Energy

---

[48]Kimura Kenjirō, interview by author, September 9, 1998.

Corp. was launched, and work began on developing the ribbon crystal technique based on the EFG technique from Mobil Tyco.

Unlike the Sunshine Project, Japan Solar Energy started to research and develop the ribbon crystal technique as a purely private sector corporation. Contrary to prior expectation, it turned out that it was not possible to manufacture silicon crystals of the expected quality even with the technology introduced from the United States. The development of ribbon crystal was extremely difficult.

According to the initial contract, the arrangement was that Sharp and Matsushita Electric would purchase the silicon crystals manufactured by Japan Solar Energy. However, since the performance was inferior to expectations, both companies hesitated to buy the silicon and use it to manufacture solar cells. So, to demonstrate its raison d'être and to keep their business going one way or the other, Japan Solar Energy decided to expand to manufacturing solar cells using their own silicon crystals. By doing it themselves, they also wanted to prove that it was possible to manufacture solar cells using their ribbon crystals.

However, this foreshadowed the collapse of the joint venture that came later. Sharp and Matsushita Electric were opposed to the actions of Japan Solar Energy and withdrew funding, leading to an impasse in the joint venture project in 1978. Japan Solar Energy was caught in a dilemma. Acting rashly to demonstrate their own importance and to try to keep the business going could also be interpreted as causing the joint venture to disintegrate mid-air. The fact that this even happened indicates that the companies that had agreed to buy the silicon crystal could not wait for the outcome of the development—a private sector business venture cannot afford to ignore the passing of time and to leave a situation unattended when results are not immediately forthcoming. In comparison, national projects have leeway in terms of both time and funding and, in this sense, they are able to wait long period of time for results. The Sunshine Project was also unable to achieve success with the ribbon crystal technique. However, before this became clear, the efforts of the private sector companies to tackle the challenge to the Sunshine Project were brought to a standstill.

5.3 Accelerating the Project and Establishing NEDO Due to the Second Oil Crisis

5.3.1 Insistence on Setting Up a Semigovernmental Corporation

For some time, MITI had coveted a new semigovernmental corporation to extend their own authority, using the Sunshine Project as a foothold. The idea of setting up a research institute or an organization in the form of a semigovernmental corporation to develop new energy had been aired already before the project was established. As previously mentioned, AIST played a central role in producing the original draft for the Sunshine Project in June 1973, which included a proposal to set up the New Energy Technology Development Center (provisional name).[49] The AIST plan assumed that due to the requirement for knowhow of specialized R&D management over the long term of the project, a research and development agency would be set up separately from the AIST to serve as an implementing agency from the beginning.[50] However, while waiting for authorization from the Industrial Technology Council, the proposal for the semigovernmental corporation submitted in December 1973 became unfeasible because in the same year MITI proposed another semigovernmental corporation. As a result, the organization could not be established in fiscal year 1974, the first year of the project. Immediately after the end of the budget negotiations, AIST submitted a document in the name of the director of AIST to the vice-director of the budget office, stating intent to set up a semigovernmental corporation in the following year, fiscal year 1975.[51]

In the new policy of June 1974, which included the draft budget for fiscal year 1975, the second year of the project, AIST changed the New Energy Technology Development Center, the first designation, to the New Energy Technology Development Agency (provisional name), a semigovernmental corporation. According to the draft, it was assumed that the agency would have a total of 120 staff posts including a director, a vice-director, a board of directors (six persons), and division heads (eight persons). The purpose of the agency was to "start up and develop comprehensive technologies in order to accomplish the established goal" of developing new energy technologies.

It was assumed that special accounts for industrial investment and subsidies for corporate operating expenses would be allocated in this budget, and plans were made to obtain a site for the corporation's development work (develop a large-scale geothermal center, a solar energy center, a coal gas liquefaction plant, a high-efficiency water electrolysis plant etc.), and that a subsidy of ten billion yen would be granted out of the industrial investment special account (100% subsidy) in the first fiscal year to cover the cost of constructing the main test building.[52] At the

---

[49]MITI Kōgyō gijutsuin (1973).

[50]One reason why the need for the establishment of an executive body for the Sunshine Project was so strongly argued was that AIST also had the following ulterior motive: "I made the suggestion that we should scrap the idea (of establishing a semi-governmental corporate entity), but was told that actually surely it was a good idea because it would be a good place for *amakudari* [i.e. re-employment of ex-government officials in private sector or private or semi-governmental posts]. They even said to me that if we didn't create an organization like this, then maybe there would be no such place for me in the future when I myself left government service. This is basically the case most of the time, whenever a government agency creates a (semi-governmental) corporation." (Suzuki Norio, interview by author, September 2, 1998.)

[51]"FY1975 New Policy Consideration Materials (Second Draft)," in "*Shōwa gojū-nendo Shin Seisaku* [*FY1975 New Policies*]", by MITI AIST Minister's Secretariat, General Affairs Division, memorandum, 1974.

[52]"FY1975 New Policy Consideration Materials (Second Draft)," in "*Shōwa gojū-nendo Shin Seisaku* [*FY1975 New Policies*]", by MITI AIST Minister's Secretariat, General Affairs Division, memorandum, 1974.

end of the new policy, there was also an emphatic reminder that "according to the findings of the Industrial Technology Council on how to promote new energy technology development, dated December 18, 1973, it is necessary to set up a semi-governmental corporation".[53]

The proposal assumed that the relationships between the Office for Developing the Sunshine Project at the AIST, the national research institutes, the private sector, and the New Energy Technology Development Agency would look like the structure outlined in Fig. 5.5. The basic plan was that AIST would give its undivided attention to formulating plans for developing new energy technologies and obtaining the budget, and that other startup functions in the implementation stages would be outsourced to the New Energy Technology Development Agency.

Based on this proposal, MITI started to negotiate the budget for the new energy set-up with the Ministry of Finance in December the same year. However, due to a policy of constraints on any increases in semigovernmental corporations consequent on budget austerities, they were once again forced to accept a zero budget and to postpone the request for funding for building the new facilities until fiscal year 1975.

In September 1975, the third year, MITI clarified the policy of establishing the New Energy Technology Development Research Center (provisional name) in fiscal year 1976.[54] The designation for the new corporation had transitioned from the Technology Center to the Technology Development Agency, and now the Technology Development Research Center.

However, the Liberal Democratic Party (LDP) Special Committee on Administrative and Fiscal Reform reorganized one third of the existing semigovernmental corporations in 1976 and 1977, and brought out a policy of approving no new semigovernmental corporations at all in fiscal year 1976. Finally, in January 1976, the negotiations were taken as far as the three key officials of the LDP and the Minister of Finance, Ōhira Masayoshi, but the negotiations fell through and the plan was postponed for a third time.

Since MITI rejected a third consecutive budget proposal for the semigovernmental corporation, AIST reluctantly started to consider plans to use the Electric Power Development Co., Ltd. or other existing organizations as an alternative in the implementation stages of the Sunshine Project. In January 1976, having received the AIST application, Electric Power Development announced that they were prepared to accept contract development,[55] but the proposal did not go past the consideration stage, and outsourcing to Electric Power Development was postponed.

---

[53]"FY1975 New Policy Consideration Materials (Second Draft)," in "*Shōwa gojū-nendo Shin Seisaku* [*FY1975 New Policies*]", by MITI AIST Minister's Secretariat, General Affairs Division, memorandum, 1974.

[54]*The Nikkei/Nihon Keizai Shimbun*, September 6, 1975.

[55]*The Nikkei/Nihon Keizai Shimbun*, December 20, 1975; December 31, 1975; and January 8, 1976.

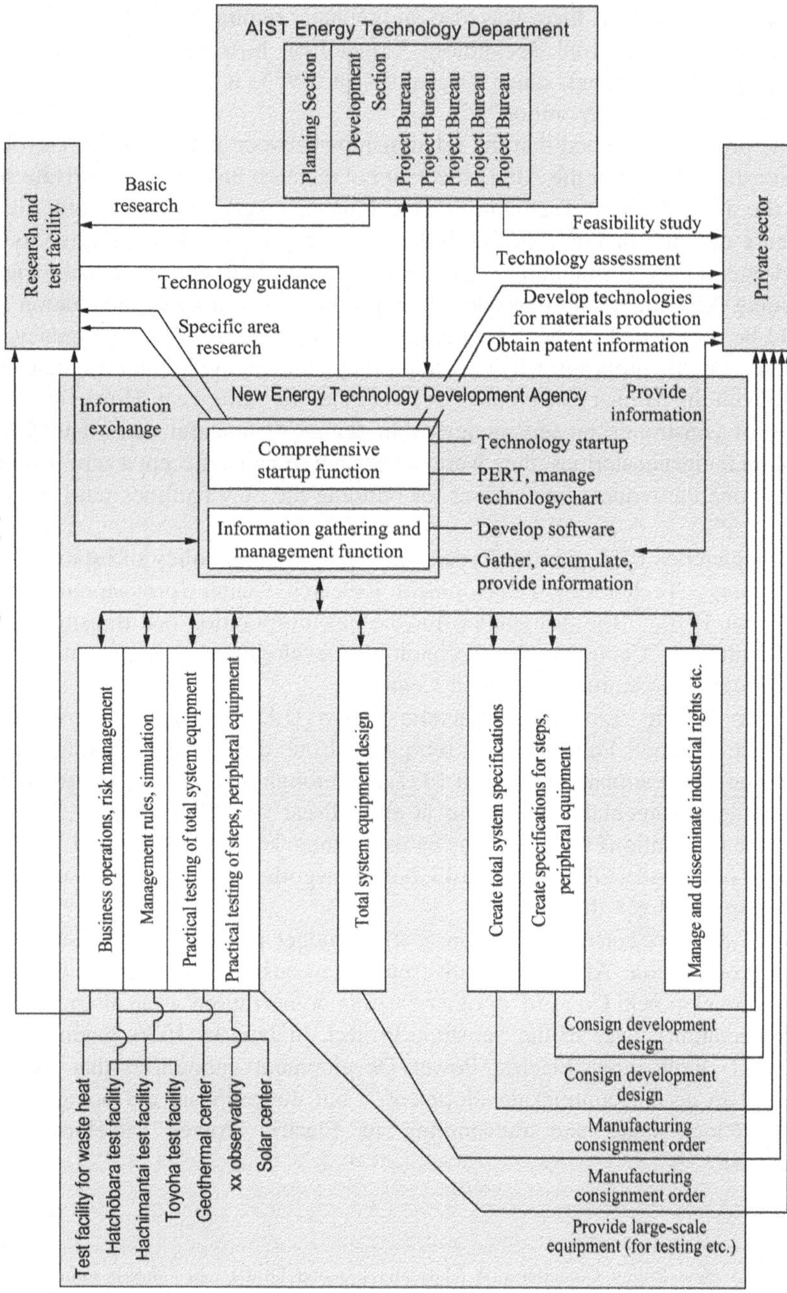

Fig. 5.5 Organizational Chart of the New Energy Technology Development Agency (June 1974 draft). *Note* Obvious misprints have been corrected. *Sources* Based on "FY1975 New Policy Consideration Materials (Second Draft)," in "Shōwa gojū-nendo Shin Seisaku [FY1975 New Policies]", by MITI AIST Minister's Secretariat, General Affairs Division, memorandum, 1974

However, in fiscal year 1977, when the fourth proposal to set up a semigovernmental corporation was postponed at the last moment, MITI sent Electric Power Development an official request in December 1976, proposing the company as the core organization for promoting the Sunshine Project.[56]

Afterwards, MITI and Electric Power Development discussed at length how to share the risks of development and how to handle themes that had no relation to power generation. Finally, in April 1977 it was decided that Electric Power Development would cooperate with the Sunshine Project.[57]

In June 1977, Electric Power Development set up the in-house Sunshine Project Business Division and started to supervise the implementation stages of the Sunshine Project on a temporary basis. In the first fiscal year, AIST promptly commissioned plant development for a total of 1.14 billion yen.[58] In September 1978, Electric Power Development established a facility for solar power research and development in Nio Town, Kagawa Prefecture; in October, a facility for research and development of water electrolysis hydrogen production in Kawasaki; in November, surveys of deep geothermal energy began in the Hōhi region of Kyūshū; and in October 1979, a coal gas liquefaction research and development facility was established at Iwaki, Fukushima Prefecture. One by one, these facilities set about plant development. Although outsourcing to Electric Power Development was a temporary measure, AIST was able to advance the project to the implementation stage and, by pointing to these endeavors, the agency was able to build the case for the necessity of setting up a new energy-related semigovernmental corporation in the future. However, when the sum total spent on the plants outsourced to Electric Power Development began to increase every year, rising to ten billion yen, it was argued that the plants were unsuitable as a supplementary business for Electric Power Development.[59]

---

[56]Ishikawa Fujio (head of AIST's Technology Promotion Division), who was seconded from AIST and became head of the Technology Office of Electrical Power Development (Denpatsu)'s Sunshine Project Division said the following: "We were essentially singled out because Electrical Power Development is the organization that springs to mind when you talk about an energy-related institution created in line with national policy. I heard that Mr. Nakasone (Yasuhiro) leafed through a long list before making the decision." (Ishikawa Fujio, interview by author, June 4, 1998.)

[57]*Nikkei Sangyō Shimbun*, June 2, 1977. According to p. 432 of *Denpatsu 30-nen shi* [*30 Year History of Denpatsu*], ed. 30 Year History Editorial Committee (Tokyo: Denpatsu, 1984), the following three conditions applied at the time of the commissioning of the plant development: (1) the government would commission Denpatsu to carry out the work under the Sunshine Project, in view of the nature of Denpatsu, as a business-conducting company; (2) the scope of the commissioned work would encompass any and all work pertaining to the construction of the Sunshine Project pilot plant; (3) Denpatsu would establish a separate organization within the company from the standpoint of ensuring that the commissioned work would proceed smoothly.

[58]*The Nikkei/Nihon Keizai Shimbun*, January 20, 1977.

[59]Comment made by Yamanaka Masami (Deputy Director-General of Technology). MITI Kogyo gijutsuin Sanshain keikaku (1984), p. 15.

In 1977, the Moonlight Project, a project to develop technologies for energy conservation, was launched alongside the Sunshine Project. In August 1977, the Special Committee for Energy Technology at the Industrial Technology Council approved the Moonlight Project, which AIST planned to establish in fiscal year 1978.[60] After obtaining approval, MITI requested funds to set up the Moonlight Project in the budget for fiscal year 1978. In the first fiscal year, the technology research and development cost came to 2.76 billion yen. Although this was a relatively small scale compared to the Sunlight Project, it resulted in a further expansion of the technology policy themes at MITI.

5.3.2 Leaning Toward Coal Energy for Reasons of Budget Justification

When the second oil crisis erupted in 1979, the situation for new energy technology development once again underwent rapid changes. Suddenly, and once again, new energy technology development was in the spotlight.

In August 1979, the Demand-Supply Subcommittee under the Advisory Committee for Energy published its interim report, *Provisionary Long-Term Energy Supply and Demand Outlook*, where they increased the target for the Sunshine Project from covering 1.6% of all energy supply by 1990, to 4.8% with a further increase to 7.1% by 1995. The breakdown was 21.4 million kiloliters from coal liquefaction and other coal energy, 5.4 million kiloliters from large-scale deep geothermal energy and other forms of geothermal energy, and 6.5 million kiloliters from solar cells, solar houses, and other forms of solar energy for a total of 33.3 million kiloliters.[61] According to these calculations, coal liquefaction and coal energy accounts for more than 60% of the whole. The focus of the Sunshine Project should have been on expectations of solar energy, but the project was reoriented toward coal at this time. What were the reasons for this anomaly?

Compared to the assumptions at the start of the project, the reason for the subsequent investment of a large share of the budget in coal energy development was due to changes to the system of budgeting for the Sunshine Project as of fiscal year 1980.

---

[60]*The Nikkei/Nihon Keizai Shimbun*, August 24, 1977.

[61]MITI Kogyo gijutsuin Sanshain keikaku suishin honbu (1980), p. 17. Incidentally, in the Long-Term Energy Demand and Supply Outlook (while the figures do include firewood and charcoal, etc.), the prediction of 5.5%, geothermal 1.0% is made for 1990 under the category "New Fuel Oils, New Energy, Other." Although there is a trend for the figures listed in the Long-Term Energy Demand and Supply Outlook to be slightly high due to the inclusion of "old" energy fuels such as firewood and charcoal, in this chapter, where no other materials are available, these figures will be treated as the demand and supply outlook for oil/petroleum alternatives.

MITI had already launched a fundamental review of the Sunshine Project in May 1979 with the idea of narrowing down the focus. At the same time, the ministry undertook a review of policy as a whole, and it became evident that a draft motion for a legislative bill that would provide a comprehensive and systematic framework for new energy policy would be presented in the summer of 1979. This became the Act on the Promotion of Development and Introduction of Alternative Energy (hereinafter, the Alternative Energy Act) enacted in May 1980.

The goals of alternative energy were (1) to expand the supply of foreign coal, hydropower, geothermal, and other alternative energy sources, (2) to promote a switch to coal and LNG in the industrial sector, (3) to promote the development and use of nuclear power, (4) to grow the market for solar systems in the private sector, and (5) to accelerate the development of coal liquefaction and other alternative energy technologies.[62] To provide a specific systematic framework, there was a switch from general budget accounts to special budgets accounts and, in terms of the organization, the decision was taken to set up NEDO.

The sources of the budget presented a problem. The fact that a large share of the budget came out of coal-related special accounts[63] had an extremely big influence on the choices for developing technologies under the Sunshine Project. To anticipate the conclusion, at this point, the Sunshine Project, which should have focused on solar energy, switched to actually developing technologies for coal energy through the 1980s.

To start with, let us look at the budget in detail. With the enactment of the Alternative Energy Act, special accounts now played a leading role in the budget for the Sunshine Project, which had previously been covered out of the general account budget (Table 5.4).

In fiscal year 1979, the budget for the Sunshine Project comprised seven billion yen from general accounts and 4.9 billion yen from special accounts, but in fiscal year 1980, 7.1 billion yen came out of general accounts and 21.5 billion yen from special accounts. By fiscal year 1984, 3.7 billion yen was covered by general accounts and 36.1 billion yen by special accounts, indicating that special accounts sank substantial investments into the project (Fig. 5.6). Based on these facts, we can surmise that the need to secure reasons for using the coal-related special budget provided the context for the switch to the focus on coal under the Sunshine Project.

As of the 1980s, and in line with the Alternative Energy Act, the budget for the Sunshine Project changed from a structure that depended on general accounts to one

---

[62] MITI Sōgo enerugī taisaku suishin honbu jimukyoku (1980), p. 10.

[63] Incidentally, these special accounts originated in 1967 as the "Special Accounts for Coal Measures". The name was changed in 1972 to "Special Accounts for Coal and Petroleum Measures," again in 1980 to "Special Accounts for Coal, Petroleum and Alternative Energy Measures," and again in 1993 to "Special Accounts for Coal, Petroleum and Alternative Energy Supply and Demand Structure Development Measures." The accounts consisted of two separate account titles: the Coal Account, and the Petroleum and Energy Demand Structure Development Account. Furthermore, in 2007, the accounts were integrated together with the Special Accounts for Promoting Power Source Development to create the Special Accounts for Energy Measures.

that was focused on the power source diversification account in the special accounts for promoting power source development (the power source account), and the petroleum and alternative energy account in the special accounts for coal, petroleum, and alternative energy (the petroleum account). Since these special account budgets were funded by the tax on electricity development promotion, the tariffs on crude and heavy oil, and the petroleum tax, it was necessary to comply with the original intent of power source diversification and coal measures even when the funds were used for developing new energy. Consequently, the power source account was divided between solar and geothermal, and almost the entire amount of the budget coming from the petroleum account, which was larger than the power source account, was allotted to the development of coal energy technologies (Figs. 5.7 and 5.8).

As a result, the proportion of coal energy technology development as a share of the budget for the Sunshine Project became extremely high. The organizational restrictions on acquiring budgets, and the social conditions behind these systems (protecting the coal industry), dictated the orientation of the technologies where funds for technology development were allocated. Here, we understand how, even in national projects, budgets are allocated to selected technologies based on the validity of the process rather than the rationality behind the technology. National projects should aim for technology development success and deliver policy goals on this basis, But, in fact, there is more focus on obtaining the budgets to maintain the project than on successfully developing technologies, and this gradually confuses the project goals, causing the goals to shift.

Let us take a look at another organization. The real decision to establish NEDO was taken in December 1979 during the budget negotiations for fiscal year 1980. At the time, there was a conflict between a policy of not approving new semigovernmental corporations, and insistence that priority should be given to the

Table 5.4 Draft budget and organizational structure for new energy Development in the Alternative Energy Act

(1) In order to secure the long-term stability of the required funding, the tax on electricity development promotion will be raised from 0.085 yen to 0.3 yen per kilowatt and, together with the petroleum tax, there will be change to how the money is spent
(2) Special accounts will be established for alternative energies. The power source diversification account in the special accounts for promoting power source development [the power source account] will be used to promote the use of power from alternative energies. Concerning other fiscal measures, alternative energy will be added to the petroleum account and the special accounts for coal and petroleum measures, and an account for coal, petroleum and alternative energy will be set up
(3) Establish a comprehensive development mechanism for new energy as the central parent organization for developing new energy

Source MITI Sōgo enerugī taisaku suishinhonbu jimukyoku (1980), p. 10

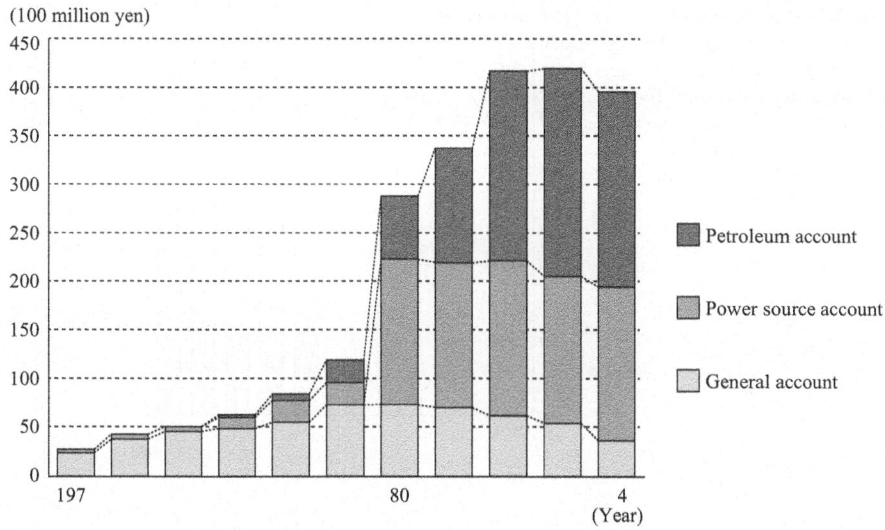

Fig. 5.6 Breakdown of the Sunshine Project Budget by Account. *Source* MITI Kōgyō gijutsuin Sanshain keikaku suishin honbu (1984), p. 12

importance of new energy. As a result, in order to establish NEDO, the condition was to dissolve existing semigovernmental corporations at MITI based on the scrap-and-build policy at the Ministry of Finance. At the end of exhaustive negotiations, the decision to set up NEDO was finally taken on December 28. A memo by one of the stakeholders describes the decision as follows.[64]

1. The need to establish the New Energy Comprehensive Development Organization (provisional name) in fiscal 1980 is recognized.
2. The organization must be concise and effective in line with the main points of administrative reform.
3. The abolition of the Coal Mining Industry Rationalization Corporation, as well as the merger of the Small and Medium Enterprise Relief Corporation and the Small and Medium Enterprise Promotion Corporation, will remove two semigovernmental corporations. These two organizations will be eliminated simultaneously with the establishment of the New Energy Comprehensive Development Organization (provisional name).
4. The required measure will now be submitted to the ordinary session of the Diet.[65]

According to the decision taken at the end of December, the production business of the alcohol monopoly would also become a NEDO department within two

[64]Ishikawa Fujio, interview by author, June 4, 1998.

[65]Material provided by Ishikawa (1979a).

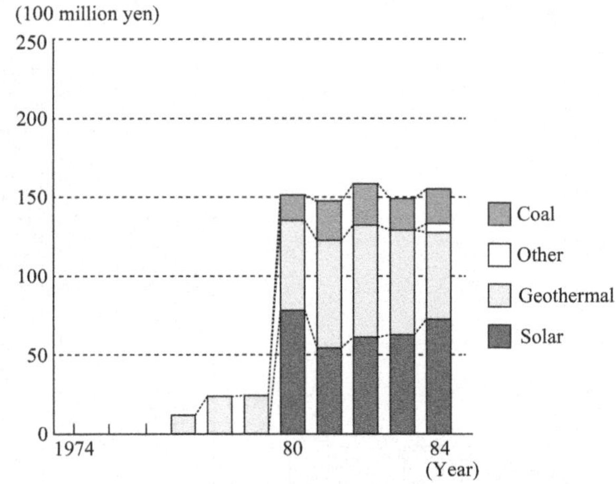

Fig. 5.7 Breakdown of Technology Themes under the Power Source Diversification Account for the Sunshine Project. *Source* MITI Kōgyō gijutsuin Sanshain keikaku suishin honbu (1984), p. 12

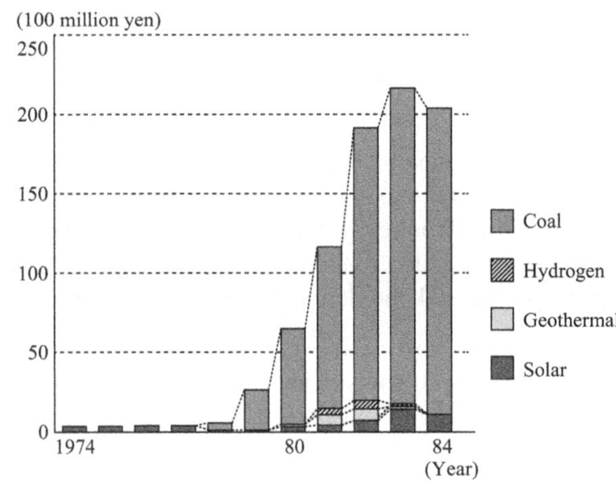

Fig. 5.8 Breakdown of Technology Themes under the Petroleum and Alternative Energy Account for the Sunshine Project. *Source* MITI Kōgyō gijutsuin Sanshain keikaku suishin honbu (1984), p. 12

years.[66] As a result, MITI accepted the bargaining points of the abolition of two semigovernmental corporations and the transfer of one specialized department in order to establish NEDO.

Therefore, when NEDO was launched, they had to set up a Coal Mining Industry Rationalization Department within the organization to take over the remaining business of the Coal Mining Industry Rationalization Corporation. This is the reason why NEDO still today coordinates coal mining parallel with new energy technology development. In October 1982, alcohol production business activities were added to NEDO, a decision taken at the end of 1979.

[66]Material provided by Ishikawa (1979b).

It is astonishing that the organizational circumstances at MITI had such a large impact on both the technology choices for the Sunshine Project and the structure of the NEDO business. The Sunshine Project should have implemented the very best plan in technology terms to deliver on its mission of developing new energies but, due to other policy-related circumstances at MITI, various other enterprises were gradually added to the project. This led to some situations that at first glance appear baffling—for example, in the name of the Sunshine Project, over half the budget was spent on coal energy development. And why is NEDO in charge of alcohol production?

5.4 Amorphous Materials Emerge

Crystalline silicon had been central to solar cell manufacturing methods and the main theme of the Sunshine Project had been to find ways to manufacture crystalline silicon economically. However, at this time a new technology was emerging, promising dramatic reductions in silicon usage amounts—solar cells using amorphous silicon. As a result, photovoltaic energy research under the Sunshine Project was changed to a dual structure where amorphous silicon was developed in parallel with conventional crystalline silicon. Now, what made it possible to incorporate amorphous solar cells in the Sunshine Project after the fact? Below, I take a look from the budget perspective at the series of processes that led to the assimilation of this technology into the Sunshine Project.

Research and development of amorphous materials began in earnest in 1968 when Ovshinsky published a paper on amorphous semiconductors. The paper caught the eye of researchers at the Electro-Technical Laboratory (Denki Shikenjo), who researched the technology in Japan at the earliest stage. Neither the Energy Department nor the Electronic Devices Department took charge of the research—the responsibility fell to the Basic Department (these departments were created after the restructuring of ETL). This coincided with the period when solar thermal energy research started at the Denki Shikenjo, but there was no point of contact between the two departments. This suggests that practical implementation of solar cells was not the goal of the amorphous materials research; rather that the research was carried out in a context that was completely separate from solar energy.

Up to this point, only a handful of pioneering researchers had worked on amorphous materials. A major turning point came in 1976 when Peter LeComber at the University of Dundee published a paper showing that it was possible to exercise PN (positive and negative polarity) control of amorphous semiconductors.[67] This paper singlehandedly broke down the limits on the application of amorphous semiconductors. The discovery indicated that it was possible to use amorphous semiconductors for large and inexpensive solar cells, and it became clear that

[67]Tanaka and Shimizu (1976).

amorphous silicon was an important material for solar cell manufacturing. Developing this method for implementation would facilitate a substantial reduction in the amount of silicon used to manufacture solar cells. This was also one of the answers to the problems confronting the photovoltaic power generation program under the Sunshine Project.

Upon receipt of this information, the Basic Department at the ETL promptly reset its goals to study approaches to using amorphous materials to make economical and high-performance solar cells. Fortunately, the Sunshine Project had already been launched by this time. A proposal on an amorphous-related theme in the form of solar cell development submitted to the project and accepted, allowing the department to take on a commission and obtain a generous research budget, would facilitate an expansion of the research on amorphous materials.

To implement this approach, the Basic Department within the ETL first had to indicate that they would research solar cells. Therefore, a public hearing about solar cell research at the Basic Department was held at ETL, and it was decided to share the responsibility with the Electronic Devices Department, which had advanced solar cell research in the past. Subsequently, hearings were also held at the AIST, which ultimately approved the inclusion of amorphous research in the Sunshine Project.

This is how amorphous materials were incorporated in the photovoltaic power generation program under the Sunshine Project. However, in the mid-1970s, the budget allocated to this technology was very small. In fiscal year 1978, the amount was 4,252,000 yen, and in fiscal year 1979 9.6 million yen.[68] These were meager amounts compared to the 20 million yen obtained by the Electronic Devices Department in fiscal year 1978 for crystalline silicon solar cells alone. The budget amounts suggest that, at this time, there were no great expectations of success with the technology for amorphous materials.

However, in the early 1980s, large amounts were suddenly allocated to the budget for the amorphous solar cell program. Rather than for some technical reason, this was due to major changes in the policies for allocating budgets to other programs in solar energy research. In fact, the reason had nothing to do with photovoltaic energy. Rather, the programs researching solar thermal energy had not delivered the expected results, which indirectly benefitted the budget for amorphous materials. Large amounts of the intended budget were instead diverted to photovoltaic power generation. This proved extremely favorable to amorphous solar cells, which appeared precisely at this time.

As already explained, at first large-scale power stations generating solar thermal energy seemed the odds-on favorites under the Sunshine Project, while photovoltaic power generation by means of solar cells was only incidental. Therefore, in the 1980s, large amounts were still invested in establishing trial plants for generating solar thermal energy. However, it gradually became apparent that plants for generating solar thermal energy did not deliver the performance expected at the start.

---

[68]Material provided by Tanaka Kazuobu (1980).

Large budgets were required for large-scale construction of plants for generating solar thermal energy, but by the time the plants entered the trial stages to verify power generation at the plants, the budget needed for the solar thermal power generation programs had been reduced by half. At this time, the budget for solar energy research under the Sunshine Project, including both photovoltaic and thermal energy, was nearly ten billion yen per year.

Once it became clear that solar thermal power generation would not produce the expected amount of energy, there was no choice but to substantially revise the program. On the one hand, there was no longer any need for the large budget amounts that had been spent on solar thermal energy but, even so, the budget framework for solar energy, including both thermal and photovoltaic, was retained. Unexpectedly, a situation developed where spending the budget for solar energy became problematic. Surprisingly, photovoltaic and thermal energy were both viewed as solar energy. Instead of revising the budget framework for solar energy, the entire amount was simply transferred to the program for generating photovoltaic energy, which was completely different in technical terms. Conducting annual reviews of the appropriate budget scales for each program was a laborious task and, if there were sudden changes in the amounts, there was the risk of censure for being unable to achieve the results. Therefore, the budget was allocated to photovoltaic energy simply because photovoltaic and thermal were both solar energies. Here as well, the association with the sun was at work.

As a result, the budget allocated to solar thermal power generation was invested in (1) trials with large-scale plants for crystalline silicon solar cells, which required large amounts of money, and (2) research into amorphous solar cells, which were raising expectations by this time. In fiscal year 1970, the budget scale for amorphous solar cells stood at a few million yen, but by 1981 it had suddenly increased to more than one billion yen. The context was the inertia that exists within such systems, suggesting that good timing was behind the allocation of a large portion of the generous budget for solar research to amorphous solar cells (Fig. 5.9). This also suggests that national projects do not settle on themes and budgets for purely technical reasons.

As discussed above, the large budgets acquired for solar thermal research had lost their purpose by the early 1980s. As a result, large amounts of money were transferred to photovoltaic research, providing amorphous solar cells with an unexpected advantage due to good timing and rising expectations.

5.5 The Slump in Oil Prices and Project Reorganization

5.5.1 The Decline in NEDO Initiatives

The situation changed in unexpected ways for the Sunshine Project when crude oil prices fell sharply in the early 1980s. As a result, NEDO was hard pressed to

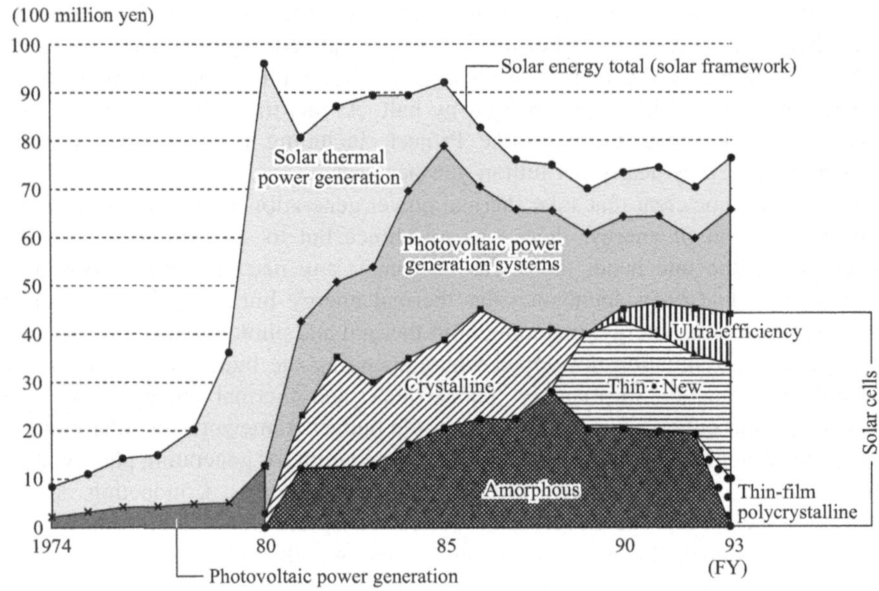

Fig. 5.9 Breakdown of the Budget for Solar Energy Technologies under the Sunshine Project. *Sources* Based on MITI (1993), NEDO (1990); and MITI Kōgyō gijutsuin, ed., *Kōgyō gijutsuin shōkai* [Introducing the AIST], editions for the respective years

achieve its own mission of developing new energy and, at the same time, felt a lack of autonomy where its own actions were concerned.

Originally, NEDO was set up as the implementing body for the Sunshine Project so it should have been possible for the organization to show some degree of initiative regarding the manner in which it moved the project forward. Critics started to say that unless NEDO was autonomous in terms of project management, the organization was simply a subcontracting agency of the AIST. The reasons for the existence of NEDO were rigorously examined.

Ōnaga Yūsaku, President of NEDO, describes criticism from other board members at a meeting of the Ad Hoc Council of Administrative Reform in 1983:

During the discussions at the Ad Hoc Council, I was kept on my toes explaining the NEDO approach in the face of criticism from two or three board members who said that NEDO was simply a channel for diverting subsidies from the AIST to the private sector, and that AIST might as well pay the money directly to the private sector.[69]

Faced with such opinions, President Ōnaga explained to the board members that NEDO was Japan's top authority on new energy technology development, that the organization had a comparatively long tenure, and was putting in place the

[69]Ōnaga Yūsaku, "NEDO de no ninen hachikagetsu [Two Years and Eight Months at NEDO]," *NEDO News*. September–October 1983, p. 4.

structures to promote technology development in the long term. He explained that whenever pilot plants were constructed, NEDO set up its own project teams, which were at the very least involved in the basic concepts and designs. Ōnaga believed that the significance of NEDO lay in bringing together expertise under these development frameworks. Consequently, he proposed to promptly establish development ideas and systems at NEDO as a means of obtaining such capable human resources.

> Since almost all of NEDO's projects are dependent on government funding in the form of subsidiaries and commissioned research funding, if we're not careful then there is a risk of falling into the state of being simply an executive body for the implementation of the government's subcontracted work (abridged). So I thought that there was a need to create a framework in which NEDO thinks and formulates policies for itself; although, of course, it is not necessarily the case that all of those will be approved by the government and the steering committee exactly as they are submitted.[70]

As a general rule, NEDO did not have its own budgetary authority, and so was bound to carry out its activities in line with the will of AIST, having received the approval of the steering committee and Industrial Technology Council. There was a significant gap between constraints such as these and the idealized concept of NEDO utilizing the power of private sector companies and advancing technology development efforts swiftly and efficiently, in a sense like a sort of private sector corporation itself. In particular, the fall in crude oil prices during the early 1980s weakened the fundamental premise for NEDO to be taking a strong approach to developing new energy technologies. At the same time, AIST too was forced to reevaluate and amend the decisions it had made with regard to making new energy the central focus of its technology policy. Despite NEDO heralding the importance of new energy and requesting to be given the initiative, the organization was instead being placed in a harsh position in which the very meaning of the project's existence—which was still failing to produce results—was being called directly into question.

During the early 1980s, NEDO repeatedly urged AIST to allow it greater flexibility in terms of both organizational structure and funding, with the aim of transforming itself into an organization that was more in line with its mission of developing new energy technologies. This was because, as far NEDO was concerned, its obligatory semi-public status was constraining the development of such technologies.

In 1983, NEDO created its own medium-to-long-term plan and attempted to assertively hammer out its own direction for the development of new energy technologies. In November 1983, Watamori Tsutomu, chairman of NEDO's board of directors, organized an internal planning committee within NEDO, and specified a set of targets that the organization should aim to achieve over the next three years, utilizing the experiences gained over the past three years since its establishment. With regard to those targets, Watamori says the following:

---

[70]Ōnaga Yūsaku, "NEDO de no ninen hachikagetsu [Two Years and Eight Months at NEDO]," *NEDO News*. September–October 1983, p. 4.

Creating a medium-to-long-term plan at NEDO is a very presumptuous thing for us to do. What I mean by that is, NEDO has neither the authority or power to have its own budget. There is no way that we should be able to do anything like that without a budget, but in any case we have created a set of working materials (for further discussion), essentially outlining the way in which we think we should proceed technologically in terms of speed and so on if the government allowed us the budget to do so.[71]

In this way, Watamori stated his views in an extremely restrained and reserved manner, in view of NEDO's official capacity. Despite this, however, during the early 1980s NEDO still had the will to cease the initiative with regard to the development of new energy technologies.

5.5.2 The Contradictory Nature of NEDO's Public Status and Policy of Utilizing Private Sector Resources

For his part, as someone who had many years of experience in originally working at a private sector company (Hitachi), Watamori was noticeably dissatisfied with the way in which NEDO was run, particularly with regard to its flexibility in terms of funding and manpower.[72]

The reason why NEDO began to demand to AIST that it be given the initiative with regard to planning the development of new energy technologies stemmed from the various constraints in the existing framework for the commissioning of research and development work.

Even after the establishment of NEDO, the budget for the Sunshine Project still had to be secured by MITI and AIST through negotiations with the Ministry of the Treasury. The established procedure was that AIST gave commissioned research funds to the various national research institutions and subsidies to NEDO; and that NEDO then managed the advancement of technology research and research outcomes by using those subsidies to commission research and development work to various private sector companies and universities.

When NEDO was originally established it was intended to be a knowledge-intensive organization based on collaboration between the worlds of industry, government, and academia; a collection of technical experts. In that sense, it was completely understandable that it should make a request to AIST to be given the initiative with regard to methods for the management and operation of technology developments, and to the direction of the project as a whole.

[71]"Energī runessansu o mezashite [Aiming for an Energy Renaissance]," *NEDO News*, January 1984, p. 3.

[72]"Rijichō taidan: Kongo no shin ene kaihatsu no hōkō [Interview with the Chairman: The Future Direction of New Energy Development]," *NEDO News*, September–October 1984, pp. 16–17.

However, the constraints placed on NEDO by AIST with regard to the management and running of the technology development were great. The following is a comment made by Watamori in July 1985. It is quite long, but I would like to quote it here as it allows to understand what Watamori's feelings were at the time.

> What I would like to ask [of AIST] is, for example, in cases where we have one budget from the special accounts for coal and petroleum measures and another from the special accounts for promoting power source development, where one source of funding has financial leeway and the other does not have much money to spare, I think that it would effective if we were allowed to use the two budgets together within NEDO as if they were one single source of funding. But sadly we are not permitted to do that. This is my biggest grievance regarding R&D and resources development. Another thing is that I would like some flexibility with regard to how we use any given budget, such as in being permitted to carry over funds to the next year, or divert funds from the same budget to other NEDO projects when we have used the budget efficiently and have some left to spare ….

Exactly the same applies when it comes to personnel. I would like it if we could get permission to transfer personnel when we are having a hard time due to personnel shortages—for example, if it was another section's allocation of 20 personnel, to transfer five of them to NEDO—under the authority of NEDO's board of directors. But we can't get permission to do that either.[73]

The use of budget funds from accounts such as the special accounts for promoting power source development or coal and petroleum measures were limited only to the specific purposes for which those accounts were intended, and NEDO could not use them together in combination with other budgets.[74] We can see from Watamori's comments—with regard to flexibility over resources and personnel also —that at the time, even with his authority as chairman of the board, he was unable to move them around as he wished. Normally, in the case of a private sector company, the number of employees would fluctuate with the advancement of a project or business effort. In the case of NEDO, however, the number of employees remained largely unchanged every year—from 1982 until 1995—at around 850. This indicates that the number of personnel was fixed at a statutory allocation, in the same way as it would be at a government agency.

As one example, seconded members of NEDO's Solar Technology Development Office transferred in and out in succession, at a typical average rate of every 2–3 years. In Fig. 5.10, we can see the change in the personnel breakdown of that office by original organization over time.

What we can see from the figure is that the proportions of personnel allocations from industry, government, and academia remained more or less constant, and that the ETL research engineering officials from ETL, who provided technical support

---

[73]"Rijichō taidan: Kongo no shin ene kaihatsu no hōkō [Interview with the Chairman: The Future Direction of New Energy Development]," *NEDO News*, September–October 1984, p. 4.

[74]As mentioned in Sect. 5.3, this was also the reason why large portions of the budget for the development of new energy technologies was allocated to the development of coal-related energy technologies.

during the early half of the 1980s, gradually left the organization, and that their place was filled with expatriate workers from MITI.

Next, let us look at the state of typical management-type departments, aside from the Technology Development Office which required specialist expert knowledge. Figures 5.11 and 5.12 show the proportion of MITI-affiliated personnel in NEDO's General Affairs and Accounting departments. Since these figures only count those personnel who had already left by October 31, 1997, they do not necessarily give a completely accurate reflection of the total number of personnel; but they do allow us to ascertain the overall trends. The charts indicate that the number of seconded MITI personnel within NEDO increased from the latter half of the 1980s into the 1990s. They also tell us that the number of full-time dedicated personnel hired by NEDO hardly increased at all over the same period.

Because NEDO was based on the concept of assembling human resources with specialist knowledge from the worlds of industry, government, and academia, NEDO's response in creating an effective framework for securing full-time dedicated personnel and maintaining them within the organization in the long term was actually delayed during the backlash of the time.[75] In this way, during the first half of the 1980s there was a difference of opinion between AIST and NEDO over the kind of organization NEDO should be and whether it should have the initiative in the development of new energy technologies. As a result of this, AIST attempted to handle the situation by sending MITI personnel to work at NEDO and quasi-internalizing the organization.[76]

As a result, NEDO's medium-to-long-term project plan did not have a significant impact on the way in which the development of new energy technologies unfolded after that.[77] Watamori often referred to himself self-deprecatingly by saying that he was "like a *min-min* cicada in the blazing sun" (a pun in Japanese playing on the word *min*, which refers to the private sector from which Watamori himself came, and also to the sound made by a certain variety of cicada. The phrase used for

[75]"Originally, what we were saying was that we would gradually hire more and more full-time dedicated employees, increasing their numbers and reducing the number of seconded personnel from other organizations; until after about 10 or 15 years over half of our employees would be full-time regular employees of NEDO. We are now nearly 20 years in, and the number of full-time employees has hardly increased. It's all very well saying that we are accumulating technical knowledge, but if the people coming to us [i.e. on secondment] go back to where they came from then although the skills and knowledge may well stick with them, it will not remain at NEDO." (Horigome Takashi, interview by author, June 13, 1998.)

[76]Around this time, Horigome (who was originally from ETL) says that Watamori would frequently call him up and ask him the true meaning of the requests made to NEDO by AIST. "[Mr. Watamori] would frequently ask me to come over and ask me questions like, 'They're telling us to do this kind of thing, but at private sector companies like Hitachi they don't do things like this, and if they did then the company would go out of business.'... 'Why do they do things like this?' I would often be called over and asked questions like that." (Horigome Takashi, interview by author, June 13, 1998.)

[77]"We discussed the medium-to-long-term plan quite a lot, and did it for about six months to a year. In the end, I'd say the medium-to-long-term plan is up in the air, isn't it? Because AIST won't delegate the authority for that." (Horigome Takashi, interview by author, June 13, 1998.)

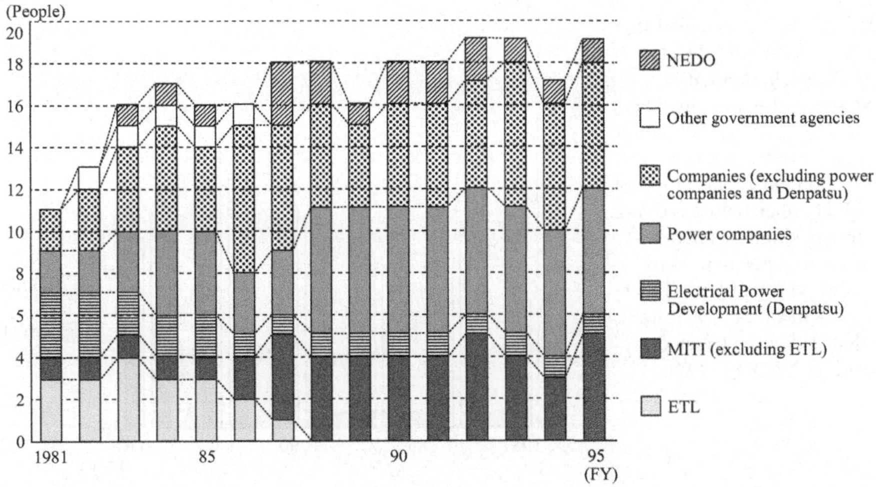

Fig. 5.10 Original Organizations of Members of NEDO's Solar Technology Development Office. *Note* This chart was created by identifying individuals who were members of NEDO on April 1 of each year by referring to the list of Solar Technology Development Office members provided in the Photovoltaic Power Generation Technology Research Association's *PVTEC gonen no ayumi* [The Five Year History of PVTEC], and identifying their organizations of origin by cross referencing between the list and a register of former directors and staff. NEDO staff also includes female office staff. *Sources* NEDO (1997); PVTEC (1996), pp. 111–113

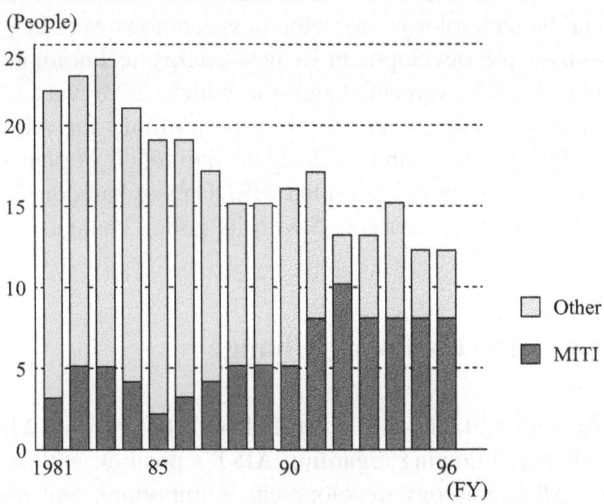

Fig. 5.11 Proportion of MITI-affiliated Personnel in NEDO General Affairs Dept. *Note* This chart counts personnel who were members of the relevant department on January 1 of the relevant year, and who held a full-time post at MITI either before arriving in that position or after leaving that position. Only persons who had left NEDO as of October 31, 1997 are included. *Source* Created based on NEDO (1997)

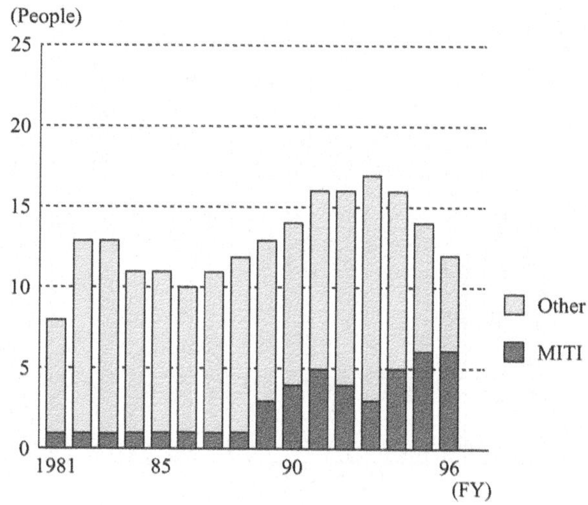

Fig. 5.12 Proportion of MITI-affiliated Personnel in NEDO Accounting Dept. *Note* This chart counts personnel who were members of the relevant department on January 1 of the relevant year, and who held a full-time post at MITI either before arriving in that position or after leaving that position. Only persons who had left NEDO as of October 31, 1997 are included. *Source* Created based on NEDO (1997)

blazing sun includes a homophone of the word *kan*, which is used to refer to government agencies.[78])

AIST's aim, on the other hand, was to broaden the scope of possible directions that it could take with regard to technology policy in the future, by sowing the seeds for various other new technologies, rather than stopping at just new energy.

As a strategy in line with this policy, there was a need for AIST to give priority to the advancement of technology developments in fields such as superconductivity and biotechnology—for which expectations were heightening during the mid-1980s—over the development of new energy technologies; which had not yielded any notable achievements, and for which there was a now a diminished sense of urgency. When doing so, it was desirable for AIST to concentrate on securing budget funding, and to delegate the actual implementation of such development efforts to an organization like NEDO (as it had done with new energy). It was in this way that AIST proceeded with its policy change.

5.6 AIST's Policy Change

At a round-table talk in the fall of 1984, Kawata Michio (then Director of AIST) said the following regarding AIST's position with respect to NEDO.

All technology development is important, and not only the latest cutting-edge technologies. Recent technology development in particular is all general

[78]"Gijutsu kaihatsu de enerugī taikoku ni [Becoming a Major Energy Nation Through Technology Development], *NEDO News*, July 1984, p. 7. A similar comment was also printed on p. 40 of the October–November 1988 edition of the same publication.

technology, not limited only to energy; so we often encounter the problem that development efforts will not advance unless we possess the basic technology required. … From our point of view, it is our intention that if we proceed with our technology development efforts and widen the scope of possible applications, we will still be able to offer a suitable response in the future, even if the direction of the wind changes with regard to cutting-edge technologies.[79]

As is also apparent from this comment, AIST's position—in response to the fall in oil prices during the early half of the 1980s—was that it wanted to respond flexibly to changes in which way the wind was blowing with regard to new cutting-edge technologies by advancing the development of basic technologies (or "technology seeds") in various areas, rather than limiting itself to the development of new energy technologies. Conversely, this was also an expression of AIST's awareness that it was a bad idea to concentrate solely on new energy as it had done until that time. AIST's position dictated that it had to continue formulating effective technology policies, even after the importance of new energy development had faded due to the fall in crude oil prices.

Initially, AIST had planned the establishment of another semi-corporate entity for the development of technologies other than new energy. AIST actually proceeded to determine the specifics of these proposals up to the budget proposal stage. However, reflecting upon the experience of spending seven years on the establishment of NEDO, it became clear that it would not be an easy task to establish another new government-affiliated corporation. AIST therefore considered using the method of again outsourcing projects temporarily to Electric Power Development (Denpatsu), as it had done with new energy projects in the past, before the establishment of NEDO. Finally, AIST arrived at the plan of giving NEDO responsibility for research and development of other technologies, in addition to new energy.

The realization of this plan would enable NEDO to handle many of the numerous projects being run by AIST, as the main implementation body with responsibility for the development of a diverse range of industrial technologies, rather than just new energy. For AIST, this meant that it could outsource other projects that it was running, aside from new energy, to NEDO by granting NEDO subsidies, while AIST itself concentrated on securing the necessary budget funding. In doing so, where necessary, it was also possible for AIST to send government engineering officials to work at NEDO on active secondment (i.e. temporarily retiring them from their government positions so that they could be loaned to NEDO).[80]

This also meant that NEDO would be freed from its mission of achieving new energy development targets, by which it had been bound until this time. If NEDO was given responsibility for various different projects, the significance of its

[79]"Nihon no gijutsu kaihatsu to sanshain keikaku no yakuwari [Japan's Technology Development and the Role of the Sunshine Project] ," *NEDO News*, September–October 1984, p. 3.

[80]Suzuki Norio, interview by author, September 2, 1998. Personnel who were seconded under this scheme could return to their original posts, and could continue normal procedures for their retirement benefits and so on.

mission (with regard to new energy) would be diluted as a natural consequence, and NEDO would be able to avoid rigorous scrutiny over having not yet achieved its targets with regard to the development of new energy technologies.

Of course, it is a fact that the development of new energy was still an important issue for AIST. However, it was becoming apparent that excessive expectations could not be placed solely on this development theme. At a NEDO round-table discussion in March 1986, Suetsugu Katsuhiko (an editorial writer for *Nihon Keizai Shimbun*, now *The Nikkei*) made the following comment: "Drawing the concern and interest of the Japanese public to the energy problem by continuing to say that that 'the next crisis is coming' is certainly one methodology. But as we have already discussed, given the current state of excess supply capacity, the 'Boy Who Cried Wolf' approach appears somewhat faded."[81]

In October 1986, a major replacement of NEDO steering committee members and senior directors took place. Major contributors such as Dokō Toshiwo and Ashihara Yoshihige, who had been involved from the initial stages of NEDO's establishment, resigned their positions. Resignations by other members of the board of directors continued between the 1985 and 1986 and, by October 1986, all ten board members who had been involved since the time of the organization's establishment—including Chairman Watamori—had withdrawn from their posts.[82]

With this transition of power, from those who had been NEDO's top-level management since its early stages to the next generation, the movement to reconsider what kind of organization NEDO should be, relative to its initial status as the main driving force in the development of new energy technologies, began to gain momentum.

For MITI, depending on how it was used, NEDO could be a very useful tool. For example, there was the following kind of usage. In the 1980s, Japanese companies were enjoying a period of strong growth, and criticism from the United States and other developed nations with regard to Japan's growing trade surplus was beginning. At this time, there was also criticism that it was unfair for the Japanese government to help private sector companies to develop technologies by way of subsidies and outsourcing fees. It therefore became difficult for MITI to take the initiative in running research outsourcing projects and research efforts by technology research associations itself via AIST.

In this situation, the organization was useful because making NEDO the main executive body for the development of not only new energy but a diverse range of other industrial technologies meant that, at least in name, the government would not be engaging in such work directly; even when active AIST engineering officials

---

[81]"Shin enerugī kaihatsu no tenbō to NEDO no yakuwari [The Outlook for New Energy Development and NEDO's Role]," *NEDO News*, March 1986, p. 3.

[82]NEDO (1990, pp. 129–131). Table of Past Steering Committee Members and Directors.

were being seconded to NEDO to conduct research. Having NEDO act as a shield in this way made it possible for the government to advance its technology development projects indirectly, while at the same time avoiding the need to clarify where responsibility lay both internally and publicly.

The backlash against the development of new energy technologies continued into the latter half of the 1980s. When the Sunshine Project budget proposal for fiscal 1988 was formulated in December 1987, the possibility of unfreezing the coal liquefaction project was discussed. The idea to unfreeze the project stemmed from the fact that MITI was lobbying the petroleum industry for a major increase in petroleum taxation, due to the drop in petroleum tax revenues—which were a source of funding for new energy development—brought about by the strong yen and a slump in crude oil prices. This move by MITI was opposed by the petroleum industry, with Tateuchi Yasuoki—Chairman of the Petroleum Association of Japan (PAJ)—saying, "MITI is putting forward a policy of allowing consumers to choose freely by encouraging competition between different types of energy. If that's the case, it doesn't make sense to be funneling petroleum taxes into the development of new energy, which is one of our rivals."[83] MITI's response was to propose the reactivation of a bituminous coal liquefaction project. However, this proposal was met with strong objections from NEDO over the sustainability of the technology. A compromise was proposed, and it was eventually determined that the scale of production at the bituminous coal liquefaction plant would be reduced from 250 tons to 150 tons per day, and that the timing of the plant's completion would be delayed by several years.

This state of affairs was not only occurring in industry and at MITI. Figure 5.13 shows the number of Sunshine Project-related articles published in the four main Nikkei newspapers (*The Nikkei/Nihon Keizai Shimbun*, *Nikkei Sangyō Shimbun (Nikkei Industrial Journal)*, *Nikkei Ryūtsū Shimbun (Nikkei Marketing Journal)*, and *Nikkei Kin'yū Shimbun (Nikkei Financial Journal)*) by year. This figure can be thought of as a proxy variable expressing the degree of attention that the Sunshine Project was receiving from the Japanese public. Public interest in the project heightened during the latter half of the 1970s. During the early 1980s there was some degree of fluctuation, but the project retained a certain level of interest. Between 185 and 1988, however, we can see that there was a rapid decline in interest. It can be said that the degree of media attention received by the project declined during this period, and NEDO's position weakened.[84]

The fall in oil prices, accompanied by a change in AIST policy and a rapid decline in public interest led NEDO to reconsider its own identity as an organization. When this happened, NEDO could no longer maintain the kind of identity that it had maintained until that time, as a knowledge-intensive think tank for the

[83] *Nikkei Sangyō Shimbun*, December 28, 1987.

[84] "From my point of view, the organization's image as a technological think tank had paled considerably, and the feeling was more that they were just doing administrative processing work 'subcontracted' to them by MITI." (Horigome Takashi, interview by author, June 13, 1998.)

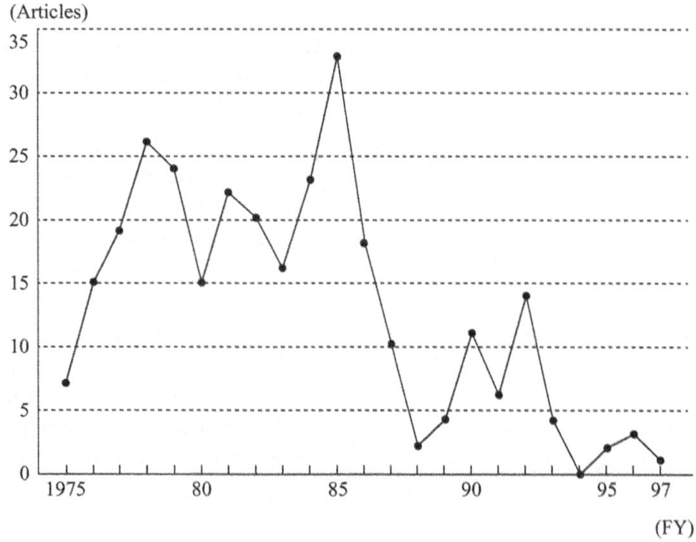

Fig. 5.13 Number of Articles Relating to the Sunshine Project in the Four Main Nikkei Newspapers. *Source* Created using data from the Nikkei Telecom database

development of new energy technologies. AIST, meanwhile, wanted to transform NEDO into an organization that would take responsibility for more urgent and important technology themes—such as those being handled under the Large-Scale Project and Next Generation (Industrial Key Technologies) Project schemes—rather than engaging solely in the development of new energy technologies.[85]

For NEDO too—for as long as the development of new energy technologies could not produce useable results in the immediate future—acting in accordance with AIST's strategy meant that the organization would be freed from the heavy burden of absolutely having to produce research outcomes in the development of those technologies, in return for relinquishing the initiative.

With Watamori's resignation as chairman, articles in which NEDO spoke out against AIST policymaking also disappeared from the pages of NEDO's organizational publication, *NEDO News*.

[85]"[NEDO's 1988 expansion] was surely because they had run out of posts. An expansion would allow them to get lots of new posts, wouldn't it? Because NEDO is a semi-governmental corporation. Directors can be posted on active secondment. In the case of normal foundations, executives cannot be seconded, but with semi-governmental corporations, active public servants can be seconded to act as executives. As a general rule, this is forbidden for non-governmental foundations." (Suzuki Norio, interview by author, September 2, 1998.)

5.7 Environmental Issues and the New Sunshine Project

The backlash against new energy development continued from the latter half of the 1980s into the 1990s. If the state of low crude oil prices continued, the Sunshine Project's sense of presence would fade further into obscurity. Amidst this state of affairs, MITI struggled with the problem of to just what extent it could maintain and ensure the survival of the project. It had become difficult for the project to achieve the pledged and widely publicized technology introduction (or installation) targets for new energy. In order for the Sunshine Project to continue peacefully without facing criticism for its failure to meet these targets, changes—such as an overhaul of the original project framework (created during the mid-1970s) and the organizational structure of NEDO (created during the early 1980s)—were required.

As mentioned earlier, NEDO had already significantly expanded the range of technological fields for which it was responsible (during the mid-1980s), and had become the organization that handled MITI's cutting-edge technology development projects, in addition to new energy. However, as the organization entered the 1990s, the reforms moved beyond this expansion, and the time finally came for a review of the actual Sunshine Project itself. Around this time, a total overall reorganization of technology policy took place, whereby the Sunshine Project was integrated with other policy programs.

During the early 1990s, global environmental issues became the subject of close-up attention in various countries around the world, becoming an international boom. The need for responding to worldwide and nationwide environmental problems, along with the existing context of natural resource and energy-related problems, provided the possibility for prolonging the life of the Sunshine Project, in the form of a response to environmental issues.

Working from the standpoint that the development of new energy and energy conservation technologies relates closely to global environmental problems, MITI decided to combine these policy programs—which until that time had been continued as separate development programs—into a single project. In January 1992, MITI compiled a plan for the integration of the Sunshine and Moonlight projects with its Research and Development Project on Environmental Technology.

The proposal for this unification of energy-related projects submitted by the Industrial Technology Council entered concrete policy discussions from May 1992 onwards. Finally, in September 1992, it was decided that AIST would launch a comprehensive program to develop energy and environmental technologies (named the New Sunshine Project) as of the following year, in fiscal year 1993.

The New Sunshine Project began in April 1993, presenting an even more ambitious grand-scale plan than the previous Sunshine Project. The plan consisted of three technology system development areas: (1) innovative technology development, (2) international large-scale joint development and (3) appropriate technology joint development; with a colossal expected total budget of 1.55 trillion yen to be allocated between 1993 and 2020, with respective budgets of 500 billion, 900 billion and 150 billion for each area. The targets set were also greatly ambitious.

The plan outlined expectations for the New Sunshine Project to provide for one third of Japan's energy needs by 2030, and to cut the country's carbon dioxide emissions by half.[86] The prediction in the past, when the original Sunshine Project was launched in 1974, was that the outcomes of the project would cover one fifth of Japan's energy needs by the year 2000, with a budget of one trillion yen. We should take note of the fact that the targets set by the New Sunshine Project were even more ambitious than those of the original Sunshine Project.

The New Sunshine Project signified the dawn of a new era, both for the development of new energy technologies and for MITI technology policy as a whole. In addition to this, with the creation of this new project, NEDO officially became the main executive body for AIST's two biggest programs: the New Sunshine Project and the Industrial Science and Technology Frontier Program.

5.8 NEDO's Forgets Its Mission, and Becomes Institutionalized as Its Existence Is Taken for Granted

So why did NEDO—the organization in charge of implementing the Sunshine Project—come to be responsible for the handling of other technology developments aside than new energy, and to significantly tone down its original mission of developing new energy technologies? From this process, we can see that a phenomenon occurred whereby the original relationship in which the organization existed because it had specific policy targets gradually changed, until the very existence of the organization itself became more important, and the policy targets were revised and adjusted to match.

At the time of the initial establishment of an organization such as NEDO, which was effectively created as a government subsidiary with a policy-based mission, promises are made with regard to achieving the realization of policy targets. However, even when it has become evident that those policy targets cannot be met, or the importance of those targets with respect to society has diminished, it is difficult to proceed in a direction that entails abolishing the organization itself. In such cases, there is a tendency for such organizations to begin to act in a manner as if they had simply existed as such from the beginning, and should therefore continue to exist as a matter of course, rather than justifying the legitimacy of their existence based on the fulfillment of their promises to achieve policy targets. Before anyone realizes what is happening, policy targets are lowered and diluted by integration with other targets; and, as personnel are periodically rotated in and out, they are eventually forgotten by everyone. This intentional disregard for policy targets institutionalizes the organization, making it possible for the organization to continue to survive without the reason for its existence being questioned.

[86] Shigen enerugīchō (1995), p. 43.

Let us look back into history. The Sunshine Project was born out of the framework of MITI's industrial technology policy. The themes relating to new energy were initially submitted in response to the call for research themes under the Large-Scale Project System in 1973. The themes were consolidated, separated and established independently, and finally brought together under the new name of the Sunshine Project. In this sense, the project was not the beginning of everything, but rather something that was derived from the parent body of the Large-Scale Project System. Without the Large-Scale Project System, the Sunshine Project would surely not have been created in the form that it was. The project did not simply appear out of thin air. Rather, it borrowed from another project system that preceded it, and was created with that as its initial starting point. The process for the selection of research and development themes was not particularly special, and followed the established procedure. By the organization carrying out routine work in accordance with established procedures, work was processed swiftly and accurately, without dependency on individual attributes. Even if members of the organization were replaced, the routine of the organization itself would continue as normal. Each individual member of the organization is expected to carry out his or her routine work unerringly as expected. In this way, AIST too carried out selection processes in accordance with established procedures.

From the end of the 1970s, MITI began to expand the range of energy policy under its control, citing the reason that a stable supply of energy was essential to the advancement of industry. This course of policy would determine the direction of energy policy centered primarily around MITI's Agency for Natural Resources and Energy, and industrial technology policy focused around AIST and the various national research institutes.

The occurrence of new social issues—such as the energy problem and later environmental problems—signified the opportunity for government ministries and agencies to expand the reach of their jurisdiction. The logic behind this trend for ministries and agencies to expand their authority is separate to that of effective national technology policy, and is related to the logic of ensuring the long-term continuity of technology research and development. In the event that a new social problem actually arises that can lead to the expansion of ministerial or agency interests and authority, the various ministries assert their positions in an attempt to make that new problem their own territory. For example, when biotechnology became the focus of attention during the 1980s, MITI, the Science and Technology Agency, the Ministry of Health and Welfare, and the Ministry of Agriculture, Forestry and Fisheries all came out with policies relating to biotechnology, each from their own unique standpoint. If a ministry or agency is able to successfully make an issue or technology its own area of responsibility from the beginning, then it is likely that it will be able to continue—at least to an appreciable extent—making that issue the target of its own policies in the long term. Herein lies the reason why we must turn our eyes towards the fact that, behind the phenomenon of long-term continuity of national projects, there is a trend for expansion of authority by making use of the inertia of organizations, which cannot be explained by technological rationality alone.

Initially, the Sunshine Project was supposed to have set concrete targets, of providing an appreciable degree of new energy solutions that would serve as an alternative to oil. Later on, however, those targets died with the fall in oil prices, and the Sunshine Project's original mission as a national project became diluted. The initial targets for the introduction of new energy technologies were eventually forgotten, and the project subsequently became a policy program under which budget funding continued to be allocated periodically to the development of new energy-related technologies.

One thing that should be viewed as a particular problem is that, amidst all of this, numerous technology development projects were carried out because organizational legitimacy over procedure was given priority over technological rationality. The term "legitimacy" here refers not to the issue of actual legitimacy in a legal context, but rather to the concept of legitimacy as discussed by the German sociologist Max Weber in his treatise on what he referred to as "legitimate rule".[87]

To give a specific example, there was the way in which—after the failure of experiments with solar thermal power generation—the massive budget for solar thermal energy was funneled into photovoltaic power generation, citing the reason that they were both concerned with the same theme of solar energy, despite the fact that the two areas were completely different from a technological standpoint. This is a classic example of organizational legitimacy being given priority over technological rationality.

While the project was given the signboard of "Sunshine," due to the fact that part of its budget was being drawn from special accounts for coal-related use, a sizeable portion of the project budget was allocated to coal-related research and development efforts. This, too, is another clear example of how organizational legitimacy (which cannot be reduced to purely technological rationality) had a major influence on technology development projects. Regarding this problem, let us examine the changes in project budget allocations by development theme throughout the entire period of the Sunshine Project.

Under the initial project proposal for the Sunshine Project it was planned for a sizeable portion of the budget to be invested in solar energy (see Fig. 5.14). However, during the 1980s, the proportion of the budget being invested in coal-related energy was several times greater than that being allocated to solar energy. As mentioned earlier, at the time of its launch, the project was based around the four pillars of solar energy, geothermal energy, coal liquefaction and gasification, and hydrogen energy.

Given the Sunshine Project's name, there is a tendency for us to be given the impression that its greatest efforts were being invested into clean and environmentally friendly solar energy. In reality, however, the area to which the greatest amount of the project budget was being allocated during the early and middle stages of the project was the development of coal-related technologies (see Fig. 5.15).

[87]M. Weber (1947).

Amounts for photovoltaic power generation, solar power systems, geothermal power generation, fuel cells and wind power generation include subsidies for expanding introduction and installations.

This figure is also included on page 24 of a document entitled Kore made no kokka projekuto no hensen (Changes in National Projects Thus Far), Material No.5 of the 31st Subcommittee for Research and Development, Industrial Science and Technology Policy and Environment Bureau, Ministry of Economy, Trade and Industry (METI) , June 2011. (After the conclusion of the New Sunshine Project, new energy research and development has been conducted under METI' s Innovation Program.)

The reason for this, as already discussed in Sect. 5.3.2, lay in the source of funding for the Sunshine Project budget. With the enactment of the Act on the Promotion of Development and Introduction of Alternative Energy in 1980, after the second (1979) oil crisis, there was a rapid increase in budget funding; and, from the 1980s onwards, the source of funding for the project was transferred from general accounting to special accounting.

These special accounts consisted of the Special Account for Petroleum and Energy Supply and Demand Structure Development Measures (part of the Special Accounts for Coal, Petroleum and Alternative Energy Supply and Demand Structure Development Measures), and the Special Accounts for Promoting Power Source Development. These special account budgets were based, respectively, on tax revenues from coal and petroleum, and the Tax for Promoting Power Source

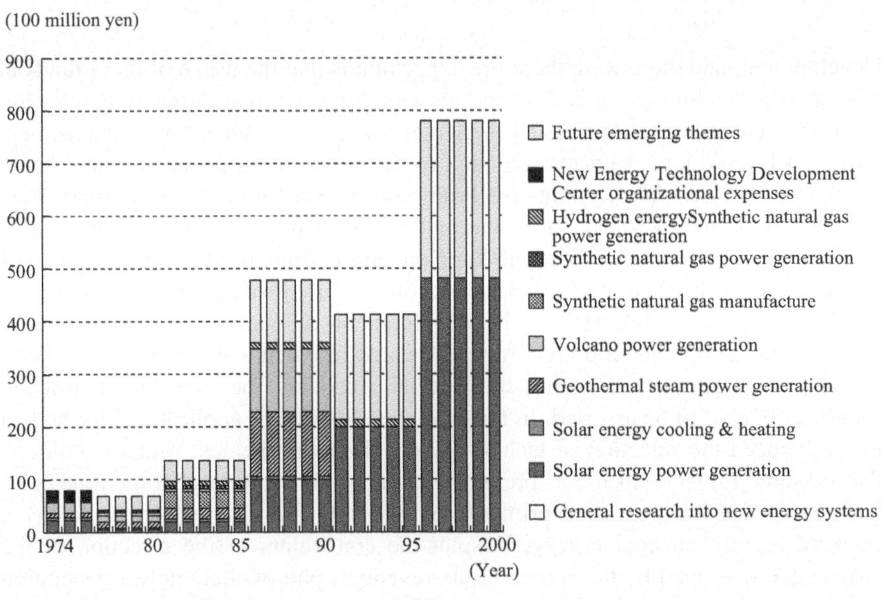

Fig. 5.14 Sunshine Project Estimated Budget (as proposed in June 1973). *Source* Created based on MITI Kōgyō gijutsuin (1973)

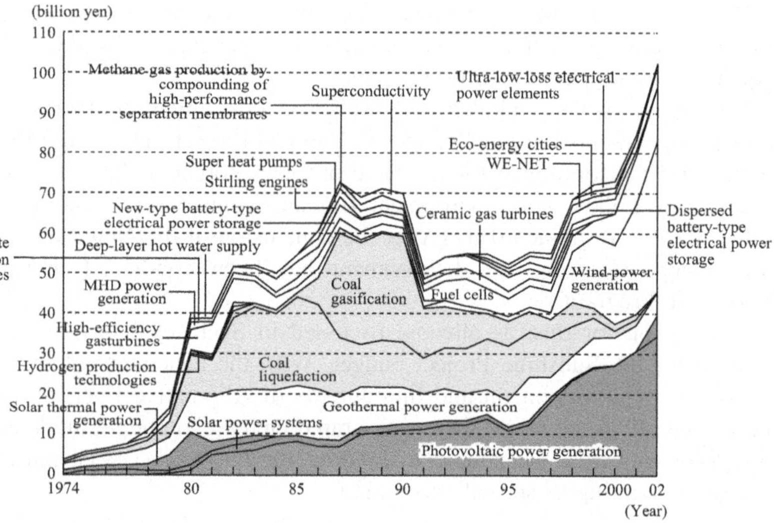

Fig. 5.15 Sunshine Project Technology Development Themes (Including Moonlight and New Sunshine Projects). *Note* Monetary amounts are converted to 2002 equivalents. *Source* Kimura Osamu, Ozawa Yoshiyuki and Sugiyama Taishi, "Seifu enerugī gijutsu kaihatsu purojekuto no bunseki—sanshain, mūnraito, nyū sanshain keikaku ni taisuru hiyō koka bunseki to jirei bunseki [Analysis of The Government's Energy Technology Development Projects—A Cost-Effectiveness Analysis and Case Study Analysis of the Sunshine, Moonlight and New Sunshine Projects]," Central Research Institute of Electric Power Industry (CRIEPI) Research Report Y06019, April 2007, p. 9

Development; and there were therefore expectations that the usage of these funds in new energy development would be in line with purposes such as the diversification of power generation resources and implementation of measures for the adjustment of the coal-based power generation (supply and demand) structure. In fact, budget funds from the Special Accounts for Promoting Power Source Development were split equally between solar and geothermal energy, while almost all of the budget that came from the Special Accounts for Coal and Petroleum Measures (which was a larger amount than the Special Accounts for Promoting Power Source Development) was allocated to the development of coal energy technologies. Utilizing large amounts of budget for coal-related research and development efforts enabled the legitimacy of using the Special Accounts for Coal and Petroleum Measures budget to be asserted. In this way, justifiability of applications for budget use influenced the selection of technology development themes. And so, although the Sunshine Project—as it was presented in the original proposal at the time of its launch—should have centered primarily on research into solar energy, in reality it centered mainly on coal energy. Despite the constraints of the direction of the project being swayed by the source of tax revenues, photovoltaic power generation would later succeed in achieving results with a comparatively small budget, thereby contributing to the project as a whole.

In project organizations, when the survival of the project is threatened by a crisis due to rapidly changing external conditions, the routine of the organization itself will try to survive by diluting its mission and watering down its targets. So, the Sunshine Project was not carried out in accordance with the judgement of government. Rather, there was a process whereby the project organization, NEDO, gradually gained autonomy, forgetting its initial mission. Technology policy became a customary action and, eventually, became semi-permanently institutionalized.

If it comes to be that the decision-making process in a technology development project is being influenced more by organizational legitimacy than by technological rationality, then it does not seem at all strange even when bizarre decisions (when viewed from the standpoint of technological rationality) continue to be made. Even where there is little or no connection in a technological sense, if the investment of budget funds into a particular research theme is procedurally correct for the organization, then the investment will be carried out. If this is the case, then we must surely turn our attention to the question of why the organization recognizes that as being the correct procedure and continues to carry it out without reform or adjustment.

Typically, organizations maintain an established routine, and detest changes that entail them having to alter that routine once it has been established. They also have a tendency to resist decisions that will reduce their size or influence. On the other hand, with regard to changes that will move them in an advantageous direction, such as by enabling them to further expand their authority, they will tend to leave things to progress as far in that advantageous direction as possible; while at the same time protecting their everyday routine.

These traits are similar to those of a living creature, naturally inclined to preserve its own life by avoiding crises that could lead to deterioration or death, and attempting to achieve expansion or growth whenever such an opportunity may present itself. If we compare any organization or organized activity to a single living organism, we can surely observe the biological manner in which the organization avoids danger, and works as if seeking to maintain and enlarge itself.

If there is some pattern that exists within organizations, which the individual members of the organization are not consciously or deliberately aware of, then unraveling the secrets of that pattern would surely be useful in preventing the failure of national projects as organization-based activities, and in reflection upon past mistakes that may lead to future successes for such projects.

References

Energy Issue Study Group at the Electric Power Division of Denki Shikenjo. (1970). Denki shikenjo ni okeru enerugī gijutsu no kenkyū kadai [Themes for energy technologies research at Denki Shikenjo]. Unpublished manuscript.

Geological Survey of Japan Editorial Committee. (Ed.). (1982). *Chishitsu chōsasho hyakunenshi [Hundred year history of the geological survey of Japan]*. Tokyo: Kōgyō gijutsuin Chishitsu Kenkyūjo Soritsu Hyakushūnen Kinen Kyōsan Kai [Geological Survey of Japan 100th Anniversary Commemoration Supporter's Association].

Material provided by Ishikawa, F. (1979a, December). Arukōru senbai jigyō no toriatsukai ni tsuite [*Regarding the handling of the exclusive sale of alcohol*]. Memorandum.

Material provided by Ishikawa, F. (1979b, December). Shin enerugī sōgō kaihatsu kikō (kashō) no setsuritsu ni tsuite [*Regarding the establishment of a general development organization for new energy (Provisional Name)*]. Memorandum.

Material provided by Tanaka, K. (1980, February 14). Amorufasu hakumaku taiyō denchi no kiso kenkyū [*Fundamental research on amorphous thin-film solar cells*]. Unpublished manuscript.

MITI. (Ed.). (1993). *Nyū sanshain keikaku handobukku* [*The handbook of new sunshine project*]. Tokyo: Tsūshō Sangyō Chōsakai.

MITI Kōgyō gijutsuin. (1973, June). *Shin'enerugī gijutsu kaihatsu keikaku (Sanshain keikaku)* [*New energy technology development project (The Sunshine Project)*]. Memorandum.

MITI Kōgyō gijutsuin Kenkyū kaihatsukan shitsu. (1973, July). *Atarashii kurīn enerugī gijutsu no kaihatsu keikaku: Sanshain keikaku* [*Development of new clean energy technology: The sunshine project*]. Self-pub.

MITI Kōgyō gijutsuin Sanshain keikaku Suishinhonbu (Ed.). (1984). *Sanshain keikaku jūnen no ayumi* [*Ten year history of the sunshine project*]. Tokyo: Sanshain keikaku jusshūnen kinen jigyō suisin konwakai.

MITI Kōgyō gijutsuin Sanshain keikaku suishin honbu. (1976). *FY1975 research outcomes report*. Tokyo: MITI Kōgyō gijutsuin.

MITI Kōgyō gijutsuin Sanshain keikaku suishin honbu. (Ed.). (1980). *Sanshain keikaku no kasokuteki suishin senryaku: Sangyō gijutsu shingikai shin enerugī gijutsu kaihatsu bukai chūkan hōkoku o chūshin toshite* [*Acceleratory promotion strategies for the sunshine project: Based primarily on an interim report by the industrial technology advisory committee's new energy technology development subcommittee*]. Tokyo: Tsūsan Seisaku Kōhōsha.

MITI Sōgo enerugī taisaku suishinhonbu jimukyoku [Energy Measures Promotion Division Secretariat]. (Ed.). (1980). *Sekiyudaitai enerugīhō no kaisetsu* [*An explanation of the law concerning promotion of the development and introduction of alternative energy*]. Tokyo: Tsūshō Sangyō Chōsakai.

NEDO. (Ed.). (1990). *NEDO jūnen no ayumi* [*Ten year history of NEDO*]. Tokyo: NEDO.

NEDO. (1997). *Yakushokuin OB meibo* [Register of Former Directors and Staff]. Self-pub.

Ōtani, T. (1995). Kenkyūkikan eno tokka [Specialization towards being a Research Institute]. *Denshi gijutsu sōgō kenkyūjo hyakunenshi* [*100 Year History of the Electrotechnical Laboratory*]. In (Ed.), Electrotechnical Laboratory 100th Anniversary Commemorative Projects Executive Committee 100 Year History Committee. Tokyo: Electrotechnical Laboratory 100th Anniversary Commemorative Projects Supporter's Association.

PVTEC. (1996). *PVTEC gonen no ayumi* [*Five year history of PVTEC*]. Tokyo: PVTEC.

Sawai, M., & Editorial Committee on the History of Japan's Trade and Industry Policy (Eds.). (2011). *Tsūshō sangyō seisakushi 1980–2000* [*History of Japan's trade and industry policy 1980–2000*] (vol. 9) Industrial Technology Policy. Tokyo: Keizai Sangyō Chōsakai.

Sharp. (1996). *Taiyō denchi no sekai kaiteiban* [*The world of solar cells revised edition*]. Self-pub.

Shigen enerugīchō. (Ed.). (1995). *Shin enerugī binran heisei shichi-nendo ban* [New energy handbook, FY1995 edition]. Tokyo: Tsūshō Sangyō Chōsakai.

Suzuki, K. (1973, May). *Shin'enerugī gijutsu kaihatsu seido no seitei oyobi shin'enerugī gijutsu kaihatsubu no secchi ni tsuite [Regarding the establishment of a new energy technology development scheme and new energy technology development department]*. Memorandum.

Tanaka, K., & Shimizu, T. (1976). Hishōshitsu handōtai ni kansuru futatsu no kokusai kaigi [Two international conferences on amorphous semiconductors]. *Journal of the Physical Society of Japan 31*(9).

Weber, M. (1947). *The theory of social and economic organization* (A. M. Henderson & T. Parsons, Trans.). New York: Oxford University Press.

Chapter 6
From the Natural System Model to the Society Development Model: Changing Perspectives II

6.1 Summary of Case Study 2

Case study 2 in the previous chapter shed light on the another side of the Sunshine Project. The factors there included the logic of procedures that worked separately from the logic of technical development. The work of allocating a budget, building an organization, choosing technologies, etc. was performed as part of public policy according to the law by following fair procedures. However, if the procedural adequacy of the work is sometimes given a higher value and priority than the new energy development itself, it is a somewhat abnormal situation. Despite this, it was a viable activity in terms of the continued existence of the project and organizations that carried out the project. There was the legitimacy of system survival that continued to be maintained, exceeding the levels of individuals' concerns.

Case study 2 suggests the possibility that the destination of policymakers' actions in the project was to continue this project as long as possible, rather than to achieve technical targets in new energy development. Particularly after the relevance of the project faded due to unexpected declines in oil prices, the initial targets of introducing new energies gradually became neglected, and NEDO started to dilute its definition of itself as an organization for new energy development by undertaking technological development projects that belonged to MITI's other policy programs. When a large-scale power plant for solar thermal power that was the image of the initial project could not achieve results, a significant budgetary allocation was given to photovoltaic power generation, on the grounds of its being another solar energy source. This had the unexpected consequence of promoting technological development in this area. For this reason, the project was maintained with the initial targets remaining unachieved, yet a phenomenon occurred in which results were achieved through technologies that had not originally been targeted.

Now, let us trace back through case study 2 to identify the key points. Solar energy development originally came about as a survival strategy for the Electrotechnical Laboratory's division for electricity transmission. Amidst the

M. Shimamoto, *National Project Management*,
Advances in Japanese Business and Economics 25,
https://doi.org/10.1007/978-981-15-3180-4_6

diminishing presence of national research institutes in research on electricity transmission, new energy development was imperative. Accordingly, solar energy research was proposed because it was difficult to develop and was too risky for the private sector. When this topic was submitted to the Large-Scale Project System, a new program that differed from the existing programs was developed, reflecting the concern that results may not be achieved in the short run. This was how the Sunshine Project came into existence. The need for winning a budget allocation made this proposed program large and prolonged. The program was approved immediately in response to the oil crisis that happened to occur in the same year. However, the plan to establish a semi-governmental corporation to implement the program was not realized at that time due to objections from the Ministry of Finance.

In solar energy development, a large-scale pilot plant to generate power from solar thermal energy was built. However, it could not achieve the expected energy production in experiments partly because the original plan was overly ambitious. Large budgets allocated every year became directionless and were then diverted to photovoltaic power generation on the grounds of being another solar energy source. This created an unintended situation of progress in the development of solar batteries, including a new amorphous solar cell system. In solar battery development, although the large-scale manufacturers headquartered in the Kanto region were expected to play an active role, they were not eager to commercialize solar batteries because they had semiconductor and other important businesses. Paradoxically, it was the manufacturers headquartered in the Kansai region that enthusiastically started working on commercialization, after not having been entrusted with important technological development in the project. MITI remained attached to developing home-grown technologies and refused proposals submitted by the manufacturers in the Kansai region to refine the foreign technologies. In response, these companies established a joint venture to counter MITI's plan but could not succeed in developing silicon substrate. Even so, with the intention of justifying its existence, the joint venture started to insist on proving that practical application would be possible with the current quality of its solar batteries and proceeding with their in-house production. As a result, the proposed joint venture business among the Kansai league collapsed.

When the second oil crisis served to accelerate the project, NEDO, a semi-government corporation, was launched. NEDO was originally established as the organization responsible for the management of the Sunshine Project with the mission of achieving goals in new energy development. The project had been proceeding with NEDO playing the central role until the mid-1980s when the project started to face criticism due to the decline in the significance of new energy development on the back of falling crude oil prices. Although NEDO asked to strengthen its initiatives, it was not permitted under the procedures already prescribed in the project. Subsequently, MITI entrusted NEDO with the development of advanced technology, rather than new energy development, thereby diluting NEDO's mission and keeping the organization in existence, while on the other hand MITI sought to facilitate new developments, by expanding into areas of technology other than new energy, and achieve risk diversification. NEDO was no longer an organization that gave top priority to meeting the goal of introducing new energy, and the Sunshine Project was to continue to exist as one of the policy programs.

In the 1990s, when environmental concerns came under closer scrutiny, MITI integrated new energy, energy-saving, and environment-related technology projects under its management into the New Sunshine Project and kept the project in existence by allocating larger budgets and long-term target plans. As a result, the Sunshine Project and NEDO were transformed into systems that would survive over the long term while the initial mission of achieving goals in new energy development was being diluted.

This chapter sheds light on the implicit theoretical assumptions through a reexamination of the study of case 2 described in the previous chapter. To anticipate the conclusion, this is referred to as the "natural system model", which depicts an organization adapting to the external environment while performing certain functions. This is also a key concept in management studies. Below, I will refer to prior studies to demonstrate that this approach carries a certain degree of validity in the sense that it explains the survival of systems. However, this chapter also highlights some of the social phenomena that are difficult to observe using this approach and suggests directions for the analysis in the next and subsequent chapters.

6.2 Organizational and Institutional Descriptions of Cases

Looking back into the history of management studies, in the flow of doctrinal history, the natural system model has existed from the early years in concurrence with the rational model covered in Chap. 4. Research into informal organizations, for example, was one of the earliest kinds of research. In the rational model, the assumption was that an organization adopts the optimum measure for achieving its goals. In contrast, in the natural system model an organization seeks self-perpetuation above all else. According to the natural system model, if achieving the goals for the entire organization were to hinder its own continued existence, it is possible that the organization would raise a positive or negative objection to achieving those goals. Assuming these premises, a method to make a national project successful must be devised based on a good understanding of these organizational characteristics.

Using an easy-to-understand example, in the rational model an organization is a machine. An organization is a tool designed to achieve a certain goal and will not violate the wishes of the user of the tool. If the user cannot make good use of the tool, the user is merely misusing it. If used correctly, the outcome desired by the user should be obtained. However, based on the natural system model, an organization is not a machine but a living being and is considered to seek its own survival and growth.

According to W. R. Scott, an organization in this model is more than a measure to achieve a defined goal. An organization is a social group that adjusts itself and attempts to survive under a certain situation.[1] In the natural system model, an organization is not a measure to achieve a goal but is an end in itself. An organization's first goal is not to achieve a goal but to survive. If that is the case, it is impossible to assume that an organization unconditionally complies with official purposes and rules to achieve goals. This is because there should be nothing odd, even when an organization has described its official structure, if the actions observed deviate from the structure. In the rational model, it was assumed that actions matched the orders issued by the organization and other norms. In the natural system model, however, because an organization's first goal is its own maintenance and survival, even its goal, of all things, can be changed.[2]

In the natural system model, humans are not a machine part, like a gear. Within an organization, an official structure arises based on personal attribution and relationships and this structure promotes or hinders the pursuit of official goals of the organization. In the natural system model, such unofficial norms and behavioral patterns also become important when an organization is analyzed. An organization was supposed to have been originally designed as a measure to achieve goals. Why and how, then, do such norms and behavioral patterns come to exist and start to exert influence on the organization, which had not been intended by the designer? Accordingly, the difference between a machine and a living creature is the central issue for consideration in this model. In other words, it may also be defined as an issue of how the dysfunction of bureaucracy starts, with this dysfunction being an embodiment of the deviation between technological optimization and organizational legitimacy (including rules of official organizations and norms of unofficial organizations).

Regarding this point, Merton pointed out that the fulfillment of and compliance with the rules that had originally existed for the pursuit of goals of an organization conversely exerted negative effects on the pursuit of the goals.[3] In a bureaucratic system, the more the rules are complied with, the more the credibility of actions in the organization increases. At the same time, however, compliance with rules itself gradually becomes its own goal and this should be considered behavior that deviates from the originally intended goals of the organization. Yet members of the organization can protect themselves against criticism from others by adhering to the rules. Hence, they devote their attention to compliance with the rules to protect their own positions within the organization. It may be that members are not even aware of this. That is, because members who belong to the sphere of indifference do not question on each occasion why the rules exist and why they must follow them. If they adhere to the prescribed rules, they can, to that extent, efficiently progress with their routine work. Intrinsically, exempting the burden of having to recognize the continuously changing external environment is the very reason for the system to exist.

[1]Scott (1998, p. 57).
[2]Watanabe (2007).
[3]Merton (1949).

The basic concept in case study 2 was to consider decision-making in a project organization not as that of a rational entity but rather as a product of the routine activities of subunits. Routines of subunits create semi-autonomous movements to certain degrees, independent from the activities for the pursuit of the goals of the project for the whole. Therefore, these movements sometimes have an unintended influence on the pursuit of the goals of the project. Based on this concept, the reasons why a project succeeded or failed are explained by the positive or negative influence exerted by the routines on the pursuit of the goals of the project for the whole.

We can learn a lesson from these perspectives. This is because a significant degree of reservations needs to be attached here to the naive supposition that an organization taking part in the project unites to achieve the goals for the whole and because the existence of a strange tendency in which the tasks of subunits are sometimes given a higher priority than the overall goals is suggested, although such tasks were initially planned for the pursuit of the overall goals. As I described in Chap. 4, Hoffmann described contradictions in government policies as "schizophrenic", while in the model for organizations presented by Allison, in activities of a large-scale organization consisting of a loose association of fundamentally diverse subunits, schizophrenic situations may well be the normal state.[4] From here comes the idea that the important role of leadership is to create a similar vector direction in such situations for the pursuit of the overall goals.

Allison presented a study that applied this model.[5] Roberta Wohlstetter raised the question why the United States government failed to do anything against the Japanese attack on Pearl Harbor, in spite of the warning signs.[6] Up until the day of the attack, the American Navy received reports of changes in Japanese encryption and of the Japanese fleet gathering in Cam Ranh Bay; there were reports from the FBI that the Japanese consul in Hawaii was burning official documents and various other information. However, the American Navy did nothing in terms of moving its own fleet outside the harbor, strengthening its aerial defense system, or deploying personnel at emergency centers. The actions of the American Navy on December 7, U.S. time, when Pearl Harbor was attacked, were the same as those on December 6. The standard output power of the organization that functioned according to entrenched routines was extremely difficult to change. There was no exchange of information between the Army and Navy units at Pearl Harbor and each unit had completely different expectations of the next enemy move at the end of 1941, based on completely different types of information from separate sources. Moreover, the meaning of the numbers of the three attack alert levels used by the Army and Navy were exactly the opposite of each other. In the Navy, "Attack 1" was the full-scale highest alert, while in the Army "Attack 1" was the lowest in the warning level —"Attack 3" was the full-scale highest alert. Consequently, on the day of Pearl

[4]Hoffmann (1965).

[5]Allison (1971).

[6]Wohstetter (1962).

Harbor attack, there was no sign of the Army acting according to a joint plan made in advance with the Navy. Although there was the overall goal of national defense, the Army and the Navy, the two subunits, acted according to their separately set routines. As a result, they could not accomplish their most important mission of appropriate defense against Pearl Harbor attack. This is an example of "schizophrenic" consequences.

This is an example where the overall goal, representing the fundamental significance of the organization's existence, was not achieved because its subunits had independent routines to follow. In addition to such unique developments among subunits and an unachieved overall goal that resulted from a lack of coöptation among the subunits, there are examples in which an organization itself seeks to maintain the routines and, as a result, the very goal that should have been pursued is replaced. The act of following the procedures for achieving goals ironically detaches the organization from realizing the expected goal, although the purpose of the organization's existence should have been to achieve the goal.

Perrow presented studies based on these perspectives.[7] One could argue that each of them are examples in which goals are "sold out" and missions are changed for the organization to survive or grow. Many of these studies reveal unintended transformation of the organization.

Messinger studied the Townsend Party, an old-age pressure group.[8] This organization laid out a radical political goal of increasing support for elderly people through economic plans. At the beginning, the political goal of supporting the elderly was the priority. As the organization started selling its members vitamins and drugs that did not need doctors' prescriptions, its funds increased due to the success of the drug sales. As a result, this organization continued to exist not as one that pursued political goals as originally intended, but as a friendship group in which elderly people got together to play cards, etc.

Gusfield conducted a study on the women's Christian temperance union and explained changes that occurred in the organization that had originally campaigned for temperance.[9] After drinking was accepted in the United States, the organization had no choice but to give up its protesting against drinking itself. As a result, the organization survived in a form that satisfied the members by changing its policy to that of the attacking manners and lifestyles of the middle class in general.

Clark conducted a study on a college in California.[10] Students of the college had insufficient academic abilities and were going through reviews of high school coursework and vocational training. Many of the teachers had been headhunted from high schools and companies and did not actually have the ability to handle a college curriculum. In these circumstances, the students still expected to transfer to regular four-year universities after two years at the college. Then the college started

[7]Perrow (1972, pp. 159–164).

[8]Messinger (1955).

[9]Gusfield (1955).

[10]Clark (1960).

to provide counseling to the students to persuade them that rectifying their English language deficiency and participating in training to acquire metalworking techniques would be more useful.

R. A. Scott conducted a study on organizations that provided rehabilitation aid to the blind.[11] Although people who needed support most were the elderly, the organizations gave preferential treatment to younger people because it was easier to collect money from those who had more chances to find jobs. The organizations started to compete for younger blind people, while little attention was paid to the elderly, who should have been given the most generous support.

Janowitz explains the reason why the politically oriented right-wing officers in the U.S. Army lost power in the late 1950s.[12] First, although many of the military officers had been from middle class families in the south, this tradition faded and the Army replenished its ranks with personnel from various areas. Secondly, around this time, many officers started to receive skills training at places other than military academies. The ability to give advice to foreign countries and weapon analysis became key factors for getting promoted to top tiers in the hierarchy—heroic acts during wartime were no longer referred to. Thirdly, as weapon systems became more complicated, officers with high-level intellectual training gained power and these officers tended to be more cautious about combat. Consequently, the politically oriented right-wing officers lost power.

Looking at these types of studies, the most classic and representative paradigms that we should refer to would be those of Michels and Selznick. The study by Michels should be marked as the foundation stone of institutionalism in organizational theories.[13] The Iron Law of Oligarchy asserted by Michels depicts the phenomenon that a political party trying to uphold democracy becomes undemocratic as it develops and becomes larger with more members. When the size of a political party increases, the administration of the huge organization becomes complicated. Ordinary members do not have the skills to manage such organization, and a few leaders are entrusted with its administration. This will turn what was originally a democratically operated organization into an oligarchy in which a small number of leaders gain enormous power. These leaders advocate democracy on the grounds of having been elected by ordinary members and attack their political enemies to protect their own positions. Even if these leaders are forced to resign after being criticized by ordinary members, they are only to be replaced by another leadership and there will be no essential change in oligarchy itself. Therefore, it is the iron law that allows no change.

Selznick's *TVA and the Grass Roots* is a monumental work that established institutionalism in organizational theories.[14] In this work, Selznick sought to

---

[11]Scott (1967).

[12]Janowitz (1960).

[13]Michels (1911).

[14]Selznick (1949). Note that the descriptions below in this chapter use the following as a reference: Kitano (1996).

indicate an organizational theory that was different from the schematic interpretation of organizations with rational objectives such as those in which objectives exist in the first place and measures are then determined to realize the objectives. His theory used a paradox to explain that measures stipulate objectives in an organization.

The Tennessee Valley Authority (TVA) is a government-owned enterprise established in 1933 by the Tennessee Valley Authority Act. However, according to President Franklin D. Roosevelt, although the enterprise had government power, the TVA had the flexibility and initiative of a private company. The aspect of having the characteristics of a private company was underlined. When the organization was created, it had the purpose of ending the argument, related to the postwar treatment of the government-owned factory for potassium nitrate used in explosives and electricity-generating dams constructed during the First World War, of whether these facilities were owned by the government or by the private sector, by declaring that they were government-owned. To justify that the company was government-owned, the purpose of its business was comprehensive regional development, something that was impossible for any private company to claim.

The TVA set up a grassroots policy to mitigate local communities' objections to the government-owned company. Specifically, the policy aimed to implement a plan democratically in a bottom-up manner, based on the collective wisdom of ordinary citizens, not in a top-down manner imposed by the federal government. The TVA was forced to pursue its business objective of comprehensive regional development, which was its organizational mission, while at the same time adhering to the grassroots democratic policy. What came out of this contradiction between democracy and bureaucracy? It was the unintended consequence of taking the teeth out of the democratic political objective by embracing local influential figures.

The TVA held up a policy of aiming to improve social welfare through the intervention by the government. This received enthusiastic support from the liberals, while the conservatives criticized it as being socialistic. For this reason, the position of the TVA as an organization was instable.

Arthur E. Morgan, the then chairperson of the TVA, was a civil engineer with a progressive outlook. He excelled in practical tasks and was considered an appropriate person to promote democracy under the New Deal. According to President Roosevelt's instruction to appoint an agricultural expert from the south to the board of directors of the TVA, A. Morgan elected H. Morgan, president of the University of Tennessee, on the recommendation of the U.S. Department of Agriculture (USDA). However, this appointment exerted decisive influence on the subsequent destiny of the TVA.

In 1933, H. Morgan and Lilienthal, another director of the TVA, selected the areas for which the three directors would have responsibility, by a majority vote of the board. Accordingly, A. Morgan was responsible for the general engineering program; H. Morgan was assigned supervision of all matters relating to agriculture; and Lilienthal was placed in charge of the TVA's power development. As a result, A. Morgan was deprived of the authority to govern the entire TVA as chairperson, while H. Morgan's political influence over the TVA's agricultural programs was strengthened.

H. Morgan and Lilienthal cooperated with each other in the TVA. In particular, H. Morgan was seen representing on the board the grassroots policy and working on behalf of the local communities. By helping H. Morgan, Lilienthal could mobilize local communities' support for the power development program. In the TVA, the power development program caused political fights and Lilienthal won the local powers over to his side through H. Morgan. In return, Lilienthal gave H. Morgan full discretion to handle the agricultural program of the TVA.

H. Morgan's significant influence posed an impediment to the implementation of the entire plan for the TVA proposed by A. Morgan. The personnel in the Agricultural Relations Department who were engaged in the implementation of the agricultural program led by H. Morgan were deployed from the University of Tennessee's organization, which provided practical instruction and demonstration on agricultural matters. They shared the same frame of mind as the rich farmers within the TVA's project areas, and that was the sense of racial discrimination and class consciousness of landlords against peasants. In addition, the head of the Agricultural Relations Department thought that it was important to protect the local systems, and the Agricultural Relations Department acted as the representative of the local communities to an excessive degree.

The local university that had exclusively been contacting the agricultural towns were vigilant about the possibility that the TVA might start a business by means such as using fertilizer plans and ignoring the university. The university attempted to impose various constraints through H. Morgan. In 1934, a coordinating committee consisting of representatives from the university, the TVA, and the USDA decided to incorporate the TVA's agricultural programs into the framework of the university's organization that provided practical instruction and demonstrations on agricultural matters. Personnel from the university became TVA officials and the USDA entrusted the university with practical instruction and demonstration on agricultural matters. The university's authority was therefore shielded.

In this manner, mutual support relationships between the TVA and the University of Tennessee were established and the TVA was viewed as supporting and standing on the side of the university. As a result, it was seen as an organization sympathetic to wealthy farmers in the class conflict between the wealthy and poor farmers.

So, the TVA sought organizational stability and continued existence in its agricultural program by obtaining the support of the representatives of the existing systems. This cooperation gave rise to, within the TVA, small groups that spoke for the representatives of existing systems such as the Agricultural Relations Department and, through these small groups, an agricultural program that was very conservative compared to the initial concept of the TVA for the whole was implemented.

In the New Deal, the existence of a grassroots policy that should have been democratic according to the initial concept brought about the consequence that the TVA supported the parties that benefited from the existing systems, at least in its agricultural program. The ideals in the policy were weakened due to unintended consequences.

This is a result of the paradoxical situation in which the localized interests of subunits of the organization were assigned priority over the overall position of the organization, and this was caused by the priority given to the interests of administrators more than the formal objectives of the policy at the time when administrative discretion was exercised.

The studies of institutionalism in organizational theories described above give us surprising conclusions about organizational phenomena that do not fit in the rational model.

6.3 National Project Research

Below, I would like to introduce studies that derived conclusions mainly relating to the natural system model from national projects in Japan. The first study was conducted by Ohtaki Seiichi, who chose nuclear steelmaking in the Large-Scale Project System as the subject matter and analyzed the relationship among the relevant organizations.[15] This study sought to clarify the factors behind the success in achieving results by the technological development organizations that consisted of various heterogeneous entities participating in the development of nuclear steelmaking.

Ohtaki considered that this nuclear steelmaking project had achieved success to a certain degree and settled on organizational factors for the success as the principal subject matter for consideration, in particular factors attributable to the relationship among the relevant organizations. Ohtaki presented the question of "what phenomena would happen when several different organizations must cooperate with each other to contribute to a larger system [project], while each organization maintained autonomy in relation to each other in terms of the purpose of each individual organization and each such organization itself". Making this the focus of interest in his paper, Ohtaki confined his attention to the nuclear steelmaking project conducted during the period between 1973 and 1980 among the technological development themes in the Large-Scale Project System of the Agency of the Industrial Science and Technology (AIST), and studied and analyzed the nuclear steelmaking project using Evan's organization-set model.[16]

According to Ohtaki, the reasons for the successful completion of the project were that there had been accumulated experience, learning and techniques in joint research with the central role played by the Iron and Steel Institute of Japan over five years preceding the establishment of the project, that there were strong project promoters in both the steel industry and MITI, and that nuclear power furnace

[15]Ohtaki (1981).

[16]Evan (1976).

development overseas became a stimulating factor for the project. It is interpreted that consensus building was easy because these factors became the "expectations set" in the areas of activities. In addition, three points are presented as factors for the success in formulating these organization sets. They are the existence of excellent project managers, the project leaders' sense of responsibility for the national interest and the fact that the relevant association educated the researchers with the intention to cultivate nuclear steelmaking supporters and sympathizers. If the reasons for the success of the project are set as a question and the stance of clarifying the conditions for the success is adopted, the method of listing conditions of the organizations (interorganizational systems) that may have contributed to the success is effective. However, the interesting point of this research paper is probably the way in which the expectations set was formulated, and it is suggested that there existed organizations categorized under a natural system.

This approach is commonly used by Sakakibara Kiyonori, who settled on management of organizations for joint research and development as the subject matter for consideration.[17] Sakakibara found the following facts from studying the case of VLSI (Very Large-Scale Integration) Technology Research Association and clarified the factors that promoted innovation in the joint research and development by the VLSI Technology Research Association based on comparison with hypotheses formulated in advance. Five propositions were found in the joint research and development. They are (1) the fact that Cooperative Laboratories, VLSI Technology Research Association became a place for information mixing through multi-level interactions of a variety of engineers whose specializations were different from one another, (2) a decentralized and flat decision-making structure, (3) massive documentation tasks, (4) frequent communication and the development of close interpersonal relationships supported by the frequent communication, and (5) the elimination of conflicts through exhaustive consultations and exchanges of unfiltered opinions. According to Sakakibara, these factors became factors to encourage innovation by Cooperative Laboratories and led to its success. In addition, quoting the definition of Selznick that an organization is called an institution when it acquires unique ideas, philosophy, and values, Sakakibara asserted that the transformation from an organization to an institution in this sense is observed at Cooperative Laboratories.[18] Sakakibara's research clearly describes the reasons for the success of joint research and development initiatives in Japan and exerted significant influence on subsequent research in this area.

The unique concept of Selznick on systems that Sakakibara quoted opened the path for institutionalism in organizational theories, leading to the new institutionalism in organizational theories that followed.[19] However, because the use of the

[17]Sakakibara (1986).

[18]Selznick (1957).

[19]Currently, Selznick is often positioned as belonging to the old institutionalism school. Refer to the following literature regarding his own concept on distinction between the new and old institutionalism: Selznick (1996).

word "institution" is different from the way it is ordinarily used, here I would like to think of the word as "organization as a tool" if no value is injected and "organization as an institution" (institutionalized organization as defined by Selznick) if values are injected. When an organization becomes an institution with injected values, the characteristics of a natural system (life of its own) appear that go beyond those of a mere disposable tool. VLSI Technology Research Association was an example of the above. According to Selznick, an organization can be an institution.

Subsequently, Selznick pursued the study of Bolshevism and found that the Communist Party of the Soviet Union built an organization that became an institution with strong norms and values.[20] His motivation behind engaging in such research was probably to show the possibility that, if such an organization can be built artificially, it can be used to achieve the goals of the organization. As suggested in the name of the book *The Organizational Weapon*, he created an insight that an institutionalized organization can be used to achieve goals as if it is a weapon. However, this concept obscures the boundary between rational system and natural system, which I refer to later in this text. Apart from anything else, the concept that persons who inject such values in an organization are the leaders forms the core of Selznick's theory of leadership.

Studies by Ohtaki and Sakakibara are common in the sense that they dealt with design methods for organizations consisting of a variety of participating entities (interorganizational systems). In his study, Ohtaki chose the Engineering Research Association for Nuclear Steelmaking as the focal organization in the nuclear steelmaking project and considered reasons why the organization set formulated by MITI, the Science and Technology Agency, the Japan Atomic Energy Research Institute, and steel manufacturers functioned well. As the subject matter for research, Sakakibara settled on the reasons why a brilliant outcome was achieved by the institution, which was an institution defined by Selznick, built in the joint research by seven private sector companies, consisting of Fujitsu Ltd., Hitachi, Ltd., Mitsubishi Electric Corporation, NEC Corporation, Tokyo Shibaura Electric Co., Ltd., Computer Associated Laboratory, Inc. (CDL), and NEC-Toshiba Information Systems (NTIS), with personnel from the AIST, MITI, and Electrotechnical Laboratory.[21] As their fundamental approach, these two studies are characterized in that they pointed out conditions in the design of specific organizations (interorganization) that were configurable. In addition, these studies asserted that the functions fulfilled by the collective formations inherent in organizations (ideas, culture, communication, leadership, etc.) were the factors for success. Given this point, these studies are based on concepts that are different from the approaches taken by Imai and Watanabe described in Chap. 4. Imai and Watanabe used, as their analytical unit, economic incentives that could be given back to the single entities. Here

[20]Selzenick (1960).

[21]Participating in this Cooperative Laboratories were Nebashi Masato from the Agency of Industrial Science and Technology as the managing director and Tarui Yasuo from Electrotechnical Laboratory as the manager.

in the studies of Ohtaki and Sakakibara, certain emergent properties are assumed. Natural systems are characterized by certain types of emergent properties.

Ohtaki and Sakakibara use organizations as their analytical unit and identify conditions for establishing structural factors that enable the organizations to achieve their goals. Behind the adoption of these approaches lies the basic recognition that micro interactions within an organization are not directly controllable but that structures can be created that enable such interactions to contribute to the achievement of the organization's goals. These studies went as far as deriving factors for success and the underlying supposition must have been that conditions for the success of an organization are controllable and that, if such conditions are met, the designer would be able to lead the comprehensive outcome from the satisfied conditions in the intended direction. There is a practical meaning to drawing implications by deriving factors that tend to set off an organization to have characteristics of a natural system in which rational policies facilitate the achievement of results in technological development.

However, the question of whether it is possible to rationally design an organization to be a natural system is basically left untouched. Selznick argues that in management there are rational, measure-oriented, and efficiency-promoting procedures and procedures of value permeation, adaptations, and reactions.[22] Essentially, these two are conceptually independent and there should be two categories. If an organization can be designed to be a natural system in a perfectly rational way, the organization can be defined by external control so that it becomes an organization falling under the rational model like a machine. In a natural system model, the point that an organization cannot be perfectly controlled by external sources must be emphasized. However, if such an organization which is like a living being, completely shuns external control, it would lead to the pessimistic conclusion that managing such an organization is impossible. Therefore, at this point, we have no choice but to give a slightly eclectic and inarticulate answer that an organization has autonomy to a certain degree but that, even so, it is not completely uncontrollable. Selznick regards rationally controlling an organization that should be a natural system as a leadership issue.[23] As an idea, this is insightful, but if I might be somewhat critical, it may be that a different label was merely put on an unsolved problem. The point is, in the words of Selznick, it is necessary to search the concrete truth of what it is to inject values in an organization that is to be implemented by leaders. Because it assumes some emergent properties that cannot be reduced to methodological individualism, conducting historical case studies may be useful to observe in what form it is developed. To see the area where the rational model and natural system model join when trying to conduct such an observation, it is essential to step into the semantic world of members of the organization. This issue of joining the rational model and the natural system model will be considered in the next chapter.

---

[22]Selznick (1948).

[23]Selznick (1957).

6.4 Aspects of Cases Lacking Sufficient Explanation

Below, I would like to look into the aspects that are not necessarily explained in full in case study 2 in the previous chapter. In case study 2, the reasons for various actions in the Sunshine Project are explained in terms of maintenance and expansion of the organizations. Although these explanations can be understood sensuously by us working daily in organizations, if we think about it further, organizational behavior surely consists of the actions of the members of the organization. The members naturally observe the organization's norms and, in many cases, they engage in daily routine tasks of the organization with probably little awareness of such matters, thinking that they are self-evident. Even so, the actions of the members are not completely without deliberation. Human beings seek to satisfy their own desires in an organization and do not always obey orders issued by the organizations in an unconditional manner. There is room for free will to a certain degree.

In case study 2, the assumptions were that the scope of the free will was almost non-existent due to the norms and systems of organizations. Accordingly, it was described that all organizational behavior was entirely for sustaining or expanding the organization itself and that people in the organization followed only its norms. In contrast to case study 1, in which governments' success stories were told using excessively fine words, a slightly pessimistic stance is seen in case study 2, in which environmental deterministic approaches were taken to describe that governments and related organizations existed only for the maintenance or expansion of their own authority and that people in such organizations do not dare to question it anew but patiently go about the daily tasks required of their roles out of necessity.

An organization responds violently when its continued existence is threatened. In this case, the organization tries to show the meaning of its existence, even if it puts itself at risk. This behavior is more a response to the expectation that individuals should play such roles pursuant to the organization's behavior rather than acting on their own intentions. It is considered that if a certain person who developed an innovation in a manner like that of an entrepreneur did not appear, another individual would have conducted the same behavior instead. The basic behavioral pattern of an organization is to protect and maintain itself according to the legitimacy of its systems at ordinary times and to seek to expand the organization within that scope if given the opportunity to do so. Accordingly, when an organization is presented by anthropomorphizing it, all organizational actions would be a matter of trial and error in adapting to the environment, as in the case of living creatures. Over the long term, an organization that experiences the selection process and succeeds in continuing to exist will survive, while one that fails will cease to exist.

However, think again, and the idea that an organization has an intention is clearly fiction. Only people within the organization can have intentions. For example, where there is more than one measure that contributes to the maintenance and expansion of the organization, what is the reason why one certain measure is selected from among all of them? Initially, who and how are the risks for an organization determined? Dealing with this question, case study 2 can only give limited answers that an organization reproduces itself and that the reason why a certain action was taken was because it was appropriate for the maintenance and expansion of the organization.

What we need to know is the individual intentions of the people within the organization. Without understanding their semantic world, we cannot explain the reason why a certain action was selected through organizational decision-making.

For example, although Electrotechnical Laboratory's division for electricity transmission was meant to survive, why then was solar energy research selected as a new research theme? There were surely many other choices for new research themes for the division for electricity transmission and it seems that there have been some accountable reasons behind the selection of solar energy research among other things. The potential of this theme could not have been completely evident. However, human beings are forced to make decisions under these uncertain circumstances. Why did solar energy become a subject of research and development? There should be separate types of explanations.

Why was the Sunshine Project conceived as a large project that should run over a long period? In the previous chapter, the reason given was that the know-how accumulated within MITI was followed. Specifically, it included making the political measure large scale to underscore the significance of the continued existence of the project and to win advantageous budgetary allocations while avoiding the risk of failure. As a matter of fact, MITI later succeeded in obtaining large budget allocations thanks to the large-scale project. However, if the large-scale project technically fails in the end, the ministry's responsibility would be significant. Did the parties involved truly believe that the project would succeed when they formulated it? On these points, it is necessary to understand what was in people's minds as they were about to develop the project.

During the period from the establishment of the project until the second oil crisis, why did MITI insist on creating a new semi-governmental corporation? In the previous chapter, the reason given was that expansion of its political authority was sought more than the technical reason for success in the development tasks of the project. However, what sort of meaning did the people who in fact tried to establish a semi-governmental corporation or the founding members wish to give to the semi-governmental corporation? Unless we listen to their explanations, we cannot understand what role was expected of this new organization.

Why could amorphous solar cell acquire large budgetary allocations at an early stage? In the previous chapter we saw that after experiments in solar thermal power generation failed, the budgetary allocations were diverted to photovoltaic power generation, and that newly emerging amorphous solar cells became the recipient of the budgets. However, even if this was the case, deciding to spend half the budget on solar call development in amorphous solar cells must have been extremely difficult without some credible grounds, given that the future outlook for amorphous solar cells was not necessarily clear. Why did amorphous solar cells win confidence so quickly? To find out the reason, it is necessary to confirm the vision of the people involved in the development of solar cells at that time.

Why did private companies continue to follow the policies of the Sunshine Project? For the people who were in favor of promoting the project, corporate earnings were not a concern and it was important to achieve results so that they would be able to explain the significance of their own organization to the public. However, private companies were likely unhappy about being involved in the national project over the long term without knowing when commercialization of the project would become possible. Under these circumstances, on what grounds did the companies participate in the project?

All the above questions can be ignored in the explanation that the organization sought to maintain and expand itself. To understand these reasons, we must go into the semantic world of the people who participated in the project by raising our resolution another notch. What is happening at the moment when an organization and system come into existence as natural systems? There exists a process by which society is built around the core of meaning. Accordingly, in the next chapter, our attention is focused on the semantic world of the people who participated in the project.

References

Allison, G. T. (1971). *Essence of decision: Explaining the Cuban missile crisis*. Boston: Little.

Clark, B. R. (1960). *The open door college: A case study*. New York: McGraw Hill.

Evan, W. M. (1976). *Organization theory: Structures, systems, and environments*. New York: Wiley.

Gusfield, J. R. (1955). Social structure and moral reform: A study of woman's christian temperance union. *American Journal of Sociology 61*(3).

Hoffmann, S. (1965). Restraints and choices in American foreign policy. In S. Hoffmann (Ed.), *Daedalus* 91, no. 4 (1962). Reproduced in *The state of war: Essays on the theory and practice of international politics*. New York: Praeger.

Janowitz, M. (1960). *The professional soldier: A social and political portrait*. Glencoe, Ill.: Free Press.

Kitano, T. (1996). *Keieigaku genron: Atarasii kachi taikei no sozō [Principles of business management: Creating a new value system]*. Tokyo: Tōyō Keizai Inc.

Merton, R. K. (1949). *Social theory and social structure: Toward the codification of theory and research*. Glencoe Ill: Free Press.

Messinger, S. L. (1955). Organizational transformation: A case study of declining social movement. *American Sociological Review 20*(1).

Michels, R. (1911). *Zur Soziologie des Parteiwesens in der modernen Demokratie: Untersuchungen über die oligarchischen Tendenzen des Gruppenlebens*. Leipzig: Klinkhardt.

Ohtaki, S. (1981), Daikibo kenkyū kaihatsu purojekuto no soshiki kan bunseki – 'Genshiryoku seitetsu projekuto' no jirei ni kansuru yobiteki kōsatsu. *Senshu Keieigaku ronsyu, 32*.

Perrow, C. (1972). *Complex organizations: A critical essay*. Glenview, Ill.: Scott, Foresman.

Sakakibara, K. (1986). Kyōdō kenkyū kaihatsu project no soshiki to management: Chō LSI gijutsu kenkyū kumiai no kēsu. In K. Imai (Ed.), *Inobēshon to soshiki*. Tokyo: Tōyō Keizai Inc.

Scott, R. A. (1967). The selection of clients by social welfare agencies: The case of the blind. *Social Problems, 14*(3).

Scott, W. R. (1998). *Organizations: Rational, natural, and open systems* (4th ed.). Upper Saddle River, N.J.: Prentice Hall.

Selznick, Ph. (1948). Foundations of the theory of organization. *American Sociological Review, 13* (1).

Selznick, Ph. (1949). *TVA and the grass roots: A study in the sociology of formal organization*. Berkeley and Los Angeles: University of California Press.

Selznick, Ph. (1957). *Leadership in administration: A sociological interpretation*. Evanston, Ill.: Row, Peterson.

Selzenick, Ph. (1960). *The organizational weapon: A study of Bolshevik strategy and tactics*. Glencoe, Ill.: Free Press.

Selznick, Ph. (1996). Institutionalism 'old' and 'new'. *Administrative Science Quarterly*, *41*(2) (40th anniversary issue).

Watanabe, S. (2007). *Soshiki shakaigaku [Organization sociology]*. Minerva Shobo: Kyoto.

Wohstetter, R. (1962). *Pearl Harbor: Warning and decision*. Stanford, Calif: Stanford University Press.

Chapter 7
The Politics of Creating New Significance

Case Study 3: The Sunshine Project as a World of Personal Significance

7.1 The Origins of the Sunshine Project

7.1.1 The Hesitancy of the Research and Development Officials

In spring 1973, in a room at the National Institute of Advanced Industrial Science and Technology (AIST) , Deputy-Director General of Development Nebashi Masato and research and development official Suzuki Ken stood in front of their boss, Councilor/Deputy Director-General of Technology Kinoshita Tōru, and reported to him that many of the Large-Scale Project proposals for that year were energy related.[1] Solar energy and hydrogen energy had been raised by some of AIST's affiliated national research institutes in the annual call for development proposals for that year, and these two themes in particular had caught their attention.

In May of that year, Nebashi and Suzuki began the task of assigning ranks to the proposed development themes, in the Office for Research and Development, which had a team of eight members in total. In view of its importance, the development officials gave solar energy the highly-favorable rank of "A". Hydrogen energy was—for the time being—given the rank of "C", since the outlook with regard to its chance of success was unclear.[2]

Later, however, when it came to the stage of deciding who would actually be responsible for the development of technology for solar energy, the research and development officials were somewhat passive in their responses. The reason for this

---

[1]An interview with Kinoshita Tōru titled "Sonotoki watashi wa [How I Acted at That Point]," published in the August 22, 1982 issue of the *Nikkei Sangyō Shimbun* newspaper. Nebashi Masato assumed the position of managing director at the VLSI Technology Research Association, which was established later. Suzuki Ken also devoted his energies to the establishment of the association. Technological counselors became known as assistant vice-ministers for engineering affairs through the revision of post titles in July 1973.

[2]Nebashi Masato and Suzuki Ken, interview by author, May 8, 1998.

M. Shimamoto, *National Project Management*,
Advances in Japanese Business and Economics 25,
https://doi.org/10.1007/978-981-15-3180-4_7

was that, while all of them recognized that research into new energy would surely be important in the future, becoming responsible for the development would be accompanied by the responsibility of producing results within a fixed period of time. There was a big difference between recognizing the necessity of the development and conducting the development themselves. Because of this, when Nebashi stood in front of the research and development officials and asked for those who wanted to do it to raise their hands, not one of them did so.[3] It was predicted that a development theme like solar energy would involve many major difficulties before any results would be achieved, and the officials were hesitant to bear the responsibility for that.

Amidst all of this, Nebashi's eyes met with Suzuki's, and he put the question to him directly. "Mr. Suzuki. How about it?" Suzuki answered, simply, "It will be too difficult."

Visions of the efforts that had been made into the development of magneto-hydrodynamic (MHD) power generation—which had yielded no favorable results even after years of work—flittered across Suzuki's mind. Suzuki believed that producing results in the development of solar energy technologies, which would require a long time to achieve, would be difficult under the current Large-Scale Project framework.

> At the time, I said that if we were going to do solar energy development within the Large-Scale Project System there was no way that I could do it. The Large-Scale Project scheme at the time typically dealt with projects which used around five billion yen over the course of between five and ten years. In the case of super-long-term projects that went beyond that, like with MHD example, in which it would not be possible to produce results in a mere five or ten years, the research and development officials responsible for them were suffering great hardships.[4]

Because Suzuki was a very busy man with multiple projects on his plate already, at first he declined Nebashi's suggestion.[5] But when Nebashi told him to think of his own way of doing it, Suzuki was placed in a situation in which he had no choice but to accept the job, and to take responsibility for the development of the technology.

Now that he had accepted responsibility for the project, Suzuki had to think of a way of producing results. He immediately began to make contact with the Electrotechnical Laboratory (ETL), the research institute that had proposed solar energy as a development theme to begin with. The individual who had proposed it was a researcher named Horigome Takashi, who was—at the time—a laboratory chief in ETL's Energy Department.

---

[3]Nebashi Masato and Suzuki Ken, interview by author, May 8, 1998.

[4]MITI Kōgyō gijutsuin Kenkyū kaihatsukan shitsu (1987, p. 51).

[5]Suzuki took charge of a project for studying and developing electric cars that was launched in fiscal year 1971. He also took part in the launch of studies of integrated automobile control technologies in the same year (Nebashi Masato and Suzuki Ken, interview by author, May 8, 1998).

7.1.2 Dreams and Cold Shoulders for the Solar Energy Researchers

Horigome Takashi was the head of a research laboratory in the Electrical Power Department at Denki Shikenjo, which subsequently became the ETL. From around the end of the 1960s onwards, Horigome had become increasingly concerned with the future of research into electrical power at Shikenjo. Specifically, he was worried about what would become of research into power systems and devices as the overall focus of Denki Shikenjo's work shifted towards electronics and information-related themes. If the research institute continued to shift further towards electronics (as it was doing at the time), he thought, power researchers like himself would find themselves feeling increasingly small and out of place there. For Horigome, having majored in electrical engineering at university, and with a track record of research results in DC (direct-current) power transmission technologies, the future of electrical power research also signified the determination of the next research theme that he himself should head towards.

By around this time, various improvements and enhancements to equipment and facilities at other research institutions meant that there were now virtually no kinds of research that could only be conducted at Denki Shikenjo. Around the same time, voices of concern also began to be raised from within the Ministry of International Trade and Industry (MITI), asking whether it was not the right time for Denki Shikenjo to start doing other kinds of research with more of a view towards the future. Because of this, Denki Shikenjo, too, was forced to leave research into power systems and devices to the private sector, and to adopt a policy of conducting research that was targeted more towards developing technologies for the more distant future. Denki Shikenjo decided that it should request its Planning Office to formulate plans for directing the reorganization of the institute's organizational structure, and to determine specific research themes for each of its departments based on discussions amongst the members of each of its departments.

Ultimately it was decided that Denki Shikenjo would be reorganized into the ETL, and the deadline was scheduled for July 1970. With the establishment of this time limit, the Electrical Power Department to which Horigome belonged was also forced to indicate specifically what kind of research it would be conducting in the future. Based on their thinking that both the Electrical Power Department and the Device Department were—in the wider sense of the term—involved in energy-related research, the Planning Office submitted an organizational reform proposal suggesting the unification of all laboratories belonging to these departments under the single name of the Energy Department. Although agreement was obtained from each department with regard to the proposal itself, the journey toward determining the specific content of research themes for the establishment of this new Energy Department was rough sailing all the way.

The Electrical Power Department's senior researchers requested that the young Horigome should be the leader in this process, giving the reason that they themselves would not be remaining at the research institute for much longer, and delegated the consolidation of opinions in deciding the department's new research themes to him. In December 1969, in search of new themes for the new Energy Department, Horigome assembled around 20 young researchers at a recreation center for civil servants at Hakone-Miyanoshita (a popular holiday destination in Kanagawa which is famous for its hot springs). The researchers stayed over at the center and debated thoroughly, from morning until night, what their new research themes should be.[6] Since Horigome did not actually have to undertake research and development efforts into the proposed themes himself, he encouraged the young researchers to speak openly based on their own free thinking; telling them that he wanted to hear every idea they had to say, even if it was something that had just come to mind. As if spurred on by his words, the young members at the Hakone brainstorming session spoke freely about their ideas for new research themes just as they came to mind; and several hundred innovative and extraordinary ideas were produced.[7] After this, Horigome and his subordinates set to work summarizing the ideas that they had brought home with them. The following year, in May 1970, they completed a report entitled *Denki Shikenjo Energy Technology Research Topics*.

In that report, Horigome and his team raised several new fields of energy research, including solar cells, wave power, fuel cells, geothermal power, and nuclear power related research.[8] It was at this time that solar energy made its first official appearance as a research theme proposal. However, it was not yet the case at this time that solar energy was receiving close attention in particular. The description given in the report, too, was only a mere few pages long, and stopped short at simply suggesting the possibilities of solar power as one possible form of new energy.[9]

---

[6]Energy Issue Study Group at the Electric Power Division of Denki Shikenjo, *Enerugī mondai ni kansuru burēn sutōmingu* [*Brainstorming on Energy Issues*], unpublished manuscript, 1969; and Horigome Takashi, interview by author, June 13, 1998.

[7]The paper mentioned above published by the Energy Issue Study Group in the Electric Power Division of Denki Shikenjo. Reference examples from the paper are as follows (figures denote item numbers): "Generate electricity by drilling holes in the earth and dropping objects in them" (use—45), "Generate electricity by coiling the earth and producing magnetic fields" (generation—17), "Collect radioactive fallout and use it as a source of energy" (use—64), and "Provide treatment by guiding electric signals into the body" (use—79). There were also smile-provoking examples that suggested the daily life of young researchers, such as "Use of rush-hour energy" (generation—46) and "Electricity generation by induction (heat and sweat from holding the hands of young women)" (generation—30).

[8]Kurokawa Kōsuke, who took part in this brainstorming session and later moved from the ETL to the NEDO temporarily at the time of the latter's establishment, noted the following: "It covered virtually all fields of research, including fuel cells, superconducting power transmission, superconducting storage, biotechnologies and medical electronics. Studies of this type were far ahead of their time. I think that all the observers were a bit surprised. But I believe that the feeling that energy technologies were pretty good emerged a little because of those studies we conducted." (Kurokawa Kōsuke, interview by author, April 20, 1998.)

[9]Energy Issue Study Group at the Electric Power Division of Denki Shikenjo, *Denki Shikenjo ni okeru enerugī gijutsu no kenkyū kadai* [Themes for Energy Technologies Research at Denki Shikenjo], unpublished manuscript, 1970.

In July 1970, two months after the completion of the report, Denki Shikenjo became the Electrotechnical Laboratory (ETL) as planned, and the new Energy Department was established.

Horigome himself, too, was agonizing over the question of what he should choose as the next theme for his own research. After much contemplation, out of all the themes raised during the brainstorming session, he eventually decided to choose solar energy as the subject of his research. In a message aimed at young researchers written by Horigome in later years, we wrote that (he had chosen it because) he thought that a new method of power generation like solar energy, which would be soft, clean and environmentally-friendly, would surely become necessary in the future.[10]

At the time, however, the solar energy research that Horigome had begun was met with cold stares from others working within ETL. This was because, at the time, the practical viability of solar energy was not clear, and it was not considered to be a very realistic option as a next-generation energy source.

Since the Large-Scale Project System itself (which was launched in 1966) was something that had come into being as a result of forceful lobbying of AIST by Denki Shikenjo, the common way of thinking at the newly created ETL was that—rather than starting another new research theme—the top priority should be produce results with some of the major projects that were already underway; such as MHD power, or the technological development of large-scale computers. Additionally, as mentioned earlier on, during this era, the higher powers within ETL had been working to advance the organization's response to the new age of electronics. There was no way that they would have excessive expectations for solar energy, which was something that differed completely from the direction in which they were heading.

"In those days, if you mentioned solar power, you'd be treated like some kind of oddball or weirdo, and brushed off lightly. You wouldn't be taken seriously," recalls Horigome.[11] The unhappiness continued further. Horigome's subordinates

---

[10]"Several collaborators and I decided to advance solar cell studies, convinced that a new power generation method that is softer, more eco-friendly and cleaner than conventional hard power generation methods (thermal power generation and atomic power generation) will be necessary in the future based on those results (*Denki Shikenjo ni okeru energī gijutsu no kenkyū kadai* [Themes for Energy Technologies Research at Denki Shikenjo]—note by the quoter). …I did not much like hard power generation methods that require ultrahigh temperatures and supervoltage in the first place. Accordingly, I chose solar power generation as the next-generation power generation method in the 21st century" (Horigome Takashi "Jinrui kyūkyoku no dennryoku energī gijutsu o motomete [Searching for the Ultimate Electric Power and Energy Technology for Mankind]," in *Energī kenkyūsha eno messēji senpai kara kōhai e: Atarashii mirai no kensetsu no tameni* [*A Message from Senior Energy Researchers to Their Juniors: For Building the New Future*], ed. Den'yūkai (1995), p. 128.

[11]Horigome Takashi. "Jinrui kyūkyoku no dennryoku energī gijutsu o motomete [Searching for the Ultimate Electric Power and Energy Technology for Mankind]," in *Energī kenkyūsha eno messēji senpai kara kōhai e: Atarashii mirai no kensetsu no tameni* [*A Message from Senior Energy Researchers to Their Juniors: For Building the New Future*], ed. Den'yūkai (1995), p. 128.

did not like the idea of solar energy research either, and even those who took a central role in the earlier brainstorming session turned their back on the theme of solar energy and moved on to pursue research in other areas.

With the loss of support from these human resources, Horigome was left with a handful of young researchers, including two members named Sawada Shinji and Tani Tatsuo, who had only joined the institute a few years earlier. "There was a feeling of inferiority in researching solar energy, in comparison with research in other areas such as fuel cells and super-conductive power transmission", reminisces Sawada, recalling attitudes towards solar energy research within ELT at the time.[12] Together with the young Sawada and Tani, and enduring the cold gaze of those around them, Horigome appropriated the remaining budget left over from his direct power transmission research (a research theme from his days in Denki Shikenjo's Electrical Power Department), and the group launched a small-scale solar energy research project starting from the 1971 fiscal year.

The outline of their research gradually began to take shape in the form of Tani's design for a solar thermal power generation system.[13] The fundamental research centered on a permselective membrane that trapped in heat using a solar energy absorption device. The results achieved with this device drew the attention of the media, and it was subsequently donated to the National Museum of Nature and Science, in November 1975.[14] Despite this, ETL did not authorize department manager demurrage expenses or any other special budget allowances for the group's solar energy research. Tani described the situation around that time as follows: "Something like solar energy will probably never amount to anything. And allowing that kind of childish, toy-like research to be carried out in our research laboratories, and assigning research funding for something like special research at ETL? Absolutely outrageous. That was the kind of attitude they had at the time."[15]

Given the response from within ETL, Sawada came to realize the importance of raising awareness and recognition of solar energy and working to achieve more widespread acceptance of the idea among the Japanese people. He hauled

[12]Sawada Shinji, interview by author, June 22, 1998; and Sawada (1998, p. 68).

[13]Horigome and Tani Tatsuo published their paper on solar energy in *Denshi to keiei* [Electronics and Management] magazine in 1972. Tani also found an article on the study of solar thermoelectric power generation systems conducted by Aden Baker Meinel at the University of Arizona in *Physics Today* magazine of the United States in the summer of 1972. Tani immediately wrote a letter to Meinel and received a draft of his research. Around the same time, Tani took an interest in an article on space-based solar power generation by Peter Glaser in *Journal of Microwave Power*. From these experiences, Tani and others came up with the idea of publishing their research findings with such research trends overseas to promote the course and potential of solar power generation. Horigome asked Tani to write a paper for *Denshi gijutsu sōgō kenkyūjo ihō* [*Electrotechnical Laboratory Journal*]. In September 1972, a paper titled "Taiyō hatsuden shisutemu no teian [A Proposal for a Solar Power Generation System]" was published in *Denshi gijutsu sōgō kenkyūjo ihō* [*Electrotechnical Laboratory Journal*] under the joint names of Horigome, Nagai Kazuyoshi, Sawada, and Tani (Horigome et al. 1972, pp. 718–729).

[14]Electrotechnical Laboratory of the Agency of Industrial Science and Technology (1991, p. 406).

[15]Tani Tatsuo, interview by author, July 9, 1998.

Horigome and Tani out to join him for an appearance on an educational TV program on NHK (Japan Broadcasting Corporation) called *Minna no Kagaku* (literally *Science for Everyone*), held seminars with the cooperation of staff from the Science and Technology Agency, and made various other efforts to increase the number of people outside of ETL who would understand and cooperate with their research.[16] However, despite these efforts, there was no change to the response towards their work from within ETL.

It was in an attempt to break the deadlock that Horigome decided, at the beginning of 1973, to propose solar energy as a Large-Scale Project theme during AIST's annual call for development proposals. Yet it seems that this was not necessarily an idea that had been met with across-the-board agreement from the higher powers within ETL. Horigome testifies that, although the proposal for solar energy research was made formally through the institute after obtaining the approval of the head of the Energy Department, the ETL Planning Department was far from supportive of the move, and was reluctant to give its consent.[17] Additionally, Tani—as explained later—testifies that the ELT's higher-level management were bemused and bewildered when they received a strong request for cooperation from AIST as a result of the proposal.[18] Once the application to the Large-Scale Projects System was determined, Horigome decided to include the solar furnace research being conducted by Noguchi Tetsuo of the National Industrial Research Institute of Nagoya in his group's solar energy research. Solar furnaces are originally pieces of apparatus used in the development of highly temperature-resistant materials, but Horigome included them because he thought that they were technically relevant to solar energy. It was in this way that Horigome's proposal came to be submitted to AIST in early 1973, under the joint endorsement of ETL and the National Industrial Research Institute of Nagoya. It was at this time that it caught the attention of Suzuki Ken.

7.1.3 Creating New Significance for New Energy

In formulating his plan, Suzuki immediately began to make contact with Horigome, who was the person at ETL responsible for proposing the theme. Suzuki told Horigome that because energy research was extremely important it was something

---

[16]Sawada Shinji, interview by author, June 22, 1998.

[17]"Generally speaking, the laboratory manager is the main official who establishes new topics. It is generally the laboratory manager who engages in frequent exchanges with the Agency of Industrial Science and Technology after obtaining the approval of people such as the division manager and the office manager as a matter of course. I had to do these things because of the position I occupied. All things, including exchanges, came to me when something happened because I proposed solar power generation." (Horigome Takashi, interview by author, June 13, 1998).

[18]Tani Tatsuo, interview by author, July 9, 1998.

that must be done as a national effort, and offered to provide support for his solar energy research. This was the beginning of a cooperative relationship between the two. Horigome later gave technical explanations in relation to solar energy as an expert advisor, at the request of Suzuki; while on the other hand, Horigome requested Suzuki's help in convincing higher-level management at ETL—who were apathetic towards solar energy—about the value and importance of his research.[19]

Suzuki was thinking about liberating his new energy technology development and making it independent from the Large-Scale Project System. In doing so, the concept that he envisaged was the creation of a comprehensive program for the development of new energy technologies that did not stop only at solar energy. Suzuki's plan was to launch a new and attractive development program by creating new meaning and significance. Solar energy research was literally the core concept at the center of this idea; and, with its clean image, Suzuki thought, they could hope to gain widespread public approval.

This kind of lobbying on the part of Suzuki was also desirable for Horigome, who had continued to endure feelings of inferiority and belittlement at ETL. If solar energy was selected as the theme for a Large-Scale Project, then surely even the higher-level management at ETL would be forced to recognize its significance. In this regard, the agendas of both the AIST development officer formulating the project plan and the ETL researchers who would actually conduct the research were in agreement. For Suzuki's part, the limitless potential power and clean image of solar energy research was essential as the central element in achieving the realization of his program for the development of new energy technologies; while, on the other hand, for Horigome, the realization of Suzuki's plan signified an opportunity for him to put solar energy research into full-blown operation at ELT.

With the formation of this cooperative relationship between Suzuki and Horigome, and with AIST making strong requests for ETL to advance the development of solar energy technologies, it was as exactly as Tani attested to in his comments (see above). The higher powers within ELT, who had never taken solar energy research seriously to begin with, were now forced—albeit in bemusement and bewilderment—to comply with the formal requests of Suzuki and the others at AIST.

There was one other thing that Suzuki needed to push ahead with. While the development of solar energy technologies would form the base of the new program, he also needed some additional new energy-related themes aside from that, as a contingency in case solar energy ended in failure. For that reason, Suzuki visited

[19]It appears that the reason for ETL submitting solar energy as a proposal to AIST was that no other promising themes were available to them at the time, and that since the theme did not clash with those being proposed by any other national research institute ETL as a whole had no choice but to submit solar energy as their proposal. Additionally, the high powers within ETL also took a skeptical view in terms of the technical aspects of the proposal, doubting whether Horigome and his team of Energy Department staff—who were originally power transmission system researchers—could really carry out research in a completely different area like solar energy.

various other agencies and research institutes in search of other themes relating to new energy.

After many meetings, he first decided to incorporate geothermal energy—an area into which research had been underway at the Geological Survey office—into the program.[20] Until then, geothermal energy had been nothing more than a comparatively plain and unglamorous research theme. But with its reevaluation from the standpoint of the development of new energy technologies, it suddenly became the focus of close attention. In an article published later (January 17, 1974) after the assignment of the budget for the Sunshine Project, the *Mainichi Shimbun* newspaper reported with regard to geothermal energy that, "Until now, geothermal (power) research had received nothing more than a pittance of survey funds equating to 'consolation money.' In comparison to that, we can surely say that (the assignment of this budget) is literally a world of difference."[21] This indicates just how small the scale of research into new energy technologies had been until that time.

Suzuki also visited the coal and mining authorities asking for suitable themes for his energy development project. It was there that he learned that a method of manufacturing synthetic natural gas, by the gasification of coal, was attracting attention. At precisely that time, a research theme relating to technologies using coal had also just been submitted by the National Research Institute for Pollution and Resources. Based on this, Suzuki also decided to incorporate technologies using gasified and/or liquefied coal into the project, as another possible type of new energy.

By the process outlined above, Suzuki arrived at a conceptual plan for the development of new energy technologies consisting of three main pillars: solar energy, geothermal energy, and coal gasification (i.e. the manufacture of synthetic natural gas): "With clean energy and solar energy alone, we wouldn't last 30 years. So I incorporated other themes, including solar, geothermal and man-made natural gas. I took the plan (to my bosses) gleefully, as a new energy development project that included all three key elements: the heavens, the earth, and men."[22]

Later, hydrogen energy—which had initially been assigned a rank of C—was also incorporated into the project plan. This was based on the recognition that if the development of solar, geothermal and synthetic natural gas energy technologies was carried out, hydrogen energy would become important as a means of transporting energy. Ultimately, therefore, the final plan consisted of four pillars instead of the initial three. Suzuki had combined several development themes that had until then been researched as completely separate themes by the various national research institutes, giving them new significance as a program for the development of new

[20]Geological Survey of Japan Editorial Committee (1982, pp. 84–85).

[21]*Mainichi Shimbun,* January 17, 1974.

[22]Testimony given by Suzuki Ken in the published interview with Kinoshita Tōru titled "Sonotoki watashi wa [How I Acted at That Point]," published in the August 22, 1982 issue of the *Nikkei Sangyō Shimbun* newspaper.

energy technologies, and created a completely new project; a new research and development framework that was entirely separate from the existing Large-Scale Project System.

In this way, Suzuki's work in connecting various new energy related research themes created new meaning and significance for the research. His vision of a project for conducting comprehensive development of new energy technologies—with a central focus on solar energy—would later have a major influence on the actual realization of the project in terms of gaining public attention.

As seen with the example of geothermal energy, until this time these individual new energy research themes had been small-scale and unglamorous and were definitely not the kinds of themes that would gain widespread attention. But binding them together and declaring them to be the ultimate plan for resolving Japan's problems created new and attractive significance that could never have been achieved with the individual themes alone.

7.1.4 An Alliance of Development Officials and Researchers

Suzuki requested Horigome's cooperation in developing this kind of project proposal, and his ETL solar energy research group also became involved in advising the creation of the proposal. That said, of course, it was not the case that Suzuki and Horigome would agree with each other on everything all of the time. For example, when Suzuki made it clear that he intended to include coal-using technologies in the new energy plan, Horigome objected; expressing concerns that the inclusion of coal-based technologies could damage the image of the project, which was the development of clean energy solutions. In response to this, Suzuki answered that because the main portent of the research was to use coal cleanly, then surely it was acceptable to include it within the scope of the project. Finally, Horigome conceded, saying that if MITI and the Ministry of Finance approved then he would accept it, too.[23]

Later on, Suzuki was envisaging a budget of around 1 trillion yen over a period of approximately 30 years for the project. But to begin with, at the startup stage, he was considering a budget of around 500 million yen. However, he had also been advised by Director of General Affairs Ōsono Hideo that he should in fact make the project an even larger-scale endeavor. Suzuki acted in accordance with this advice, making the project a large-scale program from the very beginning, and drew up an initial budget of around 1 billion yen.[24]

[23]Horigome Takashi, interview by author, June 13, 1998.

[24]Nebashi Masato and Suzuki Ken, interview by author, May 8, 1998. At the request of Suzuki, Okabe Takehisa (at the Office for Research and Development) and Suzuki Norio (at the Technology Research Section) took charge of budget integration related to solar power generation. Suzuki Norio made the following comments regarding the budget integration work he had taken charge of: "I asked for the supply of data I needed and integrated the budget all by myself in one shot because

Also around that same time, Suzuki and Nebashi's boss Kinoshita Tōru had given his approval for the project. As the budget being requested was a large sum of money (around 1 billion yen), and that it would later face the scrutiny of an advisory committee, Kinoshita suggested that perhaps they should give the project a nickname. In order to clarify the unified concept of the project, too, Kinoshita thought that it was important to give the project a familiar nickname—of the kind being given to similar projects the were being conducted in the United States—instead of following the conventional approach of simply naming the project after the technology being developed. Some people—such as AIST Director Matsumoto Keishin—were of the opinion that since it would be an internal designation it really did not matter whatever the project was called. Despite this, Kinoshita thought deeply and enthusiastically about the name of the plan. "Every morning he (Councilor/Deputy Director-General of Technology Kinoshita Tōru) would come up with something new. 'Sunrise? Sunshine? Or maybe Sunshine is no good because people will confuse it with Shoeshine.' Every morning, day after day, he would think about names for the project, and he would come to me and say, 'Hey Nebashi, how about this one?'" recalls Nebashi Masato.[25]

The initial candidates for the name of the project were "The Sunrise Project" (because of the solar energy connection) and "The Blue Sky Project" (which included the meaning of preventing pollution). However, out of consideration for local government projects with similar names, the name that was finally adopted as the official name for the project was Kinoshita's suggestion "The Sunshine Project",[26] which also carried the meaning of creating a bright and cheerful world, in which the sun poured down beams of golden sunshine from a clear and cloudless sky.[27]

Kinoshita made the existence of the project known to Yamashita Eimei, who was at the time Administrative Vice-Minister for International Trade and Industry, and informed him that they would be naming it the Sunshine Project. According to Kinoshita, upon hearing this Yamashita was very enthusiastic, saying, "It's what this era is calling for. By all means, let's put the project into action; as a connecting step to bridge the gap until the coming of the nuclear age. The name is nice and

Horigome and other people at the Laboratory were not very good at those kinds of things. It was something like five billion yen. It was an authentic-looking figure that I came up with in one shot. I said that the budget could not be that high, but they said it would pass somehow as long as it was submitted. They said that the budget looks better when it's higher. They said that a high sum remains even after cuts are made. I figured that about ten billion yen was right at that point, instead of arriving at the figure through integration, because they said that they would use one trillion yen by the year 2000." (Suzuki Norio, interview by author, September 2, 1998). In this interview, Suzuki stated that the budget was five billion yen because five billion yen out of the total budget request of 11 billion yen was appropriated for a plan to establish a laboratory as a corporation with special semi-governmental status. Suzuki subtracted this amount from the budget he compiled.

[25]Nebashi Masato and Suzuki Ken, interview by author, May 8, 1998.

[26]The published interview with Kinoshita Tōru titled "Sonotoki watashi wa [How I Acted at That Point]," published in the August 22, 1982 issue of the *Nikkei Sangyō Shimbun* newspaper.

[27]MITI Kōgyō gijutsuin Kenkyū kaihatsukan shitsu (1987, p. 51).

cheerful as well."[28] It was by this kind of process that the subtitle of "The Sun Shine Project" came to be formally added to the revised plan for the new energy technology development project written by Suzuki in June of that year (the space was subsequently removed and the name formalized as "The Sunshine Project" at the time of its public announcement that August.[29] For the AIST development officials, who were seeking to achieve the actual realization and establishment of political policy, in order to gain the understanding of the public it was important to give the project an endearing name and appeal to them about the project in an easily understandable manner.

However, the members of the ETL solar energy research group had mixed feelings when it was reported to them by Suzuki that the project had been named the Sunshine Project. Horigome testified later that when he heard the name he thought it sounded like the lyrics of a song, and Sawada said that he thought they must be joking around.[30] As far as the members of the group were concerned, regardless of what the project was called, what was important to them as researchers was the success of the solar energy development.

That being said, for them too the successful adoption of the project proposal was essential in order to secure the necessary budget for their research theme. It was Kinoshita, again, who suggested making use of the Club of Rome's *Limits to Growth* report—which had attracted widespread attention when it was published in Japan in May of the previous year—as the theoretical basis for the new energy development project. The Club of Rome argument was the ideal grounds for new energy development that would require massive funding over an extended period of time. In later years, Kinoshita made the following comment: "After bringing these [new energy related themes] together it looked as though we were going to have a pretty attractive project, so [I said] why not take the plunge and try jumping on the energy crisis announced by the Club of Rome? We don't know whether it's all true or not, but why not try coming out with a big project every now and then?[31]"

As technical development policymakers, the staff at AIST were in a position where they needed to promote the significance of the project. In doing so, they would have to gain the wide-ranging support of politicians and the public, who were not at all knowledgeable about the technology involved. Making effective use of the endearing name of the project and the Club of Rome report was one means of gaining that support. In order to mobilize the necessary budget and other resources needed for the project, it would be necessary to present new and attractive meaning and significance, and to gain the widespread agreement of society. The AIST staff—and Kinoshita in particular—recognized that winning popular support for the project

[28]The published interview with Kinoshita Tōru titled "Sonotoki watashi wa [How I Acted at That Point]," published in the August 22, 1982 issue of the *Nikkei Sangyō Shimbun* newspaper.

[29]MITI Kōgyō gijutsuin, *Shin'enerugī gijutsu kaihatsu keikaku (Sanshain keikaku)* [New Energy Technology Development project (The Sunshine Project)] , memorandum, June 1973.

[30]Horigome Takashi, interview by author, June 13, 1998; and Sawada Shinji, interview by author, June 22, 1998.

[31]MITI Kōgyō gijutsuin Kenkyū kaihatsukan shitsu (1987, p. 52).

was even more important than the purely technical problems being faced in ensuring the successful realization of the project. It was in this way that the project came to be named the Sunshine Project, in July of that year.

7.1.5 The Cooperation of Dokō Toshio

In August 1973, the Sunshine Project was publicly announced by then Minister of International Trade and Industry Nakasone Yasuhiro. In order for the project to be widely considered and discussed by experts, Minister Nakasone sought the advice of the Industrial Technology Council. Since new energy was a new theme that would not fit within the conventional framework that had existed until that time, it was necessary to create a new special committee.

But who was suitable for the role of committee chairman? In view of the importance of the position, AIST decided to appoint Dokō Toshio (Keidanren vice-president). Nebashi—who at that time was still Deputy-Director General for Development—went to see Dokō right away, and politely requested for him to become chairman of the committee. In response to Nebashi's request, Dokō immediately agreed, expressing his understanding of the importance of the task.[32] Dokō would go on to fulfill the role and expectations placed on him with the greatest integrity and sincerity.

The committee met several times between August and December of that year. Nebashi took the role of communicating with Dokō. Despite his position and standing, Dokō insisted that he would go to visit Nebashi himself whenever called upon, and actually went to visit MITI himself on numerous occasions. Out of consideration for Dokō, Nebashi said that he prepared the AIST Director's office for Dokō's visits instead of a regular councilor's room.[33]

At the first committee meeting, Sakisaka Masao—director of the Institute of Energy Economics, Japan—was singled out for chairman of the Policy Subcommittee. At first, Sakisaka hesitated to accept it, because of the heavy responsibilities of the position. It was then that Dokō drew upon him, insisting that only he (Sakisaka) could do it, and asking if he was really going to decline such an important job. Sakisaka said later that all that he could do was to fall silent. After the committee meeting, Dokō apologized politely for his discourtesy during the meeting, saying that it really was only Sakisaka who was up to the task, and that he was sorry to make such an enormous request, but he really did insist that Sakisaka should take the position. In light of this, Sakisaka finally agreed to accept the position of subcommittee chairman.[34] As demonstrated by episodes such as these, Dokō took his role with great enthusiasm, and completed a final report as chairman

---

[32]Nebashi Masato and Suzuki Ken, interview by author, May 8, 1998.

[33]Nebashi Masato and Suzuki Ken, interview by author, May 8, 1998.

[34]Nebashi Masato and Suzuki Ken, interview by author, May 8, 1998.

of this council—which was composed of the leading authorities in various fields—by winter.

Dokō recognized the importance of new energy, and there are numerous episodes that indicate his intense hard work in driving forward its development. These anecdotes indicate the reason why Dokō remained in his position, protecting the development of new energy technologies, from the time of his appointment as chairman of the NEDO steering committee in 1980 until August 1988, a mere two years before his passing.

In the formulation of the Sunshine Project, various people could be observed in forming cooperative relationships in order to fulfill their own wants and aspirations. Their ideal wishes did not always materialize just as they had envisaged, and people at times had to compromise; sometimes making use of social trends and sometimes adding names that would be easily understandable to the common people. The combination of these efforts gave new meaning and significance to the new energy project, which in turn brought forth the bud that would lead to major growth and expansion for the project itself.

If this had been a regular year, however, the project might have ended as it was, still in its budding stages. But as chance would have it, an unusual and unforeseen crisis lay ahead in the autumn of 1973.

7.1.6 Snowballing PR and Expansion of the Project PR

What kind of awareness did the Japanese government have of the energy situation during the first half of 1973, when plans for the Sunshine Project were being formulated? From today's perspective, the Sunshine Project seems to have appeared at a time that would suggest that the occurrence of the impending oil crisis had been anticipated. The Agency for Natural Resources and Energy (ANRE) was also launched in July 1973. This, again, might appear as if it was a measure taken based on anticipation of the worsening of the energy situation. Certainly, the Japanese government was surely aware of the threat of crisis in the event of an emergency with respect to the country's increasing dependency on oil. If the government had established the Sunshine Project and launched ANRE as a strategic, preemptive move against the imminent first oil crisis, it would have been a truly formidable act of foresight. However, Yamagata Eiji—who was the first Director-General of ANRE—recalls as follows:

> In around September of that year, we were just taking a breather (after having finished authorizing the electrical tariff revisions for Kansai Electric Power and Shikoku Electric Power), and I never even dreamed that six months later we would be drawn into such a terrifyingly busy state. I was thinking to myself that we could take it easy and relax now for around the next year.[35]

[35] Yamagata (1991, p. 89).

In September 1973, even the Director-General of ANRE—who stood at the very center of Japanese comprehensive energy policy—had no sense of urgency over the threat of a sudden energy crisis. Even after the outbreak of the fourth Arab–Israeli war, writes Yamagata, the possibility of the conflict developing into an oil crisis was "completely unexpected as far as the government's awareness [of the situation]".[36] Kishida Fumitake—who was head of ANRE's Public Enterprises Department at the time—also wrote: "Honestly, my first response [to the outbreak of the fourth Arab–Israeli war] was something along the lines of 'Oh, not again,' and everybody lacked the foresight to see how it would develop."[37]

At the time, Horigome—in ETL's Energy Department—had not predicted the imminent oil crisis, either: "When we drew up the budget, we had not planned for the occurrence of the oil shock whatsoever; although I was concerned that [the supply of] oil would become extremely tight. But nobody had anticipated that an oil crisis would occur that year,"[38] he recalls.

On the other hand, Suzuki Ken—who was a research and development official at AIST—said the following: "At the time, the price of oil, which had been around 2 dollars and 70 cents rose over three dollars to almost four dollars. I myself was originally an electricity guy, who also studied abroad in nuclear power, and I paid quite a lot of attention to energy-related matters, including the Putnam Report."[39] From the testimony of these individuals we can see that, although some telltale signs did exist prior to the occurrence of the crisis, it was not a state of affairs in which anyone—even experts with detailed knowledge of the energy field—could predict the timing with which a crisis would arise.

However, in October 1973, a mere two months after the announcement of the Sunshine Project (in August) the first oil crisis struck. As crude oil prices soared, the project grabbed the attention of the Japanese public.

Amidst the panic, the negation of the Sunshine Project's budget was taking place. Suzuki Norio, who compiled the budget and negotiated with the Ministry of Finance in regard to solar energy, recalled as follows:

> The Ministry of Finance, too, said something along the lines of "Don't talk stupid"… The initial appraisal didn't get us very much of a budget. Well, initial appraisals are basically set up to turn down new budget requests, anyway. But then [with the occurrence of the oil crisis in October] it was like a divine wind blew. Just at the time when we were discussing the details.[40]

---

[36]Yamagata (1991, pp. 95–96).

[37]Kishida (1991, p. 138).

[38]Horigome Takashi, interview by author, June 13, 1998.

[39]Comments made by Suzuki Ken (in Nebashi Masato and Suzuki Ken, interview by author, May 8, 1998). The Putnam Report—officially entitled *Energy in the Future*—was a report compiled in the 1950s by an engineer named Palmer C. Putnam. He was commissioned by the United States Atomic Energy Commission to write a report detailing the super-long-term global outlook with regard to energy for the next 100 years up until 2050.

[40]Suzuki Norio, interview by author, September 2, 1998.

Given the oil crisis, even the Ministry of Finance was forced to approve the budget for the project.

Amidst the panic of the crisis, the solar energy research being conducted at ETL was also drawn into the public's feverish enthusiasm for new energy. According to Tani's testimony, after the occurrence of the oil crisis, trading companies like Mitsubishi Corporation approached ETL and attempted to secure all of the patents relating to solar thermal energy research ahead of competing firms. Horigome was called before members of the Diet and had to travel around various locations explaining their research; and ETL, too, was swarmed with members of the media and numerous other visitors for days on end. At one time, Tani, too, become so tied up in dealing with them that he became effectively unable to conduct his research:

> It was kind of an abnormal time. For Dr Horigome, it must have been terrible, endless worry. At ETL he would be spoken to very harshly, and then when he would go outside in public the atmosphere was like that. Not that he was being pampered or made a fuss of, or anything like that. I think that the PR was just snowballing at the time.[41]

With the occurrence of the oil crisis, the Sunshine Project was now becoming the focus of hopes and expectations as a solution to save Japan from the energy crisis, and the project began to draw more attention from the public than either AIST or ETL had initially expected.

From the summer onwards, the AIST project proposal was discussed and considered by experts at the Special Committee for Energy Technology at the Industrial Technology Council. The solar thermal power plants presented in the project proposal at that time were an extremely large-scale undertaking.

The ETL Solar Energy Group had in fact realized—as was natural for them to do so being expert engineers specializing in that field—how unrealistic the plans put forward in the proposal were. Tani testified as follows:

> It was a very grand-scale plan. The ETL experts must have been really angry, because the proposal that they [i.e. AIST] came out with spoke of [power outputs of] 100,000 kW or 2,000,000 kW. They would say that there's no way we can do it; that this was deceitful. But from the AIST officials' perspective, there was no way that they could write 10 kW, or some low number like that.[42]

In spite of the deep concerns of Tani and the other researchers, further details of the project proposal continued to be determined by AIST.

Membership of the Special Committee for Energy Technology at the Industrial Technology Council's Solar Energy Subcommittee also included the names of Noguchi Tetsuo (the National Industrial Research Institute of Nagoya) and Horigome Takashi (ETL); under the chairmanship of Yamamura Sakae, who was a professor at the University of Tokyo. The Solar Energy Subcommittee/Panel also

---

[41]Tani Tatsuo, interview by author, July 9, 1998.

[42]Tani Tatsuo, interview by author, July 9, 1998.

included Tani, in addition to Noguchi and Horigome. They participated in the council as representatives for their respective research institutes and took part in the deliberations.[43]

Suzuki Ken offered the following testimony regarding his exchanges with Solar Energy Subcommittee Chairman Yamamura Sakae, which took place at a committee meeting in the latter half of 1973 with regard to the setting of development targets for solar thermal power plants:

> For solar energy, I had written the targets as I saw fit in order to secure the budget of eleven billion yen. So when the proposal came before the committee, I was reprimanded by Professor Yamamura. He told to me that I must not know about the essential properties of solar energy. I had written figures like 100,000 kW and 2,000,000 kW. He was angry with me, asking how we could expect to create power plants that could generate such massive amounts of power with something like solar energy, in which the energy is spread thinly over a wide area.[44]

However, in the committee's final report (during the oil crisis), the development targets for solar energy were raised even higher than in the original AIST proposal.[45] In the course of the committee's deliberations, the proposal changed to a plan demanding a design with even larger-scale development targets to be achieved at an even earlier stage. Finally, with the oil crisis continuing, the Sunshine Project budget was approved and, as a result, the project began, aiming for completion in the year 2000.

As a result of this, ETL received the solar energy research portion of the Sunshine Project budget. The Energy Department took responsibility for solar thermal power, while the Electronic Device Department took responsibility for photovoltaic power generation. Moreover, because of his involvement in the formulation of the initial proposal for the Sunshine Project, for all intents and purposes Horigome assumed the role of managing the Sunshine Project budget at ETL.

From January 1974, Nebashi took over the role of Deputy Director-General of Technology (following on from Kinoshita). Under his supervision, concrete

---

[43]Industrial Technology Council, "Shin enerugī gijutsu kaihatsu no susumekata ni tsuite [Points Regarding How to Advance the Development of New Energy Technologies]," in *Shin'enerugī gijutsu kenkyū kaihatsu keikaku (Sanshain keikaku)* [New Energy Technology Development project (The Sunshine Project)] , ed. MITI Kōgyō gijutsuin (Tokyo: Nihon Sangyō Gijutsu Shinkō Kyōkai, 1974), pp. 405 and 407. The Subcommittee on Solar Energy had a total of 11 members, including its chairman. Other participants included Suyama Junji (of the Geological Survey of Japan) in the Subcommittee on Terrestrial Heat, Kimura Hideo (of the Pollution Resources Laboratory) in the Subcommittee on Synthetic Natural Gas and Miyake Yoshizō (of the Osaka Industrial Technology Laboratory) in the Subcommittee on Hydrogen. All these participants were involved in the Sunshine Project at the respective national laboratories, like Horigome at the ETL.

[44]Nebashi Masato and Suzuki Ken, interview by author, May 8, 1998.

[45]Sangyō Gijutsu Chōsa Iinkai [Industrial Technology Council] , "Shin enerugī gijutsu kaihatsu no susumekata ni tsuite [Points Regarding How to Advance the Development of New Energy Technologies]," in *Shin'enerugī gijutsu kenkyū kaihatsu keikaku (Sanshain keikaku)* [*New Energy Technology Development project (The Sunshine Project)*] , ed. MITI Kōgyō gijutsuin (Tokyo: Nihon Sangyō Gijutsu Shinkō Kyōkai, 1974), p. 383.

preparations were carried out for the launch of the project. In February of that year, AIST established the Sunshine Project Preliminary Office, and began outsourcing work to private sector corporations for each of the various technology themes. In March, a group of project members from AIST and the various national research institutions affiliated with new energy research travelled to the United States for an observational visit. The group was led by Nebashi, and also included Ono Eiichi (Deputy-Director General of Development) from the Sunshine Project Preliminary Office and several researchers from the various research institutions, representing their respective technology development themes. In early March, the group—which also included Horigome (Director, ETL Energy Transport Research Laboratory), who was responsible for solar energy; Suyama Junji (Director, Physical Survey Department, Geological Survey of Japan) , who represented geothermal energy; Kimura Hideo (Director, First Resources Department, National Research Institute for Pollution and Resources), who represented gasification and liquefaction of coal; and Miyake Yoshizō (Director, Fifth Department, Osaka Industrial Technology Laboratory) , who represented hydrogen energy—assembled and departed for the United States. Over the duration of their trip, from March 10 until March 31, the group observed various trends and developments in new energy research.[46]

Meanwhile, Suzuki Ken—who with the establishment of the Preliminary Office had officially become the research and development official in charge of solar energy—began to approach universities regarding the outsourcing of research themes. Considering the fact that the Sunshine Project had successfully obtained a budget in the hundreds of millions (of yen), Suzuki had high hopes that universities would take advantage of large-scale funding from this budget and engage in proactive research in this field. Suzuki comments as follows: "I went to the Department of Electrical and Electronic Engineering at the University of Tokyo and said that because we had secured the budget they should use it for their research. Their response was 'Yes, by all means, please let us use it.' So I ask them how much they wanted. And they said, 'five million yen'."[47]

Until that time, there were not many universities where researchers were engaged in solar energy research. This situation illustrated the gap that lay between the suddenly and rapidly enlarged scale of the Sunshine Project and the state of progress in realistic new energy research at the time.

As we have seen above, if we raise the level of resolution and scrutinize the individual people who participated in the project, and observe the series of interactions within the various organizations involved, we begin to see a chain of actions backed up by the respective intentions and motives of the various individuals, and an image of the combined effect of these actions begins to emerge. In the Sunshine Project, amidst the panic of the oil crisis, the ideas of development officials and the dreams of researchers gathered the attention and expectations of the Japanese public; and the subsequent mobilization of resources resulted ultimately in the

[46]Investigative Committee on Industrial Technologies (1974), p. 57.

[47]Nebashi Masato and Suzuki Ken, interview by author, May 8, 1998.

creation of an epic, grand-scale project, in a form that even those individuals themselves had neither intended nor envisaged.

Despite being built around a core of ideas that nobody at the time had a clear basis for believing could be realized, the Sunshine Project gathered the hopes and expectations of various people to it like a rolling snowball. In doing so, it drew in resources almost as if it had promoted itself, and manifested itself in that form for all to see.

7.2 The Start of the Solar Energy Project

7.2.1 Backdated Rearrangement of Company Research Proposal Themes

Let us take a look at the actual state of solar energy research under the Sunshine Project, from the perspectives of various individuals who were actually involved in the process.

AIST set about determining the assignment of outsourcing themes, by first requesting various corporations to submit proposals, and then screening the submitted themes. But how did the people responsible for carrying out this task at AIST perceive this process? AIST's Suzuki Norio—who was responsible for compiling the Sunshine Project budget and was later involved in the outsourcing process—made the following statement:

> Working on such a wide-ranging project as this, there was no way that even we [at AIST] could understand everything from A to Z. So we would maintain everyday working relationships with people at various companies. Even in my case, although I do have an engineering background, I would still talk to them about various issues and ask them to teach me things. In the end, if we didn't do that then we wouldn't be able to do it [i.e. handle a project like this[48]].

Usually, when working under the name of a project, AIST would select the necessary technologies in accordance with the project and outsource them to companies. At least, that would be the typical way. If things were done according to this image, development officials and other planners with specialist knowledge would have an accurate grasp of external conditions (including trends in relevant technologies, etc.), select the most appropriate means and methods with respect to it, and outsource those tasks to the engineers at various companies in a top-down manner.

However, according to the testimony of persons who know the real situation inside AIST, the actual method being adopted was quite different. First, companies would submit proposals from their own side, in relation to technologies that they felt that they would be able to develop. AIST would then rearrange these proposals

---

[48]Suzuki Norio, interview by author, September 2, 1998.

at a later date, in a way that would make it easy for them to explain. The following statement is that of Kurokawa Kōsuke, who was seconded to NEDO from ETL: "The sorting [of the technology themes] was done later on, after the fact. First the proposals would come in. We would then categorize and rationalize them slightly for the sake of outward appearances, and rearrange them in an easily explainable way. So initially they would come to us in a completely disjointed manner."[49]

As is also suggested by comments such as this, in the Sunshine Project proposals for individual development themes were first submitted by the various corporations from their own side. Officials and AIST, ETL, and so on then set about the task of rearranging them into an overall development plan after the fact.

The process by which technologies to be developed were selected was not left to the sole discretion of AIST but depended rather upon the composition of development themes submitted by the various corporations. The nature of this arrangement meant that when companies submitted their technology development themes to the Sunshine Project, they would request themes that they felt would be commercially beneficial for them based on their own technology strategies, and that would lead to the creation of new business operations in the future. From this we can see that when it comes to national projects, it is not necessarily the case that everything is determined in a top-down manner by the project administrators. Corporations propose themes that they themselves consider to be commercially promising, and those proposals have an impact on decision-making within the project, in a bottom-up manner. In which case, the engineers at each company would surely have played a role in securing the realization of the research that they wanted to undertake at their respective companies.

What kinds of strategies did the various corporations follow in proposing their preferred technology themes and in securing contracts to develop those technologies in the outsourcing process for solar (photovoltaic) power generation technologies under the Sunshine Project? And how did AIST go about selecting the proposals made by the various companies, and determining how to divide responsibilities for the various outsourced themes amongst them?

7.2.2 Sharp: In Search of a New Method for Manufacturing Silicon

The development of new methods for manufacturing silicon—an important technology for the development of solar cells—was initially outsourced to three companies: Hitachi, NEC, and Toshiba. However, the companies that were fervently enthusiastic about the commercialization of this technology at the time were companies like Sharp, Matsushita Electric Industrial (now Panasonic), and Kyocera; a company that was to join the Sunshine Project at a later date. Let us

[49] Kurokawa Kōsuke, interview by author, April 29, 1998.

examine this outsourcing process from the various perspectives of the engineers at the companies involved.

When the outsourcing of technology developments under the Sunshine Project began in 1974, Sharp immediately declared that it would be participating in commissioned research into solar cells. This was partially due to the company's past performance in that area. Seizing this new opportunity, Sharp established a Sunshine Project Promotion Department at its central research institute, as the research department that would be responsible for solar cell research outsourced to Sharp under the project. A man named Kimura Kenjirō—who until then had belonged to the company's semiconductor business division—was appointed as the general manager of the new department.

Kimura had relevant experience from his involvement in the manufacture of the first solar cells at Sharp during the late 1950s, and it was because of that experience that he was selected as the person who would take charge of the new Sunshine Project Promotion Department. After being made general manager, Kimura promptly went to meet with Suzuki Norio—the development official responsible for solar energy technologies at AIST—and notify him of his appointment.[50] As a result, Sharp was commissioned to research and develop a new type of solar cell, from the initial year of outsourcing for R&D themes relating to solar cells under the Sunshine Project. However, as Sharp did not currently manufacture its own silicon and produce its own silicon wafers, in the assignment of themes for the initial year Sharp was tasked with the development of modular manufacturing technologies. This involved the development of packaging and modularization technologies with the aim of shortening the length of the solar cell manufacturing process, automating it and lowering manufacturing costs.[51] This division of roles was determined based on AIST's overall project policy. Recognition was given to the fact that Sharp had already been producing solar cells, and expectations were placed on Sharp producing achievements in the development of assembly processes rather than the development of materials.

However, Kimura was disappointed by this decision. This was because he believed that when it came to the mass production of solar cells at Sharp in the future, the development of methods for the manufacture of silicon—the basic raw material in the production of solar cells—would be more important than modular manufacturing technologies and other assembly-related processes. The key to reducing costs for the practical and commercially viable production of solar cells in the future lay in the development of silicon manufacturing technologies. While the fact that AIST had high expectations with regard to Sharp's experience and technical capabilities in modular manufacturing sounded good in principle, the bottom line was that Sharp had been unable to participate in the development of new methods for manufacturing silicon.

[50]Kimura Kenjirō, interview by author, September 9, 1998.

[51]MITI Kōgyō gijutsuin Sanshain keikaku suishin honbu (1984, p. 69).

To reduce the cost of producing solar cells, it would be necessary to grow crystalline silicon at a high speed (although the high level of purity used in semiconductors was not required). In order to achieve this, there was a need to develop a new silicon crystal growing method to replace the conventional Czochralski (CZ) process, in which the monocrystalline silicon used as the raw material for silicon wafers is pulled into long, cylindrical rods. Kimura, too, thought that there was an urgent need for Sharp to develop its own silicon production method, using a cheaper, more affordable raw material to replace conventional monocrystalline substrates.

Around this time, Kimura heard from one of his subordinates, Suzuki Akio, that a new ribbon crystal growing technique called EFG (edge-defined film-fed growth) had been invented in the United States. Kimura promptly made an inquiry to the inventor of this new technology—a man named Muravsky at a company called Tyco—regarding the possibility of its introduction at Sharp. Unexpectedly, Muravsky's answer was that the technology had already been introduced by Kyocera Corporation as a method for drawing out sapphire (aluminum oxide, Al_2O_3) crystals. For this reason, Tyco instructed Kimura that if Sharp wished to introduce the technology they should first obtain the permission of Kyocera.

Fortunately, Kimura was already acquainted with Kyocera founder and president Inamori Kazuo, as a result of having taken delivery of ceramic semiconductor packages from Kyocera during his involvement with semiconductor-related work in the past. For that reason, when Kimura visited the United States, he decided to have Inamori write a letter of introduction to Muravsky indicating that Kyocera had permitted the introduction of the EFG technology to Sharp.[52]

Muravsky had initially launched his own venture company, Tyco. But Tyco's technology soon drew caught the attention of the U.S.-based oil company Mobil Oil Corporation. In 1974, Mobil Oil invested in the company for business expansion, and Mobil-Tyco Solar Energy Corporation was established. Kimura was attempting to introduce Muravsky's ribbon crystal technique and begin a program of research, development, and practical commercialization with the technology at Sharp. By doing so, from Sharp's perspective, it was naturally considered desirable to secure a commission to develop the technology as part of the Sunshine Project.

Inamori at Kyocera also expressed his wishes to participate in Sharp's development of the ribbon crystal growing technology, and the two companies headed to AIST with a proposal aimed at securing a commission in relation to this development theme. AIST's Suzuki Norio described the situation at the time as follows: "If I think about it now, this small company called Kyoto Ceramic came trying to sell (the idea) to us, and at that time I thought that I'd never heard of such a company. But they came together with Sharp, and suggested trying (to produce silicon with) the ribbon crystal technique."[53] However, in the end Kyocera and Sharp's plan was rejected by AIST. Regarding the reason for the rejection, Kimura said:

---

[52]Kimura Kenjirō, interview by author, September 9, 1998.

[53]Suzuki Norio, interview by author, September 2, 1998.

> At that time, the Japanese government was taking a purist, nationalist stance, and only wanted to support purebred Japanese enterprises. The way I perceived what they were saying was that, when it came to anything that involved American capital investment, we should simply get on with it ourselves. So if we wanted to do it [on our own without government support] then that was fine.[54]

In an attempt to catch AIST's attention, Kimura argued that, hypothetically, even if Toshiba succeeded in developing its own ribbon crystal growing technique, it would eventually infringe upon the patents held by Mobil-Tyco for their technology. Engineers like Kuru Isamu and Nakagawa Masashi of the Electronic Technology Research Center, Electronics Division—who were responsible for developing ribbon crystal techniques at Toshiba—countered Kimura's argument, and the dispute between the two camps would continue for several years after that.[55] However, even with his assertions, Kimura's was not able to change the attitude of AIST. So, Kyocera and Sharp (with the addition of Matsushita) went on to form the Japan Solar Energy Corporation as a joint venture.

As we can see, the assignment of research contracts under the Sunshine Project did not necessarily match with the technology strategies of the two firms; and eventually they adopted a policy of developing the fundamental technology for solar cells themselves, separately from the project, based on their own internal decision-making processes.

The Matsushita Group, to which the Sunshine Project assigned the development of compound-semiconductor solar cells, encountered a similar situation. Matsushita, too, had the motivation to develop a ribbon crystal growing technique of its own.

7.2.3 *Matsushita Electric Industrial: Compounds or Silicon?*

The real favorite for solar cell development at Matsushita Electric Industrial at the time was compound-semiconductor cells. Staff at Matsushita's Wireless Research Institute—centering on a man called Ikegami Seiji (Research Materials Development Laboratory, Matsushita Electric Wireless Research Institute)—had begun work developing solar cells using Cu_2S (copper(I) sulfide). Matsushita was trying to manufacture solar cells that could achieve satisfactory performance at an affordable cost, using the compound-semiconductor technologies that it had already accumulated thus far.

Compound-type semiconductors differ from silicon—which produces a constant response to sunlight—in that their optical band gap can be changed depending on the material used, which meant that it was technologically possible to create more

---

[54] Kimura Kenjirō, interview by author, September 9, 1998.

[55] Kimura Kenjirō, interview by author, September 9, 1998.

responsive solar cells using compound semiconductors than with silicon.[56] However, as the research progressed, it became clear that Cu_2S cells deteriorated rapidly due to their use of copper. Later, therefore, the mainstream choice of material for compound-semiconductor solar cells (in the II–VI family of semiconductor compounds) shifted towards CdTe (cadmium telluride), which is still a prominent material used in the production of compound-semiconductor photovoltaics today.

The central figure in the research effort, Ikegami, recognized that what was necessary in order to gain the advantage in the field of CdTe solar cells (which many companies around the world were already working to develop) was an original manufacturing method that differed from those being used by other firms. Working under that premise, he began searching for a solar cell production method that would be suitable for mass production, in line with the cost reduction that was one of the key principles of the Sunshine Project.[57] What he developed as a result was a solar cell manufacturing process that made use of screen printing. Micrometer-scale particles of CdTe powder were printed onto a glass substrate in paste form. The substrate was then baked in a sintering furnace at a high temperature of several hundred degrees, in the same way as clay pottery, to produce a film. Ikegami's group eventually succeeded in producing a prototype solar cell using this screen-printing method in 1976, the third year after the outsourcing of research under the Sunshine Project began. In a newspaper article written in August 1979, after the technology was completed, Ikegami made the following statement: "In the very near future, the power generation cost [for compound-type solar cells] will become a few hundred yen per Watt… Because, if we don't do that, then there is no way that we will be able to compete with silicon-type solar cells, which can achieve high performance efficiency of 10% or higher."[58]

At the time, compound-type solar cells had conversion efficiencies of 5–6%. In performance terms, they were no match for crystalline silicon solar cells, which could achieve conversion efficiencies of 10% or higher. However, at that time, unit costs for silicon solar cells ranged from between several thousand yen to over 10,000 yen per watt of generated power. By contrast, compound-type solar cells

---

[56]Silicon solar cells do not generate electricity when light has a wavelength that is longer than 1.145 microns. Silicon solar cells generate electricity in reaction to light with a longer wavelength when this threshold wavelength is long. The material must be changed to alter the threshold wavelength because a numerical value peculiar to each material known as the optical band gap determines the threshold wavelength. The band gap of material that is most sensitive to sunlight is said to be approximately 1.5 eV. Incidentally, the band gaps for silicon and amorphous silicon are 1.1 eV and 1.7 eV, respectively. The optical band gap is a numerical value peculiar to each material that determines the threshold wavelength for power generation using solar cells. (Taiyōkō Hatsuden Gijutsu Kenkyū Kumiai [Photovoltaic Power Generation Technology Research Association] , ed., *Taiyōkō hatsuden: Sono hatten to tenbō* [Solar Power Generation: Its Development and Prospects] (Tokyo: Art Studio 76, 1998), p. 20; and Murozono Mikio, interview by author, September 11, 1998).

[57]*Nikkei Sangyō Shimbun*, August 15, 1979.

[58]*Nikkei Sangyō Shimbun*, August 15, 1979.

manufactured using the screen-printing method developed by Ikegami at Matsushita Electric were predicted to cost only a few hundred yen per watt, and so they were expected to have a significant advantage in terms of cost. The commissioned research and development of compound-type solar cells at Matsushita Electric continued from 1974 until 1981, and their original technology was met with high acclaim from both inside and outside the company.

However, due to their inferiority to silicon solar cells in terms of performance, Matsushita's compound-type solar cells did not find applications for outdoor use and, although the development had succeeded, they had not reached a state of commercialization. Small-scale development efforts at Matsushita's Wireless Research Institute would continue, even after the end of commissioned research under the Sunshine Project.[59]

While on the one hand working to develop these compound-semiconductor solar cells, the Matsushita Group also wished to take part in the project in the development of crystalline silicon. In particular, Matsushita Battery Industrial (which was the member of the Matsushita Group responsible for batteries) wanted to productize crystalline silicon solar cells for its own business domain of battery-related applications.[60] However, due to the fact that Matsushita Electric had already been commissioned to develop compound-semiconductor solar cells, Matsushita Battery Industrial—as a member of the same corporate group—was unable to achieve participation in outsourced research into solar cells during the 1970s. Matsushita Battery Industrial's Murozono Mikio explained the circumstances at the time:

> Once you're in, it's great. Like one big happy family. But getting in from the outside is quite difficult. Even though we tried to get in as silicon-type developers, there was a pretty thick wall blocking us. It was like they were saying, 'But Matsushita (Electric) is already doing compound-type solar cells, so why at this late stage does the same Matsushita Group—even though we are Matsushita Battery Industrial—suddenly want to get in on silicon?' And so we couldn't get in.[61]

In this way, the Matsushita Group was placed in a situation where if it wanted to develop crystalline silicon solar cells then it would have no choice but to advance the development alone.

So at the time, the idea put forward by Kyocera to establish a new company to produce silicon using the ribbon crystal technique matched up with the Matsushita Group's own wishes.

---

[59]Murozono Mikio, interview by author, September 11, 1998.

[60]Murozono Mikio, interview by author, September 11, 1998.

[61]Murozono Mikio, interview by author, September 11, 1998.

7.2.4 Kyocera: Opposition to the Sunshine Project

In the 1970s, Kyocera's main business was the development and production of new ceramic materials. Until then, the company's business operations had no direct connection to solar cells. However, upon hearing that Sharp wanted to introduce Mobil-Tyco's ribbon crystal growing technology, Kyocera learned that its technology could be applied to growing crystalline silicon for solar cells. During his interactions with the solar cell development staff at Sharp, Inamori began to display an interest in solar cells.[62]

In the mid-1970s, Kyocera was embarking on a proactive diversification of its business operations. By 1975, Kyocera had a widely diversified product lineup, offering products such as recrystallized emeralds, synthetic bone bio-serum, and Ceratip cutting tools, which were related in terms of the technology they used, but were aimed at completely different markets.[63]

Amidst all of this, if Kyocera's sapphire crystal technology could be applied to the growing of silicon ribbon crystals for use in solar cells, then it would signify another new market into which Kyocera could potentially enter. Kyocera's Inamori approached Sharp with the suggestion of developing ribbon crystal technology for use in solar cells as part of the Sunshine Project. Upon receiving this proposal, Sharp—which needed the Mobil-Tyco technology that Kyocera had in its possession—agreed to a joint development effort for developing ribbon crystal growing techniques in collaboration with Kyocera. However, as stated earlier, the joint proposal put forward by the two companies was rejected by AIST.

The alternative strategy taken by Inamori in response to this setback was a plan whereby Kyocera would establish a joint-venture company, together with Sharp and Matsushita Electric, for the development of silicon substrates for solar cells; based on the decision to carry out the development by themselves, as private sector companies, without joining the Sunshine Project or relying on state-sponsored commissions. For Sharp and Matsushita, too, the plan made good sense. Since the development themes for which both firms had been commissioned to carry out research under the Sunshine Project were not related to silicon crystal growing methods, and also because the plan would enable them to obtain more affordable raw materials for manufacturing their own solar cells in the future, both firms were in agreement with Kyocera's idea: to develop these technologies together through a joint venture and then buy them up and manufacture silicon solar cells later on an individual basis.

---

[62]Kimura Kenjirō, interview by author, September 9, 1998.

[63]Kagono Tadao, "Benchā Keieisha: Inamori Kazuo (Kyosera) [Venture Business Manager: Inamori Kazuo (Kyocera)]," in *Kēsubukku Nihon kigyō no keiei kōdō 4: Kigyōka no gunzō to jidai no ibuki* [*Casebook 4 on Administrative Actions Taken by Japanese Companies: A Group of Entrepreneurs and the Signs of the Times*], eds. Itami Hiroyuki et al. (Tokyo: Yuhikaku, 1998), p. 378.

In October 1975, Kyocera established a joint-venture company in partnership with Sharp, Matsushita Electric Industrial, Mobil Oil (United States), and Tyco Laboratories (United States), aiming to reduce costs in the production of monocrystalline silicon for use in solar cells.[64] That company was Japan Solar Energy Corporation. Kyocera's Inamori Kazuo was appointed as company president, and the head office was established in Kyoto, which was also the location of the headquarters for Kyocera itself. The company was established with 300 million yen in paid-in capital, and with Kyocera as the majority shareholder, owning 51% of the company's shares. Kimura—who had already been involved in the development of solar cells at Sharp's Sunshine Project Promotion Department—was seconded to the company as an executive director with responsibility for technology under the orders of his boss at Sharp, a man named Sasaki Tadashi. Japan Solar Energy formed a technology partnership agreement with Mobil-Tyco and immediately began its research into the practical commercialization of the EFG ribbon crystal technique for growing silicon crystals based on Mobil-Tyco's technology.[65] Initially, Tyco denied Kyocera's request for a partnership agreement, giving the reason that the EFG technology had not yet been perfected. Ultimately, Inamori went out of his way to travel to Tyco's offices in Massachusetts and secure the formation of the agreement in person.[66]

In this way, Kyocera—which was not initially included in the outsourcing of research under the Sunshine Project—entered into the development of solar cells on its own, without relying on research commissions. In other words, at Japan Solar Energy, another plan concerning the development of ribbon crystal technologies (which at the time were considered to show promise for the future) had begun; by way of a joint-venture between multiple private sector companies, which was entirely separate from the national project framework.

In 1975, Kimura—on secondment to Japan Solar Energy from Sharp—began to work on the development of ribbon crystal growing techniques, under the supervision of Kyocera's President Inamori. Two years later, Sharp attempted to recall Kimura. However, after consulting with Sasaki, Kimura made the decision to take the initiative and transfer permanently from Sharp to Japan Solar Energy. This was also partly for the purpose of raising morale at the new company.[67] At the time, Kimura had his back against the wall, and was prepared to commit to devote the rest of his career to the new company. The circumstances behind this move were as follows.

Initially, the main purpose of Japan Solar Energy was for staff from Sharp, Matsushita, and Kyocera to cooperate and work together with the aim of achieving the practical commercialization of their ribbon crystal technique. The respective

---

[64] *Nikkei Sangyō Shimbun*, September 10, 1975.

[65] The contract was concluded before inauguration in September.

[66] *Nikkei Sangyō Shimbun*, June 22, 1983.

[67] Kimura Kenjirō, interview by author, September 9, 1998.

firms (Sharp and Matsushita) would then purchase the ribbon crystal substrates that had been developed.

However, even when Japan Solar Energy succeeded in the mass production of ribbon crystal substrates in February 1977, Sharp and Matsushita hesitated to move forward in using them for the production of solar cells. This was partly because silicon crystals produced using this method were of inferior quality to those manufactured using other methods.[68] This situation led straight to a dead-end road for the new company, and from Kimura's point of view it was equivalent to the partners breaking their original promise. If that was going to be Sharp and Matsushita's stance, Kimura decided that the only way to break the deadlock was to actually produce solar cell modules at Japan Solar Energy and show them the level of performance that could be achieved, and so they went ahead and began producing solar cells, in breach of the initial agreement.[69] Sharp and Matsushita were angry at Japan Solar Energy for this action, and eventually they issued the company with a final ultimatum. Kimura described this situation as follows:

> So both Sharp and Matsushita came to us [with an ultimatum] saying that if we did it they would pull out and withdraw their capital, because if we did it then we would become competitors, and they couldn't be investing capital in a competing manufacturer. Well, of course, they were absolutely right. Various things happened, but in the end we said, well, it can't be helped. And so I prepared myself for the worst.[70]

In March 1978, Sharp and Matsushita Electric withdrew from Japan Solar Energy one after another. The plan to manufacture ribbon crystals through this joint-venture company had been derailed, a mere two-and-a-half years after its establishment. In contrast with a national project plan, it was difficult for a partnership between private sector companies—facing the market head on—to wait for technology development efforts to produce results. After withdrawing from Japan Solar Energy, Matsushita Electric continued its research into compound-type semiconductors. From 1980, the group also allowed Matsushita Battery Industrial to participate in the development of solar cell systems, although its involvement was limited to peripheral technologies rather than the actual solar cells themselves.[71]

As we have seen, if we take a look at the project from the perspective of the various companies, there was a clear dilemma between expectations for the long-term success of the technology development and the pursuit of profits and

---

[68]*Nikkei Sangyō Shimbun*, February 4, 1977.

[69]Kimura Kenjiro, interview by author, September 9, 1998.

[70]Kimura Kenjiro, interview by author, September 9, 1998.

[71]"There was a car battery plant near Lake Hamana in Shizuoka Prefecture in those days. Electricity generated by solar cells is direct current. At this plant, direct-current electricity was used in a manufacturing process for car batteries. …We proposed the establishment of a 100 kW system there, and the government accepted our proposal. Accordingly, members of the battery industry also took part in the Sunshine Project simultaneously with the establishment of the NEDO in 1980." (Murozono Mikio, interview by author, September 11, 1998).

other short-term benefits, among the people who were trying to achieve the realization of their own visions.

7.3 Acceleration of the Project Due to the Second Oil Crisis and the Establishment of NEDO

7.3.1 Conception of the Rainbow Project

With the occurrence of the second oil crisis in 1979 as a result of the disturbances in Iran during the previous year (eventually culminating in the Iranian Revolution), new energy development entered a new phase of acceleration. Until then, the sense of panic over the first oil crisis had calmed down to some extent and, if anything, the public's interest in new energy was beginning to fade. But when the second crisis struck it brought people's memories of the first oil crisis back to life once again and reignited the hopes and expectations of the Japanese public towards the development of new energy technologies.

AIST saw this as an opportunity. They had already created a new research and development project in 1977 for the development of energy conservation technologies, which they had named the Moonlight Project (borrowing from the image of the Sunshine Project, which focused on new energy technologies). AIST regarded these two projects as a pair: one focusing on the development of new energy technologies harnessing the power of sunlight, the other on the development of energy conservation technologies utilizing the power of moonlight.

Seizing the opportunity created by the second oil crisis, AIST conceived a further third project following on from these, the Rainbow Project. It was a project for the research and development of biomass-related technologies.

However, the hasty naming of this project would actually prove to be detrimental. In May 1975, the media reported that AIST were over-eager due to concerns over the supply of oil. The article introduced AIST's efforts in a fairly cynical manner, saying that "It is also rumored that AIST has appointed naming experts."[72] Unfortunately, the attempt to create a third project at this time—riding on the wave of this new crisis—was unsuccessful. Finally, it was decided that the development of biomass technologies, which should originally have taken place under the Rainbow Project, was incorporated as part of the Sunshine Project. From this episode, we can see that AIST at the time recognized that the energy crisis was an opportunity to expand its projects, and that skillful naming would increase the possibility of securing the necessary budget when doing so. In other words, AIST had aimed to reenact the same method they had used in launching the Sunshine Project; by giving the project an endearing name that would bring the project attractive and appealing significance in the midst of an energy crisis, ultimately

[72]*Nikkei Sangyō Shimbun*, May 30, 1979.

contributing to securing the budget for the project. However, such intentional creation of significance does not always go as planned.

7.3.2 *NEDO as a Think-Tank for Collaboration Between Industry, Government, and Academia*

In May 1980, based on the new policy of driving the Sunshine Project forward at an accelerated pace in the wake of the second oil crisis, the Act on the Promotion of Development and Introduction of Alternative Energy (otherwise known as the Alternative Energy Act) was enacted. The act stipulated provisions for the establishment of a new energy comprehensive development organization. Based on these provisions, in October 1980, the New Energy Development Organization (NEDO) came to be established. Let us examine the situation around the time of its establishment, from the perspectives of various people who were involved.

During the latter half of the 1970s, the spokesman for the Sunshine Project was a man named Sakaiya Taichi; who was an MITI official and also a well-known novelist. MITI's Ishikawa Fujio—who was also involved in the establishment of NEDO– raised Sakaiya's name first and foremost as the person exerting the most control and taking the initiative at AIST at the time: "[People like] Ikeguchi Kotarō [the real name of Sakaiya Taichi] who came to the Development Division around the time when I was at Electric Power Development. He was around at the time of the creation of NEDO. And then later on Makino Tsutomu, who took the role of Vice-Minister."[73]

Ikeguchi Kotarō (AKA Sakaiya Taichi) held the position of Director-General for General Research and Development between July 1974 and October 1978. The Sunshine Project and author Sakaiya Taichi might seem like an unlikely or surprising combination, but if we consider Ikeguchi's career and the fact that he had been catapulted into fame with the release of his novel *Yudan!* (literally meaning "carelessness" or "not paying sufficient attention") in 1975, we can begin to see the connection. *Yudan!* was a novel, set in the near future, which portrayed the crisis-like situation that would surely occur in the event of oil imports to Japan being halted.[74] As the Japanese title infers (the pair of Japanese characters used to write the word *yudan* carry the meanings of "oil" and "interruption"), the story unfolds based around the premise of oil supplies to Japan being cut off due to a crisis in the Middle East; and realistically simulates the state of affairs that would arise if oil imports to Japan were halted.

It begins with the suspension of economic activities, and goes on to depict numerous horrific situations, including people freezing to death due to fuel shortages, deaths occurring in hospitals due to power deficiencies, the complete

---

[73]Ishikawa Fujio, interview by author, June 4, 1998.

[74]Sakaiya (1975).

destruction of transport networks in times of disaster due to blockages of the transport networks, the paralysis of food production due to energy shortages, and the eruption of rioting due to social anxiety. Ikeguchi had expressed his crisis awareness in relation to energy as an MITI official in the form of this novel. The book became a bestseller that year, and its sensational content appealed to the Japanese people—who had already experienced the first oil shock for themselves—highlighting once again the importance of energy problems.

Ikeguchi was succeeded by Makino Tsutomu, who became chairman of NEDO after a stint as Administrative Vice-Minister at MITI, and remained in the role until July 1981. The post of Deputy-Director General for General Research and Development was originally created for the purpose of conducting research and development that did not fit into individual themes (such as solar and geothermal energy), and research and development that spanned multiple areas. However, in reality, during the latter half of the 1970s, with Ikeguchi, Makino and so on, the post had become a PR-type role for the project.

Kurokawa Kōsuke, who was seconded from ETL to the Office for Developing the Sunshine Project at AIST between October 1976 and December 1977, recalled Ikeguchi (who was in the same office) as follows.

> NEDO was formed after Sakaiya Taichi (i.e. Ikeguchi) left. During the period that he was there, Mr. Ikeguchi was an extremely influential person. He was involved in political policy measures, and – although I had originally thought that the role of a general development official was a petty and insignificant one – in reality it felt like he was more important than the Deputy-Director General of Development.[75]

Ikeguchi also urged that a double-sided approach to the research and development of new energy technologies—focusing on both science and engineering—was needed, and asserted that it was important to advance engineering-based research in order to validate the economic feasibility of the technologies being developed. To that end, he set about convincing the various people involved, working under the keyword of a "biaxial approach" (with the two axes of science and engineering).[76] The purpose of his efforts lay in urging the relevant parties that the development of new energy would not succeed through science-only initiatives at the various national research institutes and so on, and that semi-governmental corporate research institutes that would bear the engineering role were also necessary.

Figure 7.1 shows an explanatory diagram of the "biaxial approach," taken from a pamphlet produced by the Office for Developing the Sunshine Project at AIST. Initially, Ikeguchi and Makino had envisaged the research and technology development being conducted by semi-governmental corporate entities. When explaining that, they often said that NEDO was the new energy equivalent of the Power Reactor and Nuclear Fuel Development Corporation (PNC). Makino Tsutomu made the following comment in 1984:

[75]Kurokawa Kōsuke, interview by author, April 29, 1998.

[76]MITI Kōgyō gijutsuin Sanshain keikaku suishin honbu (1984, pp. 13–14).

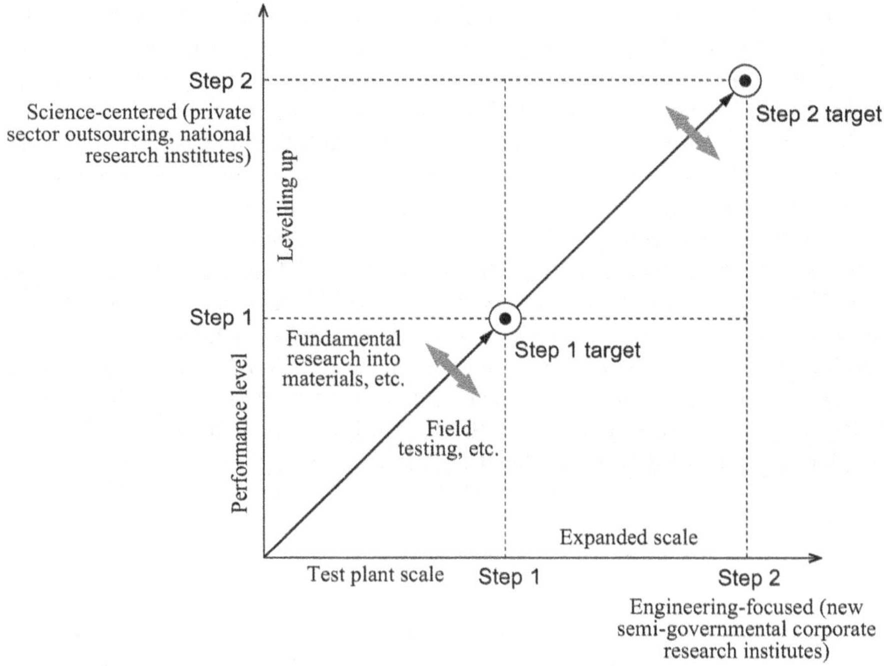

Fig. 7.1 Explanatory diagram explaining the biaxial approach (october 1974 proposal). *Source* MITI Kōgyō gijutsuin (1974, p. 23)

> Our initial aim in establishing NEDO was to create a new energy version of PNC [the Power Reactor and Nuclear Fuel Development Corporation]; a semi-governmental corporate entity that get its own hands dirty, working under the cooperation of industry, government, and academia. However, at this current time, there is a certain aspect of the organization shifting more towards a coordinating function.[77]

Despite the fact that the kind of concept described above had existed, in fact, when NEDO was launched after the second oil shock it possessed no research and development capabilities in itself, and was an organization tasked solely with project management. At NEDO, human resources dispatched/seconded from MITI and various corporations took a central role in managing the task of outsourcing research and development work to external entities, based on the framework of industrial technology policy formulated by AIST. Initially, it was conceived that NEDO would take the initiative and manage projects in an autonomous and proactive manner. However, in fact, as mentioned in Makino's comments, NEDO gradually became an organization that fulfilled the role of coordinating the project in its execution stages, within the framework determined by AIST.

[77]MITI Kōgyō gijutsuin Sanshain keikaku suishin honbu (1984, p. 16).

However, at the time of its launch, NEDO had the awareness of being an organization that was obligated to achieve the objective of developing alternative forms of energy (other than oil) through the Sunshine Project and making the proportion of Japan's energy provided by these new technologies equal to that of oil. The newly launched organization was full of grit and mettle, and the belief of actually achieving that goal. A passage that embodies the enthusiasm at NEDO around that time was recorded in *NEDO News* in March 1981.

> Putting aside the debate as to whether NEDO belongs to the public or private sector, if we consider things from the perspective that our organization was established for the purpose of achieving the government's long-term target of raising the ratio of [Japan's use of] oil to alternative forms of energy from the current ratio of 75:25 to 50:50, we can create a gauged semi-governmental corporate entity of the kind rarely seen amongst existing public corporations and corporate agencies. NEDO staff cannot afford to sit around doing nothing.[78]

In this way, during its early days, NEDO was overflowing with fervent enthusiasm and the belief that the organization was going to achieve its important mission: the development of new energy technologies to solve Japan's energy problems.

At that time—during the continuing oil crisis, when memories of the panic were still fresh in people's minds, and the threat posed by the depletion of oil resources felt quite real—a national project that could achieve the successful development of new energy technologies as one strategy for escaping from the clutches of oil dependency was considered to be of major significance.

The purpose of NEDO's establishment, according to the NEDO prospectus, was the development of alternative forms of energy (i.e. that would serve as alternatives to oil) requiring commercialization.[79] The author referred to reflections by Watamori Tsutomu after Dokō's death that Dokō voluntarily assumed the position of chairman.[80] Its organizational structure had also been cleverly contrived in order to reach this purpose. A steering committee was established at the top of the organization, consisting of individuals from the private sector. Dokō Toshio—who had served as chairman of the Special Committee for Energy Technology at the Industrial Technology Council—was appointed as chairman of the steering committee. A board of directors (consisting of ten members) was then established, under the supervision of the steering committee, as the organizational structure that would actually operate and manage NEDO and be responsible for actual

---

[78]Teramura Toshiaki, "Nedo toiu hōjin [A Corporation Called the NEDO]," *NEDO News*, March, 1981, p. 2.

[79]The NEDO was established for the objectives of (1) developing alternative forms of energy for oil that need to be commercialized, (2) assisting with the development of geothermal resources and coal overseas, and (3) promoting the development of other forms of alternative energy.

[80]The Steering Committee consisted of the following seven members: Chairman Dokō Toshio, Deputy Chairman Ashihara Yoshishige, and members Ikeura Kisaburō, Enjōji Jirō, Shiba Tadao, Nagayama Tokio, and Hiraiwa Gaishi. (Watamori Tsutomu, "Dokōsan no omoide [My Memories of Mr. Dokō]," *NEDO News*, October–November, 1988, p. 40.

decision-making.[81] The man who was appointed chairman of the board of directors, at the recommendation of Dokō himself, was Watamori Tsutomu, special advisor at Hitachi, Ltd. This choice of personnel included the intention of utilizing the thinking and energy of the private sector in the execution of the project, by making effective use of human resources from private sector companies.

According to Watamori, when Dokō appointed him as chairman, Dokō said:

> Because the Sunshine and Moonlight projects, too, have been formulated and pushed forward by [governmental agencies such as] MITI's Agency for Natural Resources and Energy [ANRE] and the National Institute of Advanced Industrial Science and Technology [AIST], there will also be many government officials and old boys amongst the staff at NEDO. In order to avoid this bias towards the government side as much as possible, we should select someone from the private sector for the role of Chairman.[82]

This choice of personnel for the top of the organization indicates the strong resolve that Dokō Toshio had the aim of creating a semi-governmental corporation that would utilize the power of the private sector. In his formal greeting in the inside cover of the inaugural issue of *NEDO News*—the organization's official news publication—in January 1981, Dokō made the following plea for the establishment of a collaborative framework between industry, government, and academia, with NEDO as its center:

> I do not think that developing alternative energy technologies as an alternative to oil will be an easy task. But if we can deepen the level of collaboration between industry, government, and academia, and mobilize the collective strengths that our country possesses, I believe that it is possible to solve this problem; and furthermore, that in the 21st century Japan will be able to stand in an even more advantageous position with regard to energy than any other country.[83]

When NEDO was actually launched, Dokō, Watamori and so on had the strong resolve that they themselves were really going to achieve these things.

Working under the concept of taking up the challenge of developing new energy technologies making maximum effective use of the networks of industry, government, and academia, as a think-tank consolidating the knowledge and expertise of all three sectors, NEDO worked hard to dispel the old image of semi-governmental corporations. Its name, too, contained the word "organization" rather than the word "corporation" or "administration/agency" as with conventional public or semi-governmental corporations. In addition, by assembling capable personnel seconded from the worlds of industry, government, and academia, it was hoped that NEDO would be able to advance its management and coordination functions—such

---

[81]The directors included the following ten individuals: Chief Director Watamori Tsutomu, Deputy Chief Director Ōnaga Yusaku, directors Matsuo Yasuyuki, Ezaki Kōzō, Ameya Masakata, Yamazaki Tetsurō, Fujinuma Rokurō, Takase Ikuya, Toyama Atsuyoshi, and auditor Teramura Toshiaki.

[82]Watamori Tsutomu, "Dokōsan no omoide [My Memories of Mr. Dokō]," *NEDO News*, October–November, 1988, p. 40.

[83]Dokō Toshio, "Kanmin no chikara no kesshū [Rallying Forces in Public and Private Sectors]," *NEDO News*, January, 1981, p. 3.

as the selection of new energy technologies for development, and the monitoring of these development activities—more effectively.

This policy of utilizing private sector resources was also intended to reduce the amount of public funds that needed to be invested by drawing on the cooperation of private sector corporations, while at the same time preemptively preventing the organization from becoming a haven for retired government officials. To achieve this, the personnel who would form the core of NEDO's full-time regular employees were secured at an early stage, and the organization was conceived as one with the independence to drive forward the development of new energy technologies autonomously. Horigome, who came from ETL at the time of NEDO's establishment to take on the role of head of the Solar Technology Development Office, said: "We thought that at the time of its initial creation it couldn't be helped, but after about 10 years NEDO would be run almost entirely by full-time, career employees. That was the organizational structure that we should promote. Everyone was in agreement with regard to that."[84]

When Horigome became head of NEDO's Solar Technology Development Office in October 1980, Kurokawa came with him, and was appointed Chief Systems Researcher. AIST had requested ETL to dispatch staff for the establishment of NEDO. It was partly due to a specific request from AIST's Ishikawa for him to take the position—in view of his history of commitment to the Sunshine Project up until that point—that Horigome accepted the post.

At that time, Ishikawa started talking to Horigome about the Sunshine Project, and requested ETL to dispatch some of its personnel, because it was ETL that was receiving the budget for the project.[85] In light of this request, Horigome took Kurokawa—who was a system expert and had a detailed knowledge of both thermal and photovoltaic power generation—with him to participate in NEDO from the time of the organization's initial establishment.

Kurokawa recalled his impressions upon visiting the NEDO office in the Sunshine Building in Ikebukuro, in October 1980: "When I went [to NEDO], there was one new desk and one box of business cards. Everyone had the same, and there was nothing else there."[86]

Starting from this initial state of having almost nothing to work with, NEDO set out on its mission, as an organized attempt aiming to successfully develop new energy technologies; and thereby to achieve the realization of a political policy to escape from dependency on oil, and to introduce and achieve the widespread adoption of new energy power generation technologies. Later on, however, NEDO would follow a different path from that initially intended by the involved parties.

As they had created and put forward the initial proposal for the Sunshine Project, the solar energy researchers at ETL themselves came to join NEDO and take on the role of project management. In doing so, as researchers, it was their wish that the

[84]Horigome Takashi, interview by author, September 16, 1998.
[85]Horigome Takashi, interview by author, September 16, 1998.
[86]Kurokawa Kōsuke, interview by author, April 29, 1998.

project would succeed with a central focus on what were originally their own areas of research: areas such as solar thermal power and, in a wider sense, power transmission systems including solar photovoltaic power generation. However, it has been pointed out that, even at this stage, MITI as a whole considered solar energy to be nothing more than a convenient slogan for the project, and that the source of energy that they had real expectations for was actually coal. Kurokawa Kōsuke, who was part of NEDO at the time, said the following: "Actually, if you look closely, for the first 10 years or so after NEDO was created, the Sunshine Project was really about coal. MITI and ANRE were not being serious about solar [energy] at all. But it was a convenient flag for them to push the project through with [i.e. gain approval for it]. It was like a banner slogan."[87]

According to this testimony, as the name suggested, solar energy was the banner slogan for the Sunshine Project. But this was nothing more than a means of appealing to the outside. As far as MITI and ANRE were concerned, with regard to the Sunshine Project, too, they were placing their hopes and expectations in what they considered to be more realistic, coal-related technologies. In fact, during the 1980s, over half of the total budget for the Sunshine Project was invested in coal-related research.

Of course, even saying that circumstances at that time meant that it was easier (in terms of the budget system) to use special account budget funding for coal-related technologies, if high expectations had not been placed in those technologies to begin with then they would surely not have been selected as a technology development theme for the sole reason of it being systematically easier to use budget funding towards them. The background to this may have been that it was MITI's true intention to place their realistic long-term hopes in nuclear power, and shorter-term expectations in coal-related technologies. However, since MITI also wanted to appeal to the Japanese public about the clean, environmentally friendly image of solar energy, it may have been the case that they wanted to make sufficient use of that image with regard to Sunshine Project itself. Even with regard to the same project, we can see that people's various positions and standpoints on technologies in which they placed their expectations actually differed. In particular, it is understandable that the researchers developing the technologies would defend and advocate for the technologies with which they themselves were involved. At any rate, despite such differences of opinion, the interests of various people who participated in the project under the image of "sunshine" were all aligned with regard to the point of continuing the development of new energy-related technologies.

---

[87]Kurokawa Kōsuke, interview by author, April 29, 1998.

7.4 The Emergence of Amorphous Materials

7.4.1 The Amorphous Researchers Group

Originally, there were two methods of manufacturing solar cells: one using crystalline silicon and the other using compound-type semiconductors. Both methods became the target of research and development under the Sunshine Project from the project's early stages. However, there was also one other additional method of producing solar cells, involving the use of amorphous silicon. How did this method emerge, and how did it come to be incorporated into the Sunshine Project? In this section, let us examine that process from the perspectives of the various individuals involved in the development of this technology.

Tanaka Kazunobu, who was studying as an intern under Kikuchi Makoto (Director, Component Fundamental Research Office) in Denki Shikenjo's Physics Department (which later became ETL's Fundamental Research Department) was one person focusing his attentions on amorphous silicon from an early stage.

After graduating from university, Tanaka Kazunobu had initially found employment with Matsushita Electric Industrial. However, at the end of the 1960s, when Matsushita cancelled the launch of its amorphous research as a result of management policy due to the worsening economy, Tanaka lost faith in the company. He left Matsushita in March 1969, and subsequently took up temporary residence at the Sophia University Graduate School. However, when Tanaka learned—prompted by the release of American inventor Stanford Robert Ovshinsky's paper in November 1968—that Kikuchi had started a program of amorphous research at Denki Shikenjo, he promptly went to visit him, and subsequently joined Kikuchi's laboratory as an intern.[88] Later, in 1971, Tanaka accepted a job offer from Kikuchi and dropped out of his courses at graduate school to join the renamed ETL. At the time, Horigome and his team in ETL's Energy Department had just started their research on solar thermal energy.

After the boom in amorphous research that begin with the publishing of Ovshinsky's paper had passed, researchers at various companies and within ETL gradually began to lose interest and moved on to other topics of research. Kikuchi, too, transferred to work at a Sony research institute a few years later, and Tanaka was left behind as the only person still conducting research into amorphous silicon in ETL's Fundamental Research Department.

There, during the early half of the 1970s, Tanaka conducted research into chalcogenide glass, and began to seek exchanges with the handful of other amorphous researchers who belonged to various universities and corporations, via the Japan Society of Applied Physics (JSAP). At JSAP, partly because amorphous silicon was a new technology, proactive discussions were being held on the subject by a subcommittee of young researchers.

[88]"Haiteku Jinmyaku [Hi-Tech Personal Contacts]," *Nikkei Sangyō Shimbun*. December 3, 1986.

After the Spring conference JSAP 1974, a researcher at Hiroshima University named Hirose Masataka came to Tanaka at ETL with a plan, to create a community of amorphous materials researchers. While there had been opportunities before then for amorphous researchers to discuss their research at JSAP, Hirose felt that it was a waste/shame that they fanned out and disbanded as soon as the JSAP conferences ended.[89] Tanaka and Hirose were in agreement about the idea, and together with Shimizu Tatsuo from Kanazawa University they set about preparing to hold a seminar. And so, in the autumn of 1974, the first Seminar on the Properties and Applications of Amorphous Materials was held in Kanazawa, hosted by Tanaka, Hirose and Shimizu.[90] After that, the seminar came to be known simply as the Amorphous Seminar. The seminar increased its number of participants each year and, several years later, even Kuwano Yukinori of Sanyo Electric (now part of Panasonic)—which would later achieve the world's first commercialization of amorphous solar cells—participated as a lecturer.[91] While the seminar had a relaxed and at-home atmosphere, with wine flowing at the evening discussions, the members gave no quarter when it came to debating technology issues.[92] Tanaka describes the harsh nature of the debates that took place at the seminar:

> When an esteemed professor or so on came along and said something, the other members would bombard him with harsh comments, saying, "That's just common sense." Then a department head from ETL would jump in and say, "That's no way to speak to an esteemed *sensei*." But the others would then say, "Well, in our field, this is the way we usually do things." That was the kind of debating that went on.[93]

Later, in 1976, Tanaka himself was invited to speak at an international conference in the United States. It was at that conference that he learned, from a report by P. G. LeComber, that amorphous silicon could be used in solar cells.

Tanaka then began to think that the research he had been conducting until that time might be of some use in the Sunshine Project. And so, in 1978, also partly due to the recommendations of Sakudō Tsunetarō, who was head of the Fundamental Research Department at the time, he decided to participate in the research and development of solar energy (which ETL had already been participating in since 1974) with amorphous silicon solar cells. To that end, in the spring of 1977, Tanaka began creating presentation materials and preparing to make his proposal.

However, until that time, ETL's Fundamental Research Department had no experience of participating in a Large-Scale Project or anything similar, and to other elements within ETL it was completely inconceivable that the department—which was concerned mainly with theoretical research into physics and so on—would

---

[89]Tanaka Kazunobu, interview by author, October 15, 1998.

[90]Material provided by Tanaka Kazunobu, *Pamphlet for the First Seminar on the Properties and Application of Amorphous Materials*, 1977.

[91]"Haiteku Jinmyaku [Hi-Tech Personal Contacts]," *Nikkei Sangyō Shimbun*. December 3, 1986.

[92]Kuwano Yukinori, interview by author, October 29, 1998. The article mentioned above ("Haiteku Jinmyaku [Hi-Tech Personal Contacts]") also contains similar statements.

[93]Tanaka Kazunobu, interview by author, October 15, 1998.

participate in this kind of project. Because of this, when Tanaka proposed the idea of the Fundamental Research Department obtaining Sunshine Project budget funding to conduct research into the research theme of creating solar cells using amorphous semiconductors, he ran into unexpected friction with other departments within ETL over the division of research responsibilities and budget funds. There was particularly intense backlash from one department within ETL that had already been responsible for solar cell research since the time of the Sunshine Project's initial launch: namely, the Electronic Devices Department.

The head of the Electronic Devices Department at the time was a man named Komamiya Yasuo. After Tarui Yasuo had moved on to the VLSI Technology Research Association, a researcher called Hayashi Yutaka in the same department had carried on his research into crystalline solar cells. Also partly because the Electronic Devices Department had also just begun research into amorphous solar cells, Komamiya objected strongly, and persistently, to Tanaka's proposal. In light of this, Tanaka submitted a request to Todoriki Itaru (Director, ETL Energy Department) and his subordinate Horigome (Director, ETL Energy Transport Research Laboratory) from the Energy Department—which was in charge of the Sunshine Project's solar thermal energy research at the time—to support his proposal for the Fundamental Research Department to carry out research into amorphous solar cells. However, even the Energy Department—which was the central entity in coordinating the Sunshine Project budget at the time—could not readily agree to Tanaka's request straight away while there was opposition from Komamiya. Horigome said the following regarding the Energy Department's response to Tanaka's request:

> To begin with, there may have been a period of time where Tanaka Kazunobu and the others were asked to garner together a certain amount of budget in the laboratory that they were affiliated with and try doing it for about a year; for example, if they were going to produce thin films, and see how much it was going to cost and so on. I think that if they hadn't done that then Mr. Komamiya and the others probably wouldn't have agreed on it in the end."[94]

Until Tanaka could obtain the understanding of Komamiya, in the Electronic Devices Department, Todoriki, Horigome and the others in the Energy Department had to take a reserved attitude towards Tanaka's proposal. Even Sakudō—the head of the Fundamental Research Department—had to admonish Tanaka's aggressive stance and convince him to abandon his proposal temporarily. Tanaka went around trying to convince the various relevant parties, but eventually learned that as long as Komamiya was opposed to the proposal then nobody could make a final judgment on the matter. In light of this, he decided to make one last gamble with regard to whether or not the development of amorphous solar cells by the Fundamental Research Department would be included in the Sunshine Project, by pitching his

[94] Horigome Takashi, interview by author, September 16, 1998. Hamakawa Yoshihiro of Osaka University reportedly visited Horigome repeatedly around this time to propose amorphous studies as well.

proposal at the Large-Scale Project Sunshine Hearing (a hearing for requesting estimated budgets for the 1978 fiscal year), which was held on May 23, 1977.[95] Although there had been cases where departmental director-class personnel had attended the hearing in the past, on this occasion Komamiya himself appeared at the venue:

> Just on that occasion, Mr. Komamiya himself came out, with a pipe in his mouth, and with a fearsome look on his face. Then, when the host of the proceedings said that we were to hear an explanation from Tanaka, who started doing amorphous research originally, Mr. Komamiya said, in an extremely threatening manner, 'Hold on a minute. Amorphous research is originally something that our Kobayashi has been doing. It isn't the case that he [Tanaka] did it first. Think carefully about what you are saying with regards to that.'[96]

However, after Tanaka's explanation, Komamiya's attitude softened. He turned to Hayashi and said, "Hayashi. You mustn't lose out to Tanaka, with his powerful drive." After he had recognized Tanaka's fervent enthusiasm for the topic, he gave his immediate consent to the Fundamental Research Department's stance on the spot, and directed Hayashi to "draw a line" and divide up amorphous research between the Electronic Devices Department and the Fundamental Research Department.[97] In this way, amorphous research cleared the difficult obstacle of ETL's internal discussions and took its first step towards becoming part of the Sunshine Project.

For Tanaka, after securing a consensus within ETL, the next thing that awaited him was the AIST hearing. It was then that Kurokawa Kōsuke—who at the time was a research and development official seconded to AIST from ETL's Energy Department—appeared as the representative for AIST.[98] A memo records that Kurokawa made the following comments with regard to Tanaka's explanation at the AIST hearing on June 2, 1977: "*Belittling (the work of) others is problematic.

---

[95]Planning Division of the Electrotechnical Laboratory, *Ō Puro, Sanshain, hiaringu nittei* [Hearing Schedule for the Large Project and Sunshine Project] , unpublished manuscript, 1977. Hearings were conducted at the Headquarters Meeting Room on May 24 and the Planning Office at the Tanashi Annex on May 20, May 23, May 24 and May 25 of that year. The hearings in question that took place on May 23 included those on studies of solar thermoelectric power generation systems (by the Energy System Laboratory and the High-Temperature Electronic Materials Laboratory) from 10 a.m., studies of solar spectral irradiance measurement (by the Applied Optics Laboratory) from 1.30 p.m., studies of new electricity generation methods (by the Energy Transportation Laboratory) from 2 p.m., basic solar cell studies by the Solid Device Laboratory from 3 p.m. and basic solar cell studies by the Semiconductor Device Laboratory from 4 p.m. After the hearings, Kazunobu Tanaka from the Basic Component Laboratory provided supplementary explanations from 5 p.m.

[96]Tanaka Kazunobu, interview by author, October 15, 1998.

[97]Tanaka Kazunobu's pocket notebook for fiscal year 1977. His note in the writing space for May 23 says, "They understood. We might be able to draw the line successfully."

[98]As described in Chap. 3, the ELT sent Kurokawa to the AIST temporarily from October 1976 to December 1977. The ELT dispatched two members to the Office for Developing the Sunshine Project at the AIST under the MITI on a temporary basis each year to work there as officers in charge of development.

(Compound-type semiconductor SCs [solar cells] have also been approved for mass-production use, and we currently know this.) *Isn't this a manufacturing device? [The question is] whether or not it will pass as an evaluation/test-use device. *Competing research is problematic, as it constitutes an overlapping investment [in the same area[99]]."

From these comments, we can surmise that the tone of Tanaka's argument at the time of his explanation to Kurokawa, too, was fairly offensive towards other methods of solar cell production. It also suggests that at this time, in 1977, Kurokawa—who was later seconded to NEDO—was already conscious of the fact that competition between researchers in different areas of component technology for solar cells would lead to overlapping investment. Finally, however, Kurokawa accepted the internal judgment of ETL and agreed to it on behalf of AIST; and, so, the inclusion of amorphous research in the Sunshine Project was officially determined.

That August, Sakudō—the head of ETL's Fundamental Research Department—promptly applied for approximately five million yen of research funds for the 1977 fiscal year, in the form of president's reserve expenses, and a budget of 1.5 million yen was granted to amorphous research for the year.[100] From the following year, even though it was true that the research had successfully acquired Sunshine Project budget funding, the sum of funds granted for the 1978 fiscal year was a mere 4.3 million yen. Funding for 1979, too, stopped short at 9.6 million yen.[101]

Tanaka—who because of these budget shortfalls was struggling even to buy experimental apparatus—began to entertain the idea of turning the development of amorphous solar cells into a large-scale project involving industry, government, and academia, in order to get the full-scale development on track. His idea involved not outsourcing research into amorphous technology on an individual basis, but instead involving universities and private sector companies to facilitate a full-scale research and development effort, from fundamental research to the manufacturing stage. Furthermore, what Tanaka had in the back of his mind was the idea of surpassing the United States, even if it was in this field alone, by concentrating resources to conduct this amorphous research as a fully collaborative project—in the true sense of the word—between industry, government, and academia.[102]

Although the slogan of collaboration between industry, government, and academia did already exist with regard to the Sunshine Project during the 1970s, in reality the outsourcing of research to universities was extremely limited due to the view that the provision of MITI funds to universities (which were under the

[99]Tanaka Kazunobu's note on the Agency of Industrial Science and Technology's hearings on the Sunshine Project, 1977.

[100]Sakudo Tsunetarō, *Shochō ryūchihi no tokubetsu sochi no irai ni tsuite* [Request for a Special Measure for Manager Retention Expenses], unpublished manuscript, August 4, 1977.

[101]Tanaka Kazunobu, "Amorufasu hakumaku taiyō denchi no kiso kenkyū [Basic Research on Amorphous Thin-Film Solar Cells]," unpublished manuscript, February 14, 1980.

[102]Tanaka Kazunobu, interview by author, October 15, 1998.

jurisdiction of the Ministry of Education, Science and Culture) constituted a jurisdictional infringement.

However, while moving technologies such as crystalline solar cells—for which the fundamental principles behind the technology were already understood—straight over to practical commercialization-oriented research (with the major goal of cost reduction) was relatively easy; for research topics like amorphous materials—for which there still remained many points (including in terms of their physical properties) that were as yet unclear—the support of researchers at universities and national research institutes in terms of fundamental theory was extremely important.

It is likely that Tanaka's fellow researchers at the Amorphous Seminar were a significant source of inspiration for him in conceiving his plan. Tanaka's plan involved absorbing these human resources—from companies and universities and so on—into the project, and driving the development forward through collaboration with regard to fundamental theory (universities), materials technology (national research institutes) and device design, and production (companies and universities), with mutual sharing of research and development outcomes between the various partners. His idea also included a vision for the future: the launch of a new energy industry, based on the pivotal technology of amorphous solar cells.[103] Tanaka argued that amorphous technology was at a different stage in its development from that of crystalline silicon research, and declared that in the future amorphous semiconductors would give birth to a new industry, in the same way as crystalline semiconductors had created the massive electronics industry. Tanaka asserted that amorphous solar cells would rival transistors. It was by displaying this vision for the future that Tanaka portrayed the possibilities for amorphous technologies in the future.[104]

From that time onwards, Tanaka worked to raise awareness and enlighten the media about amorphous technologies. When Ovshinsky visited Japan to speak at a lecture in January 1978, Tanaka acted as his interpreter.[105] Later, in March of that year, Tanaka appeared in a technology interview with *Nikkei Business* and explained in detail the future possibilities of amorphous technologies.[106]

When the second oil crisis struck in 1979, the Liberal Democratic Party (LDP) set up an Investigative/Advisory Committee on Energy and Resources. In

[103]Tanaka Kazunobu, "Amorufasu hakumaku taiyō denchi no kiso kenkyū [Basic Research on Amorphous Thin-Film Solar Cells]," unpublished manuscript, 1979, p. 8. This handwritten material for explanation is stated as additional material for Assistant Vice-Minister for Engineering Affairs Yamanaka Masami. The statement shows that by May 1979, Tanaka had already had an idea that would become a model for a figure in *Denshi gijutsu sōgō kenkyujo ihō* [*Electrotechnical Laboratory Journal*] (Fig. 7.4).

[104]Refer to Numagami (1992): 50–65 for the roles technologies play as cognitive models by functioning as focusing equipment.

[105]*The Nikkei/Nihon Keizai Shimbun*, January 30, 1978.

[106]Gijutsu taidan [Technological Dialogue], "Handōtai ni shite zairyō setsuyaku, taiyō denchi o kosuto daun [Saving Materials by Opting for Semiconductors, Cutting Down on the Cost of Solar Cells]," *Nikkei Business*, March 27, 1978, pp. 84–88.

accordance with the recommendations of Ovshinsky, who was visiting Japan at the time, the committee requested an explanation from ETL regarding their research into amorphous solar cells (which at that stage they had already begun). As we will see in the following section, in February of that year, Dr Yamano Masaru—director of Sanyo Electric's central research institute—had held a press conference where he announced Sanyo's successful development of a mass-production technology for amorphous solar cells, and the level of expectation and interest in amorphous technologies was heightening.

In May 1979, Tanaka, as the person responsible for the research at ETL, was invited to an LDP breakfast meeting. Other members in attendance included Yamanaka Masami (Deputy Director-General of Technology) and Sunami Taira (another research and development official) from AIST, Todoriki Itaru (Director, Energy Department) from ETL, and Tanaka's former boss Kikuchi Makoto (director of Sony's central research institute) . Tanaka explained to them the future prospects of amorphous semiconductors. Based on his explanation, the committee decided to establish a subcommittee, chaired by the politician Yosano Kaoru, to continue discussions regarding the technology. The subcommittee met several times, starting from June 1. Following the advice of Nakahara Nobuyuki, who at the time was a managing director at Toa Nenryō Kōgyō (now JXTG Nippon Oil & Energy Corporation), Tanaka used his connections to invite Morrel H. Cohen from the University of Chicago to attend the first meeting on June 1, to prove that the American authority was also in favor of the idea.[107] Based on Cohen's suggestions, the subcommittee prepared a report.[108] This became the draft proposal for the collaborative framework between industry, government, and academia, for the effort to develop amorphous solar cells that was launched in the 1980 fiscal year. In this way, the budget for research into amorphous solar cells increased dramatically. In the 1981 Sunshine Project budget, when universities began to participate fully in conducting commissioned research under the project, the amount of funding had risen to 1.3 billion yen, exceeding that of the project's U.S. counterparts. Tanaka's strategy had worked.

In parallel with this partnership between AIST and ETL, engineers at private sector companies were also engaged in efforts to develop amorphous technologies. One such engineer, who had been involved in the development of amorphous technologies from a very early stage, was Kuwano Yukinori of Sanyo Electric Co., Ltd.

---

[107]M. H. Cohen, *The development of amorphous photovoltaic cells in Japan*, 1979 is the record of this lecture.

[108]Subcommittee on the Development of Amorphous Solar Cells, *Report by the Subcommittee on the Development of Amorphous Solar Cells (draft)*, unpublished manuscript, 1979.

7.4.2 Sanyo Electric Pledges to Develop Amorphous Solar Cells for Practical Use

It was not only national research institutions that were focusing their attention on amorphous materials. Sanyo Electric was one company that began conducting its own independent research into amorphous materials, stimulated by the so-called "amorphous boom" that followed the announcement of Ovshinsky's paper in November 1968. Sanyo had already set about developing crystalline silicon solar cells at one point during the 1960s; and in 1963 had installed a solar cell-powered clock tower in front of its central research institute (which was established that year) as a demonstration of this technology.[109] However, as with Hitachi and other firms, Sanyo had judged solar cells to be unprofitable in regards of cost involved in producing them, and soon decided to withdraw from developing them. In 1974, when the Sunshine Project began, Sanyo participated in development of solar houses making use of solar thermal energy, but had passed on the opportunity to take part in the development of crystalline solar cells.

Going into the 1970s, Sanyo Electric's central research institute made amorphous silicon one of its official research themes. At the time, integrated circuit technology using monocrystalline silicon was not yet fully developed, and it was hoped that amorphous semiconductors would become useful as switch elements in place of conventional transistors.

At Sanyo's central research institute, Kuwano Yukinori had already begun research into amorphous semiconductors before the 1970s.[110] Making use of the expertise learned at his university, Kuwano had been conducting research into the formation of thin films, including transparent conducting films made from amorphous silicon nitride using plasma reactions, and protective insulating layers for semiconductor surfaces. From the 1970s onwards, he began to research chalcogenide amorphous semiconductors, and engaged in the development of switch elements and non-volatile memory technologies. At the time, Yamano Masaru—director of the central research institute—had two separate groups working in parallel to develop the same kinds of technologies, one using amorphous silicon and one using monocrystalline silicon.[111]

Later, however, it gradually became apparent that switching circuits made with amorphous semiconductors were inferior in terms of performance to those made with crystalline silicon, and some Japanese companies soon began to terminate their amorphous research. By around 1973, it had become clear in the development of

[109] *Nikkei Sangyō Shimbun*, May 8, 1979.

[110] Kato Katsumi, "Sanyō Denki amorufasu shirikon taiyō denchi [Amorphous Silicon Solar Cells by Sanyo Electric]," *President*, September issue, 1984.

[111] *Nikkei Sangyō Shimbun*, May 8, 1979.

non-volatile memory, too, that amorphous semiconductors were inferior to monocrystalline silicon in terms of performance.

Amidst these discoveries, Kuwano had the idea of a new application for amorphous semiconductors, as switching elements for large electric current devices. Using these, it would be possible to create solid-state fluorescent glow lamps, and to make fluorescent lights that would light up instantaneously. In the summer of 1973, Kuwano went to Sanyo's lighting division along with Ōnishi Michitoshi from his research group to try and sell them the idea. However, the person in charge there gave Kuwano a cold response, asking, "Is there really a need for fluorescent lights to come on instantaneously?" And so Kuwano and his associate's proposal was bluntly rejected.[112] Despite Kuwano's efforts, during the early half of the 1970s, Sanyo Electric was unable to find any applications for amorphous semiconductors that would lead to the creation of a commercially viable product.

Hamakawa Yoshihiro and Tanaka Kazunobu, with whom Kawano had interactions as fellow amorphous researchers, invited Kuwano to their research seminar and asked him to give a presentation about the elements that he had created.

For Kuwano, too, the presence of other researchers at universities and national research institutes developing the same kinds of technology as him gave him courage to keep going with his own development work. In later years, Kuwano wrote: "People like Professor Hamakawa from Osaka University and Tanaka Kazunobu from ETL said to me 'Kuwano, you've made some interesting [switching] elements. Please come to our seminar one time and talk about them;, and gave me an opportunity to present my work. At the time that made me really happy."[113]

In October 1974, Kuwano—who had still as yet been unable to find an application for his research would lead to a commercial productization—made a request to his boss Yamano Masaru saying that he wanted to quit his amorphous research. Yamano, however, refused to accept this; telling Kuwano thunderously, in his Osaka dialect, "We're not quitting!" Yamano not only attempted to inspire Kuwano and boost his morale, but also gave him technologically beneficial advice; suggesting at this time that he might reevaluate his amorphous semiconductors for use as an energy-related material.[114] It was not the case that Yamano already knew, at this time, that amorphous silicon could be used to make solar cells. However, with the oil crisis in the previous year, energy was a hot topic. Yamano felt that perhaps amorphous materials could be used for energy-related applications, and so he directed Kuwano to consider turning his work towards applications in that direction.[115] This would later lead to one significant coincidence.

---

[112] Kuwano (1985, p. 22). Similar descriptions appear in *Nikkei Sangyō Shimbun*, June 3, 1981, as well.

[113] Kuwano (1985, p. 22).

[114] Kuwano (1984, p. 109).

[115] Kuwano Yukinori, interview by author, October 29, 1998.

In the November 1975 edition of *Solid State Communications*, Kuwano learned that Professor Walter Spear of the University of Dundee had reported that it was possible to achieve PN-control using amorphous silicon prepared by glow discharge decomposition of silane gas. This amorphous silicon and glow discharge reaction had been Kuwano's area of expertise since long ago, and, after reading the article, he realized that the amorphous semiconductors that he himself had been working on for many years could be used to create solar cells. Kuwano immediately reported this to Yamano and requested to start conducting research into solar cells using amorphous silicon. Yamano readily agreed and, from this time onwards, Sanyo Electric began working towards the practical realization of solar cells, making use of their experiences in amorphous materials development up to that point.

In June 1976, David Carlson of RCA Laboratories announced the successful manufacturing of an amorphous solar cell. This cell had an area of 2 mm^2, and energy conversion efficiency of 2.4%. It was an extremely elementary stage device, but it was the world's first amorphous solar cell.[116] Although he had been beaten to it in creating a practical realization of the technology, Kuwano himself said that he was actually relieved to hear the announcement by RCA, because it made him feel that he had not been wrong in the direction in which he had been heading.[117] At this stage, Kuwano had still not yet succeeded in the manufacturing of his own amorphous solar cell, and was still feeling his way, uncertain as to whether or not he would actually succeed at all. Kuwano saw a ray of hope in RCA's success, and requested for Yamano to increase his research funding. Finally, in 1977, Kuwano succeeded in an experimental test of his own amorphous solar cell using new equipment. His cell was 2 mm^2 and had and energy conversion efficiency of just 1%.

Thanks to this achievement, in 1978 Kuwano's research group was promoted to laboratory status, and in September of that year they succeeded in creating a solar cell with an area of 35 mm^2 and conversion efficiency of 2.5%. Yamano announced this achievement at a press conference in Feburary 1979 and pledged that Sanyo would start to mass produce the solar cells within one year, embedding them in devices such as electronic calculators and watches. Kuwano was shocked.

Certainly, although Hamakawa's group had already announced the creation of a solar cell with area of around 3 cm^2 in 1978, at the time it was considered that the practical commercialization of the technology for use in products by a company was still three or four years away.[118] Despite this, Yamano had made a point of declaring that Sanyo Electric would be the first to bring the technology to commercialization and by as early as 1980. Kuwano reflected upon Yamano's declaration:

---

[116]Sharp (1996, p. 113).

[117]Kuwano Yukinori, "Waga kaihatsu monogatari: Amorufasu shirikon taiyō denchi ni toritsukareta otoko [My Development Story: A Man Obsessed with Amorphous Silicon Solar Cells]," *WiLL*, October issue, 1984, p. 110.

[118]*Nikkei Sangyō Shimbun*, May 9, 1979.

> When I first tried to make a calculator [using the technology], the thin film would peel off little by little; and when I tried to attach the film to glass it would tear straight off, just like a piece of dried squid... Well, since it was something invented in '75 we just thought it was just going to be that way. Coming up with a practical commercialization in just two or three years was an outrageous proposition.[119]

However, Yamano's announcement catapulted amorphous solar cells to the center of attention, and because of their superiority in terms of production costs they soon became the target of great hopes and expectations; that they would contribute to resolving Japan's energy problems.

After the press conference, securities companies began to visit Sanyo Electric to conduct surveys, and the company received a stream of phone calls from power companies, electrical manufacturers, and trading companies requesting explanatory materials.[120] When Tanaka attempted to incorporate the development of amorphous solar cells into the Sunshine Project, these achievements at Sanyo Electric had already been made public. Seizing the opportunity provided by the press conference in February 1979, Sanyo Electric moved to create a joint project, shifting amorphous solar cell development from its previous status as a regular research theme to being a mass-production-oriented research theme.

The announcement of amorphous solar cells by RCA's Carlson in 1978 also stirred up interest in amorphous technology among other Japanese companies, aside from Sanyo. Fuji Electric began its amorphous research program in 1978. At Fuji Electric, from 1979 onwards, Uchida Yoshiyuki (a member of the company's central research institute) took charge of the development as manager of the company's amorphous solar cell research group.[121] Seeking to differentiate itself from Sanyo, Fuji Electric narrowed down its development policy in the direction of aiming to achieve not only improved energy conversion efficiency but also large-scale surface area. That year, the company succeeded in producing a prototype 7 cm^2 solar cell with 2% conversion efficiency. When Fuji Electric participated in commissioned research under the Sunshine Project in 1980, large surface area was the company's distinguishing characteristic.

That same year, Kanebuchi Kagaku Kogyo (now Kaneka Corporation) also began developing amorphous solar cells. In April 1980, the company dispatched Tawada Yoshihisa (a member of its central research institute) to work under Hamakawa Yoshihiro at Osaka University. There, Tawada would produce remarkable achievements in improving conversion efficiency with heterojunction techniques utilizing methods from organic chemistry, which was his field of specialization.[122] Around this time, various companies seeking to launch their own

[119]Kuwano Yukinori, interview by author, October 29, 1998.

[120]*Nikkei Sangyō Shimbun*, May 8, 1979.

[121]Uchida Yoshiyuki, "Sekai ni sakigaketa amorufasu taiyō denchi no shōhinka [Commercialization of Amorphous Solar Cells Ahead of Competitors around the World]," *OHM*, November issue, 1984.

[122]Tawada Yoshihisa, interview by Tomae Hisao, August 9, 1995, transcript, Tomae (1996, pp. 71–74); and Konagai Makoto, interview by author, October 12, 1998.

amorphous research programs dispatched engineers to work in Hamakawa's laboratory—which at the time was regarded as a kind of Mecca—and to absorb his expertise and know-how. However, in spite of Kaneka's efforts, AIST and NEDO did not approve the company's participation in the Sunshine Project at the time of its launch, citing the reason that Kaneka was a chemical manufacturer.[123] Despite setbacks such as this, researchers from the industrial sector like Fuji Electric's Uchida and Kaneka's Tawada—and led by Sanyo's Kuwano—continued to deepen their interactions and exchanges as representatives for amorphous research at private sector companies.

Moreover, in 1981, a stream of private sector companies began to enter the development, production, and sale of solar cells. This development was not only limited to amorphous solar cells. Companies like Kaneka and Taiyo Yuden launched their own efforts to create solar cell businesses; and Showa Oil formed an agreement with U.S. firm ARCO Solar and began importing and selling solar cells in Japan.[124] Mitsubishi Electric's LSI Research Institute also announced achievements in its research into compound semiconductor solar cells. A solar cell "boom" was taking place.[125]

The reasons behind this boom included the rapid expansion of the solar-powered electronic calculator market that came with the appearance of amorphous solar cells, and the resulting increase in everyday people's recognition of solar cells. This emergence of amorphous silicon solar cells between the end of the 1970s and early 1980s had a massive impact on the development of crystalline silicon solar cell technologies that had taken place up until that time under the Sunshine Project. The engineers who had been researching and developing other methods of solar power generation had mixed feelings with regard to these amorphous solar cells, which had made their entrance with such a great and thunderous fanfare. They regarded the appearance of amorphous solar cells as a challenge to them. By as early as August 1979, Ikegami of Matsushita Electric Industrial—who was developing compound semiconductor solar cells—was saying the following: "[Amorphous solar cells are] the technology of the future so as someone not an expert in that field I can't make complaints. But I would like us to take this method of solar power generation [i.e. compound semiconductor solar cells] to the commercial stage too, and without fearing criticism against amorphous."[126] Hitachi's Saitō Tadashi, too, made the following comments in 1983, from the perspective of an engineer in the crystalline silicon camp:

> With amorphous [solar cells], in short, there are many cases where they are well received by amateurs, or people who have no real knowledge of solar power; since they can be placed on top of glass, or on steel plate, and because they do not consume a lot of energy in production as they can be made at temperatures of between 200 and 300 degrees. But I think that these people simply haven't done a purely technical and economical evaluation.[127]

[123]Tawada Yoshihisa, interview by Tomae Hisao, August 9, 1995, transcript, Tomae (1996, p. 76).

[124]*The Nikkei/Nihon Keizai Shimbun*, September 14 and December 21, 1981.

[125]*The Nikkei/Nihon Keizai Shimbun*, October 22, 1981.

[126]*Nikkei Sangyō Shimbun*, August 15, 1979.

[127]*NEDO News*, January 1983, p. 23.

Sanyo Electric's Kuwano reflects on these kinds of reactions, from the time of the emergence of amorphous solar cells:

> Well, [those kind of reactions are understandable] because this new emergent force [i.e. amorphous solar cells] had appeared on the scene, and we started saying what we wanted to say. We all had big mouths on us, too: Mr. Tanaka and myself, and Professor Hamakawa and so on. Well, I'm sure that made everyone else angry. Because although we weren't actually saying that they (and their technologies) were bad, we were saying that ours were the best. I think they did well to endure it the way they did. The crystalline people, I mean.[128]

Around that time, the future possibilities of amorphous solar cells were not something that anyone could judge with any certainty. However, even while this was going on, in March 1981, Sanyo Electrics's Kuwano announced at JSAP that his group had achieved an energy conversion efficiency of 6.91% with their amorphous solar cells. This achievement, too, received significant media coverage.[129] Furthermore, during that same month, it was also reported that ETL had developed a new method for manufacturing silicon thin-films for use in amorphous solar cells.[130] This continuous stream of achievements with amorphous silicon also led to the NEDO steering committee declaring its expectations for the material. The Sunshine Project, too, had already launched a program of amorphous research, but at the time it had not yet reached the practical commercialization stage that NEDO was responsible for.

While the engineers at the various companies had a cooperative relationship working with the support of the various universities and national research institutes with regards to the development of solar cells, in terms of the aspect of selecting the elemental technologies that would be used to achieve the realization of this goal they also had fierce competitive awareness based on the technology strategies of their respective companies. In particular, towards the end of the 1970s, amorphous semiconductors—which until that time had no relevance to the energy field—heralded the possibilities of their applications in the manufacture of solar cells; and became the sudden focus of intense interest, with claims of potential savings on silicon raw materials, easy manufacturing methods, and the promise that cost reductions down to one-hundredth of current costs would soon be achievable. The engineers at the various companies who had been engaged in the development of solar cells based on crystalline silicon and compound-type semiconductors met this arrival with great skepticism.

In the face of this, using their network of connections stemming from the Amorphous Seminar that had begun during the mid-1970s, the amorphous researchers from the worlds of industry, government, and academia formed an

[128]Kuwano Yukinori, interview by author, October 29, 1998.

[129]*Nikkei Sangyō Shimbun*, April 2, 1981.

[130]*Nikkei Sangyō Shimbun*, March 26, 1981.

organization, which could be commonly referred to as the Amorphous Family. On the university side, there were researchers such as Osaka University's Hamakawa Yoshihiro, Hiroshima University's Hirose Masataka, and Kanazawa University's Shimizu Tatsuo; all Amorphous Family researchers commissioned to conduct research on behalf of the project.[131]

Of course, it was not the case that universities were only backing amorphous research. There were also researchers engaged in the development of various other technologies. We cannot forget the sizeable role that the universities played in solar cell research with regard to technologies other than amorphous silicon.

In November 1979, these university researchers held the First Symposium on Fundamentals and Applications of the Photovoltaic Effect in Semiconductors. This was the first university-hosted symposium on the subject, and solar cell engineers from various companies got together; including the aforementioned Ikegami of Matsushita Electric, Kimura of Japan Solar Energy, Kuwano of Sanyo Electric, and Nakagawa of Toshiba.[132] Konagai Makoto of the Tokyo Institute of Technology said the following with regard to the role played by the universities: "The significance of doing it together in collaboration between industry, government, and academia is that we do research together, and also that we are in a position to state our opinions on the direction in which we should take this kind of research and development in the future, since universities are neutral and impartial institutions."[133]

It can be said that the role of universities with regard to the development of solar cells was in the hosting of symposiums such as this, and later—via the NEDO steering committee—in advising the direction that the development of solar cells should take overall.

As we have seen, national research institutes, private sector corporations, and university researchers created a family concerning itself with the development of amorphous solar cells, appealed to others about their significance, worked their way into the Sunshine Project and advanced their research and development efforts with the aim of achieving the successes that they had predicted.

[131]MITI Kōgyō gijutsuin Sanshain keikaku suishin honbu (1987).

[132]The Japan Society of Applied Physics, *Daiikkai handōtai no hikari kidenryoku kōka no kiso to ōyō ni kansuru shinpojium* [1st Symposium on the Basics and Application of the Photoelectromotive Force of Semiconductors], 1980. Engineers from NTT, Japan Silicon, NEC, Fujifilm, Central Research Institute of Electric Power Industry, Mitsubishi Electric, Sharp, and Fujitsu also took part in the Symposium.

[133]Konagai Makoto, interview by author, October 12, 1998.

7.5 Falling Crude Oil Prices and Reorganization of the Project

While on one hand amorphous solar cells were suddenly thrown into the limelight, experiments on crystalline silicon solar cells were also underway. From the end of the 1970s onwards, discussions had begun towards the construction of automated, fully integrated plants that would handle the entire manufacturing process, from the preparation of the silicon raw materials to the assembly solar cell modules themselves, with an annual output equivalent to 500 kW. The development of these experimental plants began in the 1980 fiscal year and continued until March 1985.[134]

Around the time that preparations were being made for testing the operation of these 500 kW plants, engineers from the private sector companies involved in the construction and operation of the plants were holding discussions with members of NEDO's Solar Technology Development Office with regard to how they should proceed with the plan to construct even larger-scale experimental plants, as the next stage of the development. However, manufacturers who had been carrying out commissioned research into crystalline solar cells from the initial stages under the Sunshine Project (with expectations being placed on them by AIST because of their technological capabilities) began to express reservations with regard to these larger-scale Sunshine Project plants. The companies expressed their wishes to NEDO —from their own point of view as commercial entities—for the project not to end simply as an experiment, saying that they wanted to link this development on to something that would generate profits. With regard to this point, there was a difference in awareness and opinion between NEDO and the companies, due to the difference of their respective organizational standpoints. Below, let us focus on the dialog that took place between NEDO and these companies with regard to the manner in which the project should be advanced. This was a debate in which both sides argued and tried to convince the other– from their own respective standpoints—what the project should aim to achieve.

The January 1983 edition of *NEDO News* contained an article describing what took place at a formal debate entitled "Photovoltaic Power Generation Systems: Issues Towards Practical Realization". Participants included engineers from various companies involved in the construction of the integrated crystalline solar cell production plants, including Saitō of Hitachi, Murozono of Matsushita Battery Industrial, Nakagawa of Toshiba, and Motomiya Tatsuhiko of Shin-Etsu Chemical (who was chief engineer in his company's R&D Department), as well as Takeda Yukihiro, of the Central Research Institute of Electric Power Industry (CRIEPI) . Horigome and Kurokawa of NEDO's Solar Technology Development Office were also in attendance. These were the key individuals in the photovoltaic power generation component of the Sunshine Project during this era. For this reason, the

[134]MITI Kōgyō gijutsuin Sanshain keikaku suishin honbu (1984, p. 73).

record of this debate enables us to learn what kinds of thoughts the central figures in the Sunshine Project had regarding the direction the project should take.

The main theme of the debate was what should happen in future with regard to the 500 kW crystalline silicon solar cell plants, which were in test operation at the time. The main topic of discussion with regard to this was how the NEDO technology development should be conducted moving forward. Specifically, the key focal points of the discussion were: (1) concerns for the future and solutions with regard to the practical realization and commercialization of solar cells and photovoltaic power generation; (2) the division of labor and responsibilities between the companies when increasing the size of the plants to large-scale facilities; and (3) the disclosure of technology-related information by NEDO.[135]

In the first part of the debate, Toshiba's Nakagawa and Shine-Etsu Chemical's Motomiya expressed their concerns over the fact that, when viewed from a business standpoint, solar cells were lacking a suitable market.

Nakagawa voiced his company's concerns, saying:

> We [Toshiba] have built a 250 kW-scale ribbon crystal plant at our Himeji Plant, but now that we have reached the stage of moving into actual production we are feeling quite anxious. Basically, while we think there will certainly be demand for solar cells if they become cheaper, we are concerned about the question of how many years it is going to take before we reach that stage.

Motomiya voiced similar concerns:

> Chemical plants have to be kept running on 24-hour operation. In other words, we have the unusual operational constraint of having to keep three groups of personnel working on rotation, watching over the plant 24 hours a day. So, we have the problem of whether the products that we make there are actually going to find their way onto the market. If it turns out that they are not, then that means we need to be developing that market from now onwards.

From the point of view of the companies, it was all well and good that they had built the AIST and NEDO-led integrated solar cell plants, but the most compelling problem was whether or not there was actually a market for the solar cells they would be manufacturing. In addition to these comments, Murozono of Matsushita Battery Industrial—which had participated in systems development—also stated that there was more of a need for development of applicable fields and demand-stimulating measures at the national level in order to grow and expand the market. For the companies involved, unless the government took to stimulate demand and expand the market (such as by developing areas in which the technology could be applied), even if they were to build large-scale solar cell plants in a state where there was still no market for the products then the cost of maintaining them would simply increase (with no returns). The companies requested that the national government should take responsibility for developing a market

[135]The following discussion is based on statements made in "Taiyōkō hatsuden shisutemu—Jitsuyōka eno kadai [Solar Power Generation Systems—Issues for Their Practical Application]," *NEDO News*, January 1983.

commensurate to the scale of the facilities being developed. Alternatively, where they were forced to develop and carve out this market for themselves, then they did not want to have any restrictions imposed on them by NEDO. Additionally, the companies that were specialized towards producing crystalline solar cells were concerned about the trend towards amorphous solar cells, which at the time were making a remarkable breakthrough, and for which commissioned research under NEDO was due to begin as of the 1983 fiscal year.

However, as far as NEDO was concerned, in order to display those achievements, it was important to design large-scale plants from as early a stage as possible; and to lower costs in order to facilitate that, it was considered desirable to eliminate overlaps in the development. This was not only the opinion of NEDO. When MITI requested budget funding for the development of amorphous solar cells, the Ministry of Finance had also complained, expressing their concern that it was an overlapping investment.[136] At that time, AIST avoided the Finance Ministry's criticism, by explaining—based on technical advice from NEDO and ETL—that crystalline and amorphous solar cell technologies had different applications. Despite this, however, AIST was still forced to make the firm promise that eventually the target of the research would be narrowed to just one of the two technologies, rather than continuing to develop the two technologies in parallel.

For that reason, it was NEDO's policy firstly to restrict free competition between various companies, and to create a large-scale plant with their cooperation, narrowing to one specific technology (either crystalline or amorphous) that they would adopt. It was because of this policy that exchanges like the one below took place.

> [Saitō] In order to reduce costs it is very important to introduce the principle of competition. So I think that if you don't allow us to compete and narrow down to just one of the two choices—crystalline or amorphous—in 1985, then the costs will definitely not go down.
>
> [Motomiya] I think in the end that even if we come up with ideas in the course of the manufacturing research process, where we think that utilizing this idea would lower costs further, under the current NEDO system we won't be able to do it.

In response to these comments, NEDO's Kurokawa answered as follows:

> Technology development funding is extremely limited. If we consider that, then we are faced with the question of whether or not competition is really possible… [Allowing competition means that] we would be allowing two intrinsically similar types of development to run in parallel. It means that, even if we do see some beneficial effects from the competition, for the most part the development requires twice the amount of funding. So there is the feeling of whether or not that will be tolerated.

In response to this, Saitō continued to hold on, saying, "I am doubtful as to whether we can produce anything efficient through cooperation between companies when have narrowed it down to just one technology." To this Kurokawa responded, "Hypothetically, even if we allowed competition, I think that it would be fine to do so once a base of [consumer] demand has been established. There isn't even a

[136]Horigome Takashi, telephone conversation with author, November 19, 1998.

market, right?" When Saitō replied, "Not right now, no", Kurokawa came back immediately saying, "Yes, so that's why I'm saying, surely it's more efficient to concentrate the wisdom and know-how that is currently dispersed (across various different companies) in one place. Don't you agree?"

Kurokawa went on to say, "One thing that always concerns me when listening to the voices of manufacturers is that it sounds like you all approach research and the development with the attitude that it will all be wrapped up and completed within two to three years." From Kurokawa's point of view, the words of the companies must have given the impression that their attitudes were too profit-centered to achieve the major objective of photovoltaic power generation. As a systems specialist and researcher, too, Kurokawa may have had hopes for the early-stage completion of large-scale production facilities. However, from the point of view of the corporate engineers, the development of solar cells was already a minor concern within their respective companies to begin with; and it was only natural for them to hope to move the development in a direction that would lead to profits for their companies in some way, shape or form.

Ultimately, Horigome decided that they would continue the parallel development of both crystalline and amorphous technologies for the time being, but that they would set the deadline of the 1985 fiscal year, when they would select one of the two solar cell manufacturing methods based on an evaluation by NEDO. Six months later, in July 1983, Horigome made the following declaration to newspaper reporters.: "We will continue the parallel development of both technologies [crystalline and amorphous] until the 1985 fiscal year. Then, at the end of that year, we will evaluate the results of the development and then continue with prioritized research for the practical commercialization of the more promising of the two technologies from 1986 onwards."[137]

In response to this comment by NEDO's Horigome, the media reported that "A face-off between the two materials—amorphous versus polycrystalline—is set to play out on the big stage of national project-level development."[138] For his part, Horigome's decision was most likely based on the belief that selecting one of the two technologies, crystalline or amorphous, based on the results of a fixed period of tentative competitive development would partially satisfy the assertions of both Kurokawa, who wanted to consolidate the project, and Saitō, who insisted on the importance of competition between companies, and at the same time fit in with the wishes of AIST.

[137]*Nikkei Sangyō Shimbun*, July 20, 1983.

[138]*Nikkei Sangyō Shimbun*, July 20, 1983.

7.6 Project Outcomes

7.6.1 Sharp and Kyocera's Response

Sharp and Kyocera, which until that time had been developing crystalline silicon solar cells, had already begun to display a swift response to the series of breakthroughs being achieved with amorphous solar cells by companies such as Sanyo Electric.

By as early as July 1979, Sharp had formed licensing agreements with Ovshinsky's Energy Conversion Devices (ECD), Inc. and Burroughs Corporation and had begun introducing their amorphous semiconductor memory technology.[139] Sharp had judged amorphous semiconductors to have great technological potential for the future, and had made the decision to begin conducting its own amorphous technology development program. Sasaki—who was head of Sharp's Semiconductor Business Division—had the idea to first introduce the technology at the company, and then later to undertake commissioned research into amorphous-related themes under the Sunshine Project. Horigome, who had joined NEDO in 1980, said the following with regard to Sasaki's approach to NEDO: "Sharp was using its crystalline [solar cells] as a power source for satellites, but they wanted to develop amorphous solar cells too; and not using their own technology, but using the patent of a man called Ovshinsky. So Mr. Sasaki, who was vice-president of the company at the time, came to NEDO with a very enthusiastic proposal."[140]

As a general rule, NEDO's policy was to restrict the development themes that could be handled by any one company. But Horigome could not turn down such a strong and enthusiastic request from Sharp. And so later, in 1983, Sharp participated in amorphous research for the project under the name of "reliability-increasing technology".[141]

At Japan Solar Energy, too, where Kyocera had become the central driving force behind the company after the withdrawal of Sharp and Matsushita Electric, it was felt that the momentum of the amorphous boom was too great to ignore; and research had begun into amorphous technologies, out of fear of being left behind and too slow to jump on the bandwagon if amorphous silicon became the favorite in the future. With regard to these efforts, Japan Solar Energy's Kimura said the following.

---

[139]*Nikkei Sangyō Shimbun*, September 1, 1979.

[140]Horigome Takashi, interview by author, September 16, 1998.

[141]"We accepted the extremely strong request made by Mr. Sasaki. He told us that amorphous looked promising, and Sharp really wanted to join the Project and perform research in competition with companies like Sanyo." (Horigome Takashi, interview by author, September 16, 1998).

> I thought that I couldn't simply be doing no research at all into amorphous solar cells, because there was a risk that the age of amorphous solar cells might come (and we would be left behind), and so as a safety measure I decided to organize a small research group, and to check out the future possibilities [of amorphous solar cells] myself.[142]

At one point, towards the end of the 1970s, Japan Solar Energy also dispatched engineers to accumulate amorphous development know-how under Professor Hamakawa at Osaka University. In response to the appearance of amorphous solar cells, both Sharp and Kyocera began to research and develop amorphous technologies in addition to the crystalline technologies that they had been developing up until that time.

Although Kyocera had been unable to take part in the Sunshine Project during the 1970s, during that time, too, the company had been continually applying to participate in the project each time the AIST research and development official changed. AIST's initial judgment, too, gradually changed as research and development officials changed every few years and, from 1980, Kyocera itself was finally able to participate in commissioned research into amorphous solar cells under the Sunshine Project, separately from Japan Solar Energy. The research theme at that time was the development of amorphous solar cells using ceramic substrates—ceramics being Kyocera's main business—and, using this as a pretext, Kyocera was finally able to succeed at getting its foot in the door of the Sunshine Project.[143] As a result, Kyocera established an arrangement whereby the company would conduct amorphous research itself, and ribbon crystal (i.e. polycrystalline) research through Japan Solar Energy.

7.6.2 Technology Development Policies of Sharp, Kyocera, Sanyo and Matsushita

In June 1983—around almost the same time as Hitachi, Toshiba and the other Kantō-based (eastern Japan) manufacturers that participated in the plant development were engaging in the debate with NEDO—Kyocera's Inamori, Sharp's Sasaki and Sanyo Electric's Yamano participated in a round-table discussion hosted by Hirono Tadashi (previously a research and development official at AIST) entitled "The Dawn of the Solar Energy Era".[144] Kyocera, Sharp, and Sanyo would later become the top three companies in solar cells and photovoltaic power generation but, in fiscal year 1983, Sanyo Electric had the largest market share in terms of solar cell production volume in Japan, at 40%. This was followed by Fuji Electric and Sharp, both at 16%, while Kyocera's share accounted for only 8% of total solar cell

[142]Kimura Kenjiro, interview by author, September 9, 1998.

[143]MITI Kōgyō gijutsuin Sanshain keikaku suishin honbu (1981); and Kimura Kenjirō, telephone conversation with author, November 10, 1998.

[144]Hirono and Seiichi (1985, pp. 61–87).

production, and Daido Hoxan and various other companies accounted for the remaining 20%.[145]

At the discussion, both Sharp and Kyocera criticized the major Kantō-based manufacturers for the poorness of their efforts towards solar cell development. Sharp's Sasaki said: "In Kantō there are many heavy electrical companies that manufacture generators. Those who make generators have a tendency to turn their attention towards concentrated facilities with proportional energy, so I think that many of them took developing solar cells in a derivative manner out of necessity, when they were forced to do so."[146]

Kyocera's Inamori, too, said the following, emphasizing he and his firm's own fervent enthusiasm and efforts towards the development of solar cells in comparison with Kantō-based manufacturers: "Only Kansai-based [western Japan] companies have created dedicated business divisions and they are all operating in the red. First off, Kantō-based companies can afford to sit and watch which way things go. I think it's like they are sitting on the fence waiting to come down [into the business] if things go well."[147]

As far as Sharp and Kyocera were concerned, companies like Hitachi and Toshiba were merely dodging the issue by participating in the commissioned research funding bracket. On the other hand, the two firms were confident that they themselves were fully committed to the project, to the extent that they were investing large amounts of their own business resources into trying to create a market.

In response to crystalline-type solar cell manufacturers who were dependent on AIST and NEDO's outsourced research and development programs, wherever there was an opportunity, companies like Sharp and Kyocera were looking proactively to enter into the development of different technologies (such as amorphous solar cells) under the Sunshine Project; and, furthermore, to establish new markets by themselves, without any relation to outsourced research under the project. Except for a small portion of applications such as consumer-use solar cells and calculators, the majority of these required the development of new market, and great effort on the part of the companies.

From the latter half of the 1970s onwards, Sharp developed a wide range of products using solar cells, focusing primarily on calculators and watches, but also including products for use in outer space, communications relay stations, rainfall gauges, pump systems, and power generation modules. For Sharp, which had originally begun the commercial productization of solar cells ahead of other companies, space was already an existing market. Sharp maintained this and other existing markets, while at the same time targeting the calculator market as the primary focus of its solar cell business.

Meanwhile, Kyocera was ravenously advancing the development—in cooperation with other firms—of any and all products into which there was even a slight

[145]NEDO (1996).

[146]Hirono and Seiichi (1985, p. 69).

[147]Hirono and Seiichi (1985, p. 71).

possibility of embedding solar cells. Kyocera attempted to create new markets for the application of solar cells in all manner of different fields, wherever it saw an opportunity. This included the development of solar cells for radios (January 1981), consumer-use portable power supplies (May 1981), solar energy lamps (May 1982, in partnership with Misawa Homes Institute of Research and Development; then later in October 1982 with the agricultural machinery manufacturers Iseki & Co., Ltd), battery chargers (September 1982, in partnership with Laox), car batteries (May 1983, in partnership with Maruenu Corporation), bicycle tail lamps (November 1983, in partnership with Bridgestone), solar cell-powered water heaters (March 1984), and solar-cell-equipped fee-charging toilets (June 1984, in partnership with Nikken). Kyocera was also proactive in its donation of equipment overseas—donating electrical power systems using polycrystalline silicon solar cells to Thailand, Pakistan, China, Indonesia, and various countries in the Middle East—all independent of NEDO projects—between 1983 and 1985.[148]

Meanwhile, around this time, Sanyo Electric's Yamano was saying the following in response to the point made by Kyocera's Inamori that there was still some concern in terms of stability with regard to using amorphous solar cells (for which the market for use in products such as calculators and watches was steadily growing) for photovoltaic power generation:

> If we postpone putting [amorphous solar cells] out in the open and exposing them to the harshest sunlight until last, there is currently no problem using them under milder conditions, such as in calculators and watches. The idea is that if they gradually change [and improve] and we gradually move them outside little by little, then surely we can get there somehow eventually.[149]

Although amorphous solar cells display excellent conversion efficiency rates when used in calculators and other similar applications indoors, under fluorescent lighting and so on; there were problems with their deteriorating quality and low energy conversion efficiency in comparison with crystalline-type solar cells when used in outdoor applications. At this time, although both Kyocera and Sharp were producing amorphous solar cells in parallel (i.e. along with their other solar cell technologies) for use in calculators and other similar applications, their attitude with regard to photovoltaic (PV) power generation systems (the development of which was the original objective of the Sunshine Project) was that there was no other option but to use crystalline-based technologies.[150] On the other hand, by continuing to run ahead of the pack in terms of the development of amorphous solar cell technologies, Sanyo Electric was envisaging a scenario for the future in which it would also be able to offer cheaper products in the PV power generation field, based on these amorphous

---

[148] *Kyōto Shimbun*, October 12, 1985.

[149] Hirono and Seiichi (1985, pp. 80–81).

[150] The following is a statement made by Inamori: "I don't know how long it will take for amorphous solar cells to become extremely reliable, but solar cells exposed to intense sunlight must be new crystalline cells no matter what that takes. I feel that reducing their cost is the only major area left for development" (Hirono and Seiichi 1985, p. 81).

technologies. When doing so, Sanyo would be able to make use of the research outcomes achieved through its commissioned research into amorphous solar cells under the Sunshine Project (which had just begun at that time).

While Sanyo Electric was working entirely under the concept of making amorphous solar cells the dominant technology in the field, firms like Sharp and Kyocera were treating amorphous solar cells as a tentative solution, to be used as and when needed depending on the application; while maintaining crystalline solar cells as their true favorite for the future.

Also around this time, Murozono of Matsushita Battery Industrial was searching for a market for the compound-semiconductor solar cells developed by Ikegami, of the same Matsushita Group. To that end, he released a newspaper article explaining that one of the distinguishing characteristics of compound-semiconductor solar cells is that, although their conversion efficiency is not good under fluorescent lighting, it is extremely good under red light.[151] Someone at U.S. firm Texas Instruments who read the article took an interest in this property of compound-semiconductor solar cells and made an inquiry to Murozono. Texas Instruments had been searching for solar cells to use in its calculators, which could be used under comparatively dim conditions, such as in aircraft cabins and during meetings when an overhead projector was being used. In response to this, Murozono promptly held a discussion with Ikegami, who was concerned about the risk of the project being discontinued after the completion of commissioned work under the Sunshine Project, and it was decided that the Matsushita Group would develop and manufacture compound semiconductors for calculators.[152]

And so the Matsushita Group developed a market for compound-semiconductor solar cells to be used in calculators, and was thus able to link the outcomes of its commissioned research and development efforts under the Sunshine Project on to the establishment of an actual business.

As a result, Matushita Battery Industrial started production of compound-semiconductor solar cells for calculators and—as explained in the next section, also partially due to the withdrawal of other companies from the amorphous solar cell market—rose to a position in which it essentially controlled half of the market, with the other half being controlled by Sanyo Electric.

7.6.3 Company Interests Become More Evident

From the mid-1980s onwards, NEDO rapidly began to lose the initiative in the development of new energy technologies. As a consequence, it became difficult once again for NEDO to involve itself in making decisions regarding the direction in which the Sunshine Project was heading. In 1983, NEDO had established its own medium to long-term project plan, and was still trying to take the initiative in the

[151]Murozono Mikio, interview by author, September 11, 1998.
[152]Murozono Mikio, interview by author, September 11, 1998.

development of new energy technologies. However, in 1985, Watamori complained about the lack of freedom afforded by NEDO, and in the following year all of NEDO's initial directors resigned en masse. Additionally, 1985 was also the start of the period when the public's level of interest in the development of new energy rapidly began to decline. Around this time, AIST also began to diversify the key focal points of its technology policy, and a plan began to emerge under which NEDO would also be given responsibility for the practical realization stages of other projects, under the Large-Scale Project and Next Generation Project schemes. With the fall in oil prices, key technology policy fields became dispersed, and AIST came to adopt that way of thinking that it could respond better to future changes in direction by diversifying and expanding the range of its technology development projects.

Also partly due to this change in circumstances, ultimately NEDO was unable to make a decision to narrow the development of PV power generation technologies under the Sunshine Project based on the selection of a single solar cell technology. In the end, no choice was made between crystalline and amorphous solar cells in 1985; and, as a final result, the proclamation made by NEDO's Horigome ended as a bluff, at least from the point of companies involved in the project. Competition between the companies in their commissioned research over which technology would become the dominant one had produced results in terms of improving the performance of solar cells overall; but ultimately the decision that should have been made by NEDO in 1985 as to the selection of a single technology—which was the original premise for such competition—was never made. After that, NEDO became even less able to select and discard technologies. Kyocera's Kimura had the following to say with regard to this tendency towards the unlimited continuation of commissioned research under NEDO: "When they say, let's hold a meeting and assess the [research and development] achievements of the Sunshine Project, they should do an assessment to determine which companies should be dropped, and drop them. But what actually happened was they just delayed, and on, and on."[153]

During the mid-1980s, Sharp and Kyocera—which had until then developed and produced both amorphous and crystalline solar cells—began to recognize that they had reached their technological limits, and, predicting a future slump in demand for solar cells for use in calculators and watches, established a policy of reducing their production of amorphous solar cells. As a result, Sharp began to narrow the focus of its attention onto conventional monocrystalline solar cells, while Kyocera began to focus on polycrystalline silicon solar cells, manufactured using the ribbon crystal technique. Sharp's Sasaki recognized that research into amorphous semiconductors in fact constituted research into thin-film technologies, with possibilities for practical applications that did not stop simply at solar cells; so Sharp diverted the thin-film techniques that it had accumulated through its amorphous research efforts towards liquid crystal technologies.[154] Kyocera, too, later made PV power

[153]Kimura Kenjirō, interview by author, September 9, 1998.

[154]Hirono and Seiichi (1985, p. 80); and Mr. X at company S (name withheld by request), interview by author, September 10, 1998.

generation systems utilizing the polycrystalline silicon technologies developed at Japan Solar Energy one of its strategic business targets. As companies like Sharp and Kyocera—which until that time had conducted multiple technology development efforts in parallel—began to realize that they could expect to see no further growth in markets for consumer-use amorphous solar cells, they saw that it was now necessary to turn—on all fronts—towards the original objective of PV power generation systems. According to data from 1998, around 80% of amorphous solar cells were being used for applications such as calculators and watches. If other portable devices are included, this figure reaches 90%.[155] The main application for amorphous solar cells was in small-scale devices such as these. Along with this change in policy direction, Kyocera's Inamori began to feel more strongly that the main emphasis of solar cell development should no longer be on technology development, but rather on research into production technologies and market development. He preached to the research and development officials at AIST and members of NEDO's Solar Technology Development Office about the necessity of measures for encouraging the introduction and widespread installation of these technologies. Kimura, also of Kyocera, gave the following account.

> Mr. Inamori told the government to hurry up and invest its energy into researching private sector production methods and market creation, otherwise the industry would never get going; that the government should cooperate with creating a new industry. It's good for the government to try very hard to develop things like amorphous [solar cells] or some other absurd thing that we are not sure is actually going to amount to anything in the end. But we are done. Our company is done with that. Now, it's just a question of how MITI is going to make use of these things. Surely that's the role of MITI? He told them that very strongly, again and again.[156]

NEDO's Horigome has a similar recollection: "At that time, Kyocera and so on were saying quite a lot that amorphous solar cells were no longer going to equal to anything, and that there was no reason to invest so much effort into them.[157]

In October 1984, Kyocera invested around two billion yen in total construction costs in completing the Solar Energy Center at Sakura, in Chiba Prefecture. There, Kyocera began designing and manufacturing large-scale PV power generation systems for power companies and overseas entities, using the polycrystalline silicon solar cells being manufactured at Japan Solar Energy.[158] The center was equipped with facilities for carrying out foundation-level research, practical experiments and testing, and manufacturing and demonstrations of solar energy systems. Also, partially for promotional purposes, the entire surface of the building's south face was covered with solar panels.[159]

---

[155]Fuji (1997, p. 81).

[156]Kimura Kenjirō, interview by author, September 9, 1998.

[157]Horigome Takashi, interview by author, September 16, 1998.

[158]*Nikkei Sangyō Shimbun*, October 25, 1984.

[159]*Nikkei Sangyō Shimbun*, October 25, 1984.

Since the withdrawal of Sharp and Matsushita in 1978, Japan Solar Energy had been producing solar cells using its own original ribbon crystal technique. However, despite its development efforts, the company had not been successful in producing any solid achievements with this ribbon crystal method. Due to technological restraints such as ruffling of the base and the slow speed of crystal growth, the company was still having difficulty in manufacturing solar cells that would surpass the quality of those produced by other methods. A similar situation existed at Toshiba, which had been carrying out commissioned research under the Sunshine Project, and ultimately none of the ribbon crystal techniques developed by these companies proved to be the key to effective mass production of solar cells.

For this reason, Japan Solar Energy had fixed its attention on an alternative method: the casting method (a polycrystalline silicon growing technique) that was being marketed by the German firm Wacker. Taking this opportunity, Japan Solar Energy—which until then had advanced its development while conducting discussions with its partner Mobil—decided to abandon the development of ribbon crystals and shift all its efforts to polycrystalline silicon. To that end, Kimura travelled with Inamori to Mobil's head office in New York to inform them that Japan Solar Energy—which until then had been paying royalties to Mobil—would now be abandoning the ribbon crystal EFG (edge-defined film-fed growth) method and switching over to polycrystalline. After this, Kimura flew immediately to Germany, together with Moriyama Shingo (president of Japan Solar Energy) and held negotiations with Wacker for the introduction of their casting technique at Japan Solar Energy. When they did so, Kimura was surprised to see that an MITI-affiliate came out to greet them upon their arrival and arranged the details of their stay.[160] This was because Moriyama had previously been Commissioner of the Agency for Natural Resources and Energy.

From his experience of having been unable to participate in the Sunshine Project for many years, Inamori had felt keenly the importance of having a pipeline to MITI.[161] It was for this reason that he had appointed Moriyama—who was not only an ex-Commissioner of the Agency for Natural Resources, but also hailed from the same hometown as Inamori—as a vice-president of Kyocera, and president of Japan Solar Energy, upon his retirement from government office. Just as it had introduced the ribbon crystal technique directly after the launch of the Sunshine Project, Kyocera was now introducing the casting method, a foreign technology, and attempting to make it the fundamental technology behind its solar cells. In the past, with the ribbon crystal technique, AIST had refused Kyocera permission to participate in the Sunshine Project. However, when Kyocera requested to take part in the project on this occasion, even with this German casting technique, there were no complaints from AIST. This was thanks, in part, to the influential presence of Moriyama.[162]

[160]Kimura Kenjirō, interview by author, September 9, 1998.

[161]Kimura Kenjirō, interview by author, September 9, 1998.

[162]Kimura Kenjirō, interview by author, September 9, 1998.

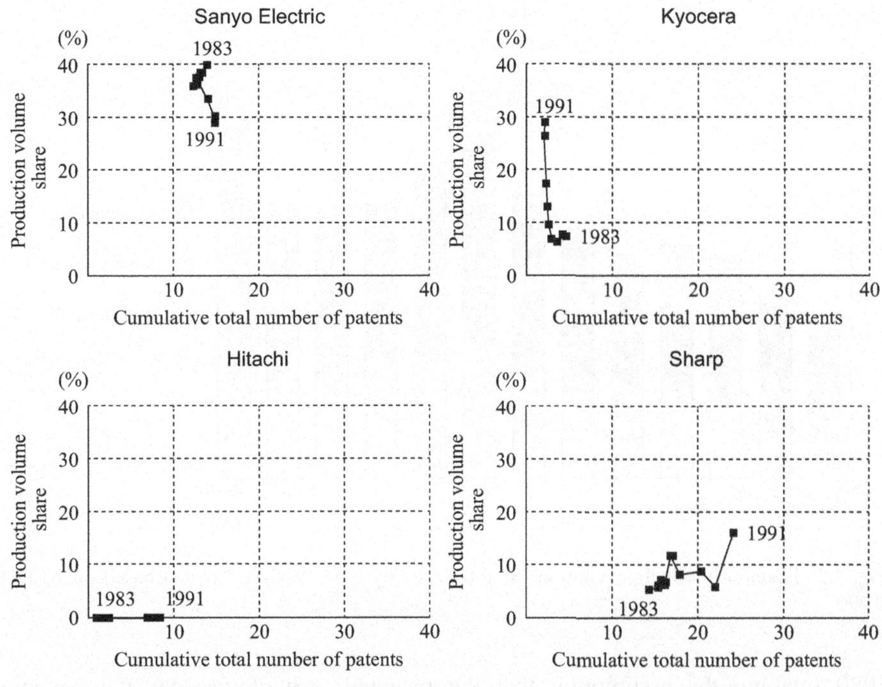

Fig. 7.2 Production volume shares and cumulative numbers of patents by company. *Sources* Created based on *The Classified Index of Publicly* Announced Patents compiled by the Japan Patent Information Organization (JAPIO) for each fiscal year; Fuji (1997), and NEDO (1996)

Later in 1986, Inamori likened Kyocera's technology to polycrystalline silicon. He recognized that the strength of his company lay—even in the case of solar cells—in introducing core technologies from overseas without hesitation and utilizing other surrounding technologies to improve upon them:

> When making a great company, there are two basic patterns: there are companies that have particularly fantastic technology and come out with those at the head of their advance; and then there are companies that skillfully arrange and organize common technologies. Our company [Kyocera] is the latter. Let's say our technologies are like polycrystalline silicon: wrapping up many small, core technologies like grains of sand with common technologies and know-how.[163]

In fact, Kyocera had very few patents for solar cells at that time in comparison with other companies like Sanyo and Sharp. Figure 7.2 is a set of graphs plotting cumulative number of solar-cell-related patents (classification H01L31—04) against production volume share for each company between the year 1983 and 1991. The graphs assume a lag of three years before patents are made public, plotting the cumulative number of patents along the horizontal axis starting from

[163]Takeuchi et al. (1986).

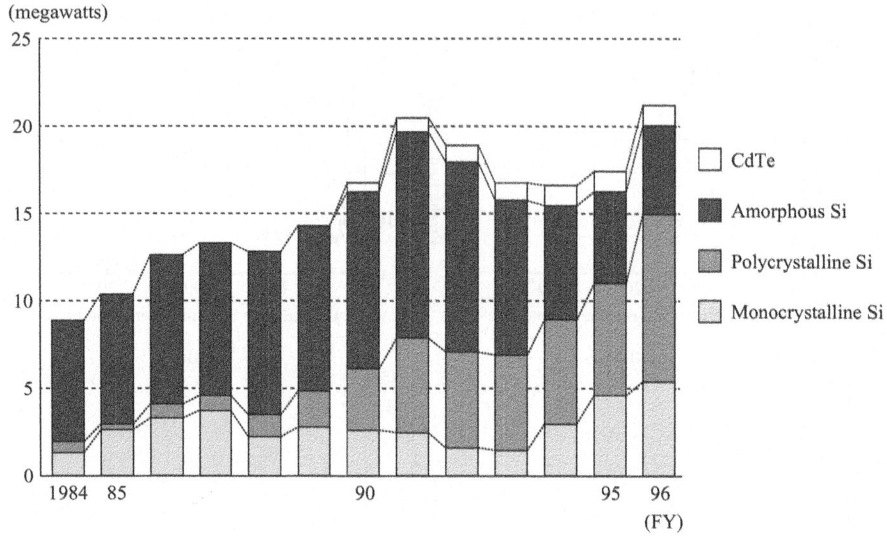

Fig. 7.3 Japanese production volumes of solar cells by type. *Source* Created based on NEDO (1996)

1980 (making the assumption that the patented technology does not become obsolete), and plotting production volume share along the vertical axis. Accordingly, the further to the right-hand side a point is located the greater are that company's technological capabilities; and the higher up a point is located the greater the volume of products that company is actually delivering to the market.

Sanyo Electric gradually decreased its market share due to a slump in demand for amorphous solar cells. As we can see from the way the line of the graph curves, Sanyo began to invest its efforts in technology development once again from the latter half of the 1980s, but at the beginning of the 1990s those efforts had not yet been sufficiently rewarded. Sharp proceeded along the route of linking its technological capabilities to an expansion of its production volume share; while Hitachi, despite maintaining a certain degree of technological capabilities, was not actually producing products to a significant extent.

We can see from the graph that, while this was going on, Kyocera was rapidly expanding its production volume share, despite having reduced its number of patents to a relatively low number in comparison with other firms. In particular, Kyocera's production volume share shows significant growth from the time of its introduction of Wacker's polycrystalline technology and onwards. From that time onwards, the proportion of polycrystalline silicon solar cells being produced relative to the production volumes of other types of solar cells in Japan gradually increased. By 1995, polycrystalline silicon solar cells had eventually surpassed amorphous silicon solar cells in terms of production volume (Fig. 7.3).

In 1985, after cutting its ties with Mobil, Japan Solar Energy finally became a wholly owned subsidiary of Kyocera. Later, in 1988, the company was absorbed into Kyocera.

Also during 1985, the test operation of the 500 kW crystalline solar cell plants came to an end and, during the next few years, Hitachi, Toshiba, and NEC all voluntarily withdrew from commissioned research into crystalline silicon solar cells under the Sunshine Project. Saitō, who subsequently left Hitachi to become a professor at Tokyo University of Agriculture and Technology, said: "I came to the university around the time when the Emperor Hirohito passed away [in 1989]. It was apparent that they weren't going to do any business [with solar cells] at Hitachi even if I stayed there, so I just thought there was nothing for it and decided to move on."[164]

As a result of this withdrawal, one of the three poles in this tri-polar market structure disappeared, and Kyocera and Sharp—which now actually produce these products—took their place as the dominant forces in the crystalline-type solar cell market, competing against the amorphous solar cell manufacturers.

In this way, the Kantō-based manufacturers that AIST had initially envisaged would drive the development of Japanese-made technologies left the Sunshine Project, and companies that were originally not permitted to participate in the project at the time of its original launch came to produce achievements in the development and manufacture of solar cells, using methods that were not permitted at the time of the project's initial launch.

7.7 Environmental Issues and the New Sunshine Project

During the latter part of the 1980s, too, the development of both crystalline and amorphous solar cell technologies continued, in parallel, under the Sunshine Project. At the same time, measures for encouraging the introduction and widespread installation of PV power generation systems for general household use gradually became a focus of all manufacturers.

Around this time, a technology research association for PV power generation was also being created. In 1990, the Photovoltaic Power Generation Technology Research Association (PVTEC) was established. After this, a new system was adopted whereby PVTEC handled the commissioning of all research to both private sector companies and universities, and conducted negotiations with NEDO. In a manner of speaking, NEDO was outsourcing the task of commissioning research and development work to a research association consisting of private sector companies, due to the increasing cumbersomeness of the commissioning process.

[164]Saitō Tadashi, interview by author, September 7, 1998.

It is said that the companies were initially opposed to this proposal by AIST.[165] This was because, while the creation of a technology research made NEDO's job easier, the private companies would be forced to bear the burden instead. For that reason, the companies considered that NEDO should handle the task using its own funds and personnel, instead of relying on a research association such as PVTEC, which would force private companies to cover the cost themselves.

However, partially due to NEDO's personnel constraints, ultimately AIST handled the situation by moving to establish PVTEC. Sanyo Electric president Iue Satoshi was appointed as chairman of the association, and a total of 24 companies and two other organizations participated as members.[166]

Introducing and expanding installations of PV power generation systems for general household use not only required solar cells. It was also essential to conduct research into the systems that would support them. Research into systems for PV power generation under the Sunshine Project began when Kurokawa Kōsuke—who was on secondment from ETL to AIST—submitted a research proposal concerning these systems to the project in fiscal year 1978. Kurokawa had argued strongly from an early stage that the development of such systems was necessary to the development of PV power generation systems in the same way as solar cells.[167] Also partly due to the fact that the Sunshine Project was created primarily through the efforts of ETL's Energy Department, which had originally been conducting research into power transmission systems, Kurokawa had from early on been appealing to the relevant parties about the importance of these systems with regard to PV power generation. And so, from the early 1980s onwards, the Sunshine Project began conducting proof-of-concept (demonstration) research into various types of systems, with test operation of these systems beginning in the latter half of the decade. Murozono of Matsushita Battery Industrial—which participated in systems development efforts under the Sunshine Project from fiscal year 1980—said:

> Looking back at the feelings we had at the time, it was unthinkable that we could install solar cells as part of our factory systems. Because, there was no point in installing power generating equipment that cost 10,000 or even 20,000 yen per watt [of generated power]. But when the government made its best efforts to try it out ahead of industry, the engineers – including ourselves – got a real feeling for what PV power generation was, and at the same time for what kind of things we needed to do with it.[168]

Since it was not possible for private sector companies to cover the cost of this proof-of-concept research, engineers like Murozono gained opportunities to learn about techniques and technologies for when solar cells were embedded and used in actual systems by participating in commissioned research.

Towards the end of the 1980s, coupled with advancements in the development of such systems, solar cells reached a stage of practical commercialization in the form of PV power generation systems for general household use. With this, the need

---

165 Horigome Takashi, interview by author, September 16, 1998.

166 PVTEC (1996, p. 8).

167 Kurokawa Kōsuke, interview by author, April 29, 1998.

168 Murozono Mikio, interview by author, September 11, 1998.

arose for the manufacturers of solar cells and other related products—which during the early half of the 1980s had been in competition over the development of these technologies—to cooperate in order to achieve the common objective of the widespread introduction and popularization of these systems.

Power grid interconnection-type systems had already been mentioned in an explanatory document entitled *Research and Development of Photovoltaic Power Generation Systems*, written by ETL's Kurokawa a mere three years after the launch of the Sunshine Project.[169] In this document, Kurokawa explained in detail about methods of grid interconnection for widespread installation and popularization of residential-use systems. Later, Kurokawa made the following comment: "From that time, I had been engaged in research into the reverse flow of generated energy [i.e. back onto the grid]. When I presented my research in the United States I was laughed at, but now it's the global mainstream standard."[170]

And so Kurokawa had to wait more than ten years before his idea became reality. However, it was thanks to numerous achievements in the development of solar cell modules and other systems over those ten years that the long-anticipated introduction of a PV power generation system for general household use finally came at the end of the 1980s. The first stage was the revision of the Electricity Business Act in June 1990. The act was amended in such a way that it became sufficient to simply submit a notice to a regional Bureau of Economy, Trade and Industry when installing a system with power output of less than 500 kW, and a system was established whereby the maintenance of such systems would be the responsibility of local electrical safety inspection associations.[171]

With the establishment of the Demand–Supply Subcommittee under the Advisory Committee for Energy's revision of its Long-Term Energy Supply and Demand Outlook in October 1990, the handling of measures for the future introduction and widespread popularization of PV power generation systems became a problematic issue for MITI. Until this time, AIST had been responsible for technology policy. However, the creation of policies for the introduction and widespread popularization of these technologies was an energy-related issue and, as such, came under the jurisdiction of the Agency for Natural Resources and Energy. The agency's Ōtsu Yukio (Director, Energy Efficiency and Oil-Alternative Energy Division) took charge of negotiations with the various parties to achieve the realization of grid interconnection, which was the bottleneck to achieving widespread adoption of PV power generation systems.

Ōtsu approached Horigome (who had already left NEDO for a university posting) and ETL's Kurokawa and asked for their cooperation, collaborated with people like Sanyo Electric's Kuwano, company representatives of firms like Sharp and Kyocera, and the Japan Electrical Manufacturers' Association (JEMA), and made his best efforts to achieve the realization of grid interconnection. One thing

169 Kurokawa Kōsuke, "Taiyōkō hatsuden shisutemu no kenkyū kaihatsu [Research and Development of Solar Power Generation Systems]," unpublished manuscript, 1977.

170 Kurokawa Kōsuke, interview by author, April 29, 1998.

171 Kuwano (1992, p. 113).

needed in order to achieve grid interconnection was the permission of MITI's Technology Division. It was also necessary to obtain the approval of the power companies and to establish a support framework for grid-connected PV power systems. However, there was intense opposition from the power companies with regard to incorporating PV power generation systems installed at individual homes into the power grid. Kuwano describes the situation as follows:

> The power companies thought of electrical power as being something that was supplied, and so for us to come along and say that power was going to be flowing backwards onto the grid [was outrageous to them]. For example, there were issues like who will guarantee the quality, what will we do when the systems break down, if power is going to be returned to the grid then what do we do about (option) price, how do we determine prices… MITI couldn't just decide these things by itself.[172]

Initially, negotiations with the power companies ran into difficulties. In particular, the decision over price setting for the purchase of electrical power generated by individual households was delayed significantly. However, at the time there were strong calls for the relaxation of government approval and licensing schemes under the slogan of deregulation, and in the end the power companies accepted a same-price scheme for the purchase and supply of electricity, and it became possible for individual members of the public to connect the PV power generation systems installed at their homes to the power grid at a favorable price.[173] In March 1990, guidelines for connection with low-voltage power lines in order to achieve actual realization of the grid interconnection concept were determined, and it became possible for PV power systems to be connected to power lines owned by power companies. Then, in April 1992, the Federation of Electric Power Companies (FEPC) made the formal decision that power companies would purchase surplus electrical power generated by household PV power generation systems. As a result, it became possible—through the installation of two electric meters (one for buying electrical power and one for selling it)—for regular households to both receive a supply of power from the power companies, and to supply power to the power companies themselves. In 1992, Sanyo Electric's Kuwano himself took the initiative and became the first person in Japan to install a reverse-flow-enabled, grid-connected PV power generation system at his own home.

The third issue standing in the way of PV power generation systems, aside from the Electricity Business Act and the issue of grid interconnection, was the need for the introduction of government subsidies to promote their widespread popularization. Reducing the cost of PV power generation systems would require manufacturers to utilize the advantages of large-scale productivity achieved through mass production. To achieve this, it was considered that firstly lowering the prices of each individual system though the dispensation of government subsidies during the introductory period—and thereby increasing the motivation of general users to purchase them—would be a promising method. Development efforts up until that

---

[172]Kuwano Yukinori, interview by author, October 29, 1998.

[173]PVTEC (1996, p. 49).

time had produced a certain degree of technological achievement. The energy conversion efficiency of solar cell modules had risen, and their cost had also fallen. Despite these improvements at the beginning of the 1990s, a general household-use 3 kW PV power generation system still carried a price tag of six million yen, and fierce negotiations took place between the public sector companies and the Agency for Natural Resources and Energy over their request for national government subsidies to the sum of half of that price tag. Kuwano said:

> At the time [the systems] cost six million yen, so if you installed one on your own house it would cost around three million yen. But giving a three million-yen subsidy to an individual private citizen was completely unheard-of. From the government's point of view, I mean. And who would be installing these systems? Well, surely anyone who buys a car or anything else for six million yen must be rich. So of course their logic was, why on earth should we hand out those kinds of subsidies to rich people?[174]

This lobbying on the part of the private sector companies and support for this from NEDO—boosted by the close attention being given to environmental problems at the time—spurred the government into action. From 1992, preferential taxation measures were implemented for installed equipment and, from 1994, the Residential PV System Monitoring Program was launched. This was a scheme whereby the government would dispense subsidies equivalent to half the installation costs for the installation of PV power generation systems at individual residences. In return, the homeowner would become a monitor, with the aim of gathering data to ascertain consumer needs and enable manufacturers to improve the performance of their devices in order to better match those needs. As a result of these developments, in 1994 539 PV power generation systems were installed. In 1995, this number rose to 958, and in 1996 to 1,866.

Before the attempts to develop new energy technologies—spanning more than 20 years since the launch of the Sunshine Project—led to the establishment and development of the solar cell industry as an independent industry, and the widespread introduction and popularization of PV power generation systems, there was a process in which AIST, NEDO, and various private sector companies pledged to achieve common goals in the form of a national project, while at the same time taking measures to achieve their own individual objectives.

When that happened, the project took the role of a framework that the various players could conform to in determining the strategies that they could take. Although project proposals are formulated based on participant-driven agreement, where there is asymmetry in the information available about certain technologies, or where it is necessary to make predictions about future possibilities, it is impossible to obtain a unified decision—even between researchers and engineers—over the selection of those technologies. Under such conditions, the players may move to restrict the actions of others, or to secure the possibility of achieving their own objectives; sometimes even by manipulating the very framework of the project itself. Accordingly, it was not at all the case that the Sunshine Project produced the achievements that it did because AIST's plan was realized just as it was initially conceived.

[174]Kuwano Yukinori, interview by author, October 29, 1998.

7.8 Conclusion

What can we learn from this case study? The interpretation in Case Study 1, which was examined in Chap. 3, was that the Japanese government played the leading role, and mapped out and executed the Sunshine Project from a viewpoint that was technologically and economically rational. The case study focused on how the government perceived external conditions and implemented policies toward goals set for the project as a whole. In that sense, the interpretation of the history of the project in Case Study 1 was that the government did its utmost to study and develop new energy technologies, but that changes in external macroeconomic conditions prevented it from achieving the initial targets. Based on this interpretation, in Case Study 1, changes in oil prices and the likelihood of success in technological development were asserted as important external conditions that the government may face.

In the meantime, the way the procedures for the Sunshine Project themselves expanded was made clear in Case Study 2 in Chap. 5. In organizational behaviors, practically the same things are predicted to occur even if the members are not individuals with specified names, as long as the routines continue. Provided that organizations originally exist for playing the expected roles outside the control of human attributes, routines rather than individuals become important. Accordingly, I changed the way of thinking in Chap. 5, and described the Sunshine Project by observing changes in organizational routines themselves, instead of directing my attention to individual intentions and actions. These descriptions made it clear that the organizations for executing the project were oriented toward expansion. A chain of actions comprises a project, if we call routines such a chain. A project works to maintain various activities undertaken in its name as long as it exists. A project is not a machine left to the mercy of humans as described in Case Study 1. It is an entity that appears to have a life of its own. In this case, organizations appear to take actions that are favorable for their survival that differ from the official, overall targets adopted initially when they are studied from the outside. Organizational actions are understood to be dysfunctional for target achievement in the presence of their divergence from the overall targets.

However, why a given chain of actions starts and which direction chains of actions that have become routine take when this particular chain undergoes changes were not made sufficiently clear in Case Study 2 either.

Case Study 3, which was examined in this chapter, focused on the moment when such a new chain of actions starts. Case Study 3 described the parties involved and their intentions when the Sunshine Project was launched and when it underwent significant changes. In many cases, new meanings were given to existing phenomena, technologies that emerged anew and the like. The observed phenomenon was that a new chain of actions started with a cluster of such new meanings. As I explained earlier in Case Study 2, the attachment of new meanings does not always cause a new chain. However, the start of actions linked to such meanings causes a chain of actions connected to the meanings to emerge ex post facto. Call to mind the popularity of a certain product or a certain trendy word, for example. It is not always the case that many people began to use the product or the word because they had high inherent functions of some kind. They became popular because many people shared their meanings and followed this up with related actions in response to a call to society made by a specific individual. Once started, a chain of actions divides actions into related ones and unrelated ones with itself as a yardstick. The continuation of related actions produces a loop system of chains of actions that has a certain degree of stability. Case Study 2 did not clearly show actions that are related and others that are unrelated. People do not agree with the appropriateness of their actions after judging the appropriateness of each of them. They determine ex post facto whether or not all the meanings, including those of connected actions, are related. In other words, a meaning is established ex post facto after an action of some kind is taken, and a chain of actions is confirmed on the basis of the meaning established in that way. Subsequent actions are related, rather than related actions following.

In recognition of such a structure, the types of actions that produce a chain of actions with the next new cluster of meanings were examined in Case Study 3. A concept secured by the future was discovered as a result. The occurrence of this concept itself in the future is not definite. However, a chain of actions appears in society as a system when many people believe in its occurrence and continue taking actions based on the assumption of its emergence. Once it has appeared, the system of a chain of actions becomes a reality in society as an organization or system that appears to have an independent life of its own as long as actions that follow those based on the system continue. However, phenomena such as a major change in a project occur when new meanings secured by the future in which the system is reconsidered cause the loop of a chain of actions to change and produce a new loop based on another cluster of meanings. In this way, we can come out with meanings secured by the future and move existing systems onto new tracks based on the agreement of others or their tacit consent regarding such meanings. In that case, nothing in the real world can secure an expectation for the future. In other words, an expectation for the future cannot be verified as the rational optimal solution in advance. An expectation for the future becomes such a solution in a self-fulfilling manner if others agree to or approve it ex post facto. Unless such agreement or approval exists, an expectation for the future is not connected to subsequent actions, and the expectation disappears without becoming a system of actions. In that sense, many people proposed meanings for the future in the national project case. A subjective agreement among people who chose to follow a leader prescribed the future of the national project in reality. What appear to be environmental adaptation actions for survival on an organizational level are activities for persuading the surrounding people, giving new meanings to the environment and changing the environment if possible on an individual member level.

As a concrete example of this, Tanaka Kazunobu of the Basic Research Division of the ETL maintained that amorphous silicon would achieve results comparable to

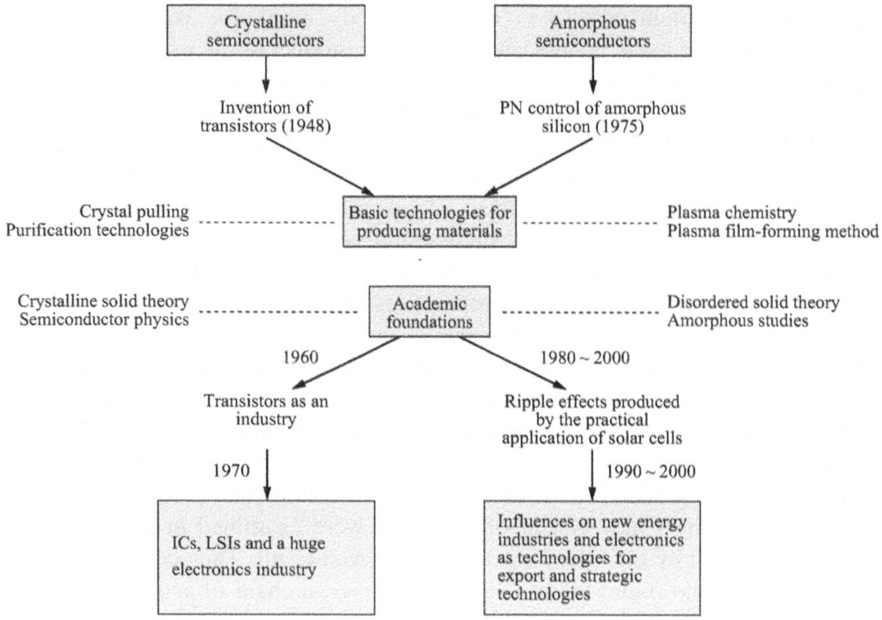

Fig. 7.4 Forecast for the development of amorphous silicon by analogy with crystalline semiconductors. *Source* Tanaka (1983, p. 30)

those produced by crystalline semiconductors when the material arrived on the scene (Fig. 7.4). Amorphous researchers at universities and companies gathered around Tanaka and formed a group nicknamed the Amorphous Family. The NEDO also had high expectations for the future of this technology. As a matter of fact, the NEDO achieved remarkable improvements in sunlight conversion efficiency in the 1980s, opening the way for amorphous silicon use in pocket calculators and other products. Amorphous solar cells made the presence of small, practical solar cells known to the world through their commercialization. Heterojunction with intrinsic thin layer (HIT) solar cells that are now marketed by Sanyo Electric are also the result of the application of amorphous technologies.

There were researchers and business managers who held up a vision and attempted to persuade the people around them in almost all national project cases where a big move occurred. As a matter of course, not all of them succeeded, but there is a process by which a forecast is realized ex post facto with the display of technological probability secured by the future as the first step toward success and the mobilization of resources to that end. Needless to say, these are not actions that all people can take easily, because risks are involved. However, researchers and business managers took a chance on these and, assuming the risks, attempted to prove what they could do here and now—or, as Aesop would have it, "Here is Rhodes, jump here." It was these researchers and business managers who ultimately led innovations. We call these people entrepreneurs.

References

Den'yūkai (A social gathering at the electrotechnical laboratory under the AIST) (Ed.). (1995). *Energī kenkyūsha eno messēji senpai kara kōhai e: Atarashii mirai no kensetsu no tameni [A message from senior energy researchers to their juniors: For building the new future]*. Tokyo: Powersha.

Electrotechnical Laboratory of the Agency of Industrial Science and Technology. (1991). *Denshi gijutsu sōgō kenkyūjo saikin no junen shi soritsu hyakushūnen kinen shi [The Last 10 Years at the Electrotechnical Laboratory in the Journal Commemorating the 100th Anniversary of the Laboratory's Establishment]*. Self-pub.

Fuji, K. (1997). *'97 Taiyō denchi kanren shijō no zenbō [An overall picture of markets related to solar cells in 1997]*. Tokyo: Fuji Keizai.

Geological Survey of Japan Editorial Committee. (Ed.). (1982). *Chishitsu chōsasho hyakunenshi [Hundred year history of the geological survey of Japan]*. Tokyo: Kōgyō gijutsuin Chishitsu Kenkyūjo Soritsu Hyakushūnen Kinen Kyōsan Kai [Geological Survey of Japan 100th Anniversary Commemoration Supporter's Association].

Ishizaka S., & Hirono, T. (Eds.). (1985). *Nijūisseiki eno enerugī: gijutsu ga hiraku shin enerugī [Energy for the 21st century: New energy developed by technologies]*. Tokyo: Tsūshō Sangyō Chōsakai.

Horigome, et al. (1972). Taiyō hatsuden shisutemu no teian [A proposal for a solar power generation system]. *Denshi gijutsu sōgō kenkyūjo ihō [Electrotechnical Laboratory Journal] 36*(10).

Investigative Committee on Industrial Technologies. (Ed.). (1974). *Gijutsu kaihatsu josei seido to tsūsanshō no ōgata purojekuto [System for assisting technological development and large projects by the ministry of international trade and industry]*. Tokyo: Sangyō Kagaku Kyōkai.

Kishida, F. (1991). Kiki o norikoete [Overcoming the crisis]. In Denki Shimbun (Ed.), *Shōgen daiichiji sekiyu kiki: Kiki wa sairaisuruka? [Testimony from the first oil crisis: Will there be another crisis?]*. Tokyo: Nihon Denki Kyōkai Shimbunbu [The Newspaper Division of the Japan Electric Association].

Kuwano, Y. (1984). Waga kaihatsu monogatari: Amorufasu shirikon taiyō denchi ni toritsukareta otoko [My development story: A man obsessed with amorphous silicon solar cells]. *WiLL*, October issue.

Kuwano, Y. (1985). *Amorufasu: Fukashigi na hishōshitsu busshitsu [Amorphous: A mysterious amorphous material]*. Kōdansha: Tokyo.

Kuwano, Y. (1992). *Taiyō denchi o tsukaikonasu: Taiyō denchi ga hiraku shinjidai [Using solar cells efficiently: A new era of solar cells will unfold]* (p. 113). Tokyo: Kodansha.

MITI Kōgyō gijutsuin. (Ed.). (1974). *Shin'enerugī gijutsu kenkyū kaihatsu keikaku (Sanshain keikaku) [New energy technology development project (The Sunshine Project)]*. Tokyo: Nihon Sangyō Gijutsu Shinkō Kyōkai.

MITI Kōgyō gijutsuin Kenkyū kaihatsukan shitsu [MITI AIST, Office for Research and Development]. (Ed.). (1987). *Ōgata Purojekuto nijūnen no ayumi: Wagakuni sangyō gijutsu no ishizue o kizuku [The course taken by large projects in the last 20 years: Laying foundations for industrial technologies in Japan*. Tokyo: Tsūshō Sangyō Chōsakai.

MITI Kōgyō gijutsuin Sanshain keikaku suishin honbu. (1981). *Shōwa gōjugo-nendo seika hōkokusho [FY1980 Annual Report]*. Self-pub.

MITI Kōgyō gijutsuin Sanshain keikaku suishin honbu. (Ed.). (1984). *Sanshain keikaku jūnen no ayumi [Ten year history of the sunshine project]*. Tokyo: Sanshain keikaku jusshūnen kinen jigyō suisin konwakai [Sunshine Project 10th Anniversary Commemorative Projects Promotion Committee].

MITI Kōgyō gijutsuin Sanshain keikaku suishin honbu. (1987). *Shōwa rokujūichi-nendo seika hōkokusho [FY1986 Annual Report]*. Self-pub.

NEDO. (1996). *Shin enerugī gijutsu kaihatsu kankei dētashū sakusei chōsa: Taiyōkō hatsuden [Research for preparing collections of data related to the development of new energy technologies: Solar power generation]*. Self-pub.

Numagami, T. (1992). Ninchi moderu to shiteno gijutsu: soshiki to nettowāku no keiei kōsōryoku [Technology as a cognitive model: The powers of management conception of organizations and networks]. *Business Review 40*(2).

PVTEC. (1996). *PVTEC gonen no ayumi [Five year history of PVTEC]*. Tokyo: PVTEC.

Sakaiya, T. (1975). *Yudan!* [*Negligence!*]. Nihon Keizai Shimbun: Tokyo.

Sawada, S. (1998). Horigome Takashi Hakushi no Ayundekita Michi [The course taken by Dr. Horigome Takashi]. *Kōgyō gijutsu* [*Industrial Technologies*] *39*(4).

Sharp. (1996). *Taiyō denchi no sekai kaiteiban* [*The world of solar cells revised edition*]. Self-pub.

Takeuchi, H., et al. (1986). *Kigyō no jiko kaikaku: Kaosu to sōzō no manējimento* [*Corporate self-reforms: Management of Chaos and creation*]. Chuōkōron Shinsha: Tokyo.

Tanaka, K. (1983). Amorufas shirikon: Hakumaku taiyō denchi shinzairyō [Amorphous silicon: A new solar cell material]. *Denshi gijutsu sōgō kenkyujo ihō* [*Electrotechnical Laboratory Journal*], *47*(7).

Tomae, H. (1996). Kyōdō kenkyū ni okeru ritateki shinrai [Altruistic trust in joint research and development] (PhD dissertation). Hitotsubashi University.

Yamagata, E. (1991). Gekidō no hibi [Turbulent days]. In Denki Shimbun (Ed.), *Shōgen daiichiji sekiyu kiki: Kiki wa sairaisuruka?* [*Testimony from the first oil crisis: Will there be another crisis?*]. Tokyo: The Newspaper Division of the Japan Electric Association.

Chapter 8
Organizational Analysis from Multiple Perspectives: Conclusions

8.1 Summary of Case Study 3

In Case Study 3, which was examined in the previous chapter, we directed our attention to the realm of meanings for people involved in the Sunshine Project through testimonies and accounts provided by the individuals concerned. Starting from the perspective of the individual participants, the Sunshine Project can be seen as a macrocosm of activities that bureaucrats, researchers and businesspeople affiliated with academia, the government, and industry undertook to fulfill their respective desires. Those individuals sought to develop new energy technologies, introduce them, and popularize them in fields such as public policy, corporate management, and research and development, utilizing their own expertise. Participants in the project tried to persuade others by creating new meanings. Such attempts, as often observed in the case studied in the previous chapter, were particularly memorable in those processes. The politics of meaning construction existed in national projects. The politics were attempts to fulfill individual intentions and realize desirable plans by persuading others, including members of the general public. The respective individuals worked to realize their own ideals based on a variety of intentions. Naturally, however, the Sunshine Project did not live up to the expectations of all the participants. This was the case because of a complex chain of actions taken by diverse individuals involved in the project.

Let us review the main points of Case Study 3 here. A certain researcher proceeded with a study for developing solar energy at the Electrotechnical Laboratory (ETL, currently the National Institute of Advanced Industrial Science and Technology (AIST)) based on his personal ideal. However, he was treated coldly at the laboratory based on the assessment that solar energy had low feasibility. In these circumstances, the researcher came up with the idea of winning a budget for his research by proposing research for a large project undertaken by the Ministry of International Trade and Industry as a whole. When submitted, this proposal attracted the attention of a member of the Office for Research and Development at

M. Shimamoto, *National Project Management*,
Advances in Japanese Business and Economics 25,
https://doi.org/10.1007/978-981-15-3180-4_8

the AIST for a different reason. The member of the Office for Research and Development suggested formulating a new energy development project separately from the existing large projects as a new policy for that year. Their opinions differed regarding the name of the project and the selection of a development theme, but the researcher at the ETL and the member of the Office for Research and Development sought to make the project more appealing to the public and increase the budget it could obtain by making the project larger through their cooperation in the fields of technology and policies. The project received stronger praise from the public than they had expected because the first oil crisis occurred in the same year. The project surpassed the expectations of the individuals concerned and began to develop a life of its own. As a result, it reached a stage where those individuals could no longer stop it themselves.

Subsequently, the NEDO was established as part of a policy for accelerating the project with a think tank based on the concept of industry–government–academia cooperation when the second oil crisis broke out. This organization abounded in eagerness to develop new energy successfully by expanding its staff of career researchers over time and managing the project as a central player. Experts in the fields concerned assembled at this organization from academia, the government, and industry.

Solar power generation gradually became the mainstream in solar energy studies. Amorphous solar cells based on a new method made an appearance at the end of the 1970s. Regarding this new method, researchers at national institutes, universities, and private sector companies cooperated with each other and strongly requested their participation in the Sunshine Project by forming the so-called Amorphous Family. The future of amorphous studies was still slightly visible at the beginning. However, researchers at national institutes promoted amorphous semiconductors as having similar potential to crystalline semiconductors. Researchers at universities supported this promotion to achieve results. Researchers at companies put solar cells into practical use in pocket calculators and other products. Thanks to the solidarity displayed by the family, amorphous solar cells received a huge budget and their presence became more strongly felt in the Sunshine Project. However, this achievement caused friction with crystalline solar cell developers. As a result, fierce competition for developing technologies occurred between the two camps. Amorphous solar cells had a dominant position in small-size applications, such as pocket calculators and watches. However, high-performance crystalline solar cells had an advantage later on, when the use of solar power generation systems began to spread. Be that as it may, amorphous technologies were subsequently combined with crystalline technologies in the form of Sanyo Electric's heterojunction with intrinsic thin-layer (HIT) system. Their combination led to the development of solar cells with higher sunlight conversion efficiency.

Steep declines in crude oil prices in the 1980s acted as a headwind for the Sunshine Project. The NEDO did not become the core manager for the Sunshine Project as people had expected. The mission for the NEDO weakened as the Ministry of International Trade and Industry commissioned topics other than new energy from the organization. The NEDO also became unable to spell out strong project orientation in other projects, such as those on solar power generation. Technological development in fields where profits as a business may be achieved in the future, such as solar power generation, advanced more as a result of the weakening of the project orientation at the NEDO.

With the arrival of the 1990s, the Ministry of International Trade and Industry changed its course regarding solar cells from technological development to introduction and popularization, and advanced legal revisions, the provision of subsidies, and the construction of infrastructure. Budgets for introduction and popularization support began expanding faster than those for research, development, and verification experiments in the mid-1990s.[1] Companies that had competed fiercely with each other in technological development now acted in concert to realize common interests, including the construction of infrastructure. As a result, the volume of solar power generation systems produced, and the number of such systems introduced, began increasing in the mid-1990s, ultimately turning Japan into the global leader in the production and introduction of solar power generation systems.

In this chapter, I clarify a theoretical perspective assumed in the case study discussed in the previous chapter through reexamination. To state my conclusion first, the theoretical perspective was to search for the individual realm of the meanings and clarify a mechanism woven by a chain of actions based on that realm. In doing these things, we must pay attention to the presence of political movements that aim to realize their plans by persuading others with the use of meanings, and by mobilizing resources. I will discuss these movements in detail in the next chapter.

As the final conclusion in this book, I would like to discuss in this chapter the fact that the analysis of organizations from multiple perspectives brings the advantage of visualizing things that are invisible from a single perspective. Such analysis makes our viewpoint on society reflective. It becomes the first step toward destroying the perception that people are beings that simply react to external conditions or only follow norms and institutions. Such analysis also makes the forecast for the social phenomena expected of social science less reliable. However, the analysis achieves the profound understanding of social phenomena and offers suggestions that are needed for moving society in a better direction for new innovations, at least subjectively, even if this point is taken into consideration.

8.2 Explanations for the Case from the Perspectives of Politics and Social Construction

Accordingly, which specific previous studies analyzed decision-making within organizations on the basis of the theoretical perspective described above? According to Graham T. Allison, politics is bargaining players positioned in a

[1]Kimura (2007), p. 8.

hierarchy engaged through regular routes.[2] The action-takers in decision-making from a policy point of view amount to two or more. A large number of players comprise this process. Organizational actions are politically derived results that depend on the strengths and abilities of supporters and opponents. The ability to persuade others plays an important role in these actions. For a start, I would like to cite examples of this political model in the following.

Richard Neustadt argues that not even the president of the United States forces everyone to obey him with his mighty authority.[3] In the first place, secretaries and assembly members who assist the president are obliged to perform their responsibilities based on their own ways of thinking. They do not blindly follow the commands of the president regarding all matters. For that reason, in reality, the president works very hard to make those participants understand that following his command absolutely is in their best interest in order to ensure that the participants perform their respective duties. The very ability to persuade, flatter, and encourage other government members to perform appropriate actions becomes a criterion for evaluating the president. This fact shows that even individuals in high official positions are unable to accomplish their intentions as they wish within the government. Conversely, no one among the subordinates who is willing to offer their advice to the top leader performs that challenging role believing that someone else will do it. Unexpected results can occur. This is also a viewpoint peculiar to political models that are difficult to grasp with the fields of vision based on rational models and organizational models. (I will discuss these models later in this book.)

Gabriel A. Almond maintains that the pluralistic observation of foreign policy decisions is important.[4] Charles E. Lindblom also explains that incremental policy decisions offer greater advantages than complete selection when determining policies.[5] Both of these opinions suggest that decisions, such as complete selection by a single entity assumed in rational models, are not made in the actual process of decision-making from a policy point of view.

Warner R. Schilling describes how decisions are formed as a political process through round-robin bargaining with the secretary of defense at the center, using budget compilation by the U.S. government as a source.[6] Each of the players in this process has expectations of his or her own. The players form an alliance on a certain issue or have their alliance declined on another based on their expectations. According to Schilling, organizational decisions are made as the sum total of such political bargaining.

Samuel P. Huntington asserts that decision-making processes related to the level of U.S. military forces and the overall scale of U.S. arms after the Second World War were not products of plans made by experts based on his study of those

[2]Allison (1971).

[3]Neustadt (1960).

[4]Almond (1950).

[5]Lindblom (1965).

[6]Schilling (1962).

processes.[7] In fact, the processes were the results of disputes, negotiations, and bargaining among public officers and groups whose interests and viewpoints varied. Huntington divides the decision-making processes for policies into administrative processes and legislative processes. In administrative processes, the power of the participating units is distributed in a stratified manner. There is no conflict regarding the basic targets and values. The scope of the available options is limited. In such processes, policies are decided on the basis of prescribed administrative procedures. There is another view, however. This view is that in legislative processes, relatively equal participating units determine the available options by a majority vote in the presence of a major disagreement of opinions regarding policy goals. Checking these views against the case of the Sunshine Project studied in this book, we can say that administrative processes generally correspond to Case Study 2 and legislative processes generally correspond to Case Study 3.

Roger Hilsman describes policy decisions as battles. As a factor affecting final decisions, Hilsman points to the relative strength of the different groups involved in the decisions in the same way.[8] A consensus for supporting a policy must be secured to realize that policy. Groups with conflicting opinions compete with each other to seek support for their policies. All techniques for alliances, including persuasion, compromise, and bargaining, are used in such competitions. An organizational decision on a policy is based on a consensus reached at the end of such competitions. Accordingly, what factors are essential for persuasion and consensus-building? As these factors, Hilsman cites the attractiveness of the targets that groups seek to achieve and the persuasiveness and good sense of discussions held by groups. Understanding the inside of the realm of meanings for participants at the point of analysis is indispensable for vicariously experiencing the battles of the participants regarding attractiveness, good sense, and persuasiveness cited by Hilsman. Case Study 3 examined in this book showed a series of processes, such as the presentation of the attractiveness of targets, persuasion, consensus-building, and resource acquisition, which also existed in various battles that occurred regarding the Sunshine Project.

Next, as a variation of the political models described above, I would like to refer to cases where collective actions of some kind take place with the natural consent of others, instead of against their intentions, by means including compensation and violence, as social construction models. I would like to examine these models further in this chapter. In political models, there were leftovers from rational models in the sense that individuals make decisions that are economically rational in a broad sense by calculating carrots and sticks, and that joint actions take place as long as they do not go against those decisions. However, the possible change of criteria for making such rational decisions through dialogues and persuasion is originally assumed in social construction models. Whether or not a certain action is

[7]Huntington (1961).

[8]Hilsman (1967).

rational is assumed to change depending on the meaning an individual attaches to that action as well.

I think that a famous episode in *The Adventures of Tom Sawyer* by Mark Twain is an easy-to-understand example of this situation. As a punishment for mischief, Tom's fearsome aunt ordered him to paint a fence. Initially, the other children made fun of Tom for painting the fence. However, before they realized what had happened, the painting had turned into a pastime that the children wanted to try, even in exchange for money, because Tom portrayed it in an intelligent manner. As a result, Tom got his friends to work in his place while also earning pocket money from them. The labor as a punishment became an amusement that the children wanted to try by paying money for it once they perceived it as something that looked like fun. In cases like this, we cannot ask the children what the painting means to them without understanding their realm of meanings. This example is from a trivial episode in a novel. However, it portrays the way people build social reality.[9]

Likewise, if we think that the development of renewable energy has social meanings, we must elucidate a process through which meanings are attached; in other words, how people began to interpret this hard-to-realize large national project that requires a huge budget as something desirable that must be undertaken at all costs beyond the perspective of mere short-term profits and losses. In fact, Case Study 3 discussed in the previous chapter focused on clarifying the changes in meanings attached to the Sunshine Project and the reasons for the project's continuation for a long period of time from viewpoints including the naming method for the project, the assertion of the significance of individual development themes, and activities for gaining social support, and acquiring funds and other resources. The meanings of renewable energy development changed with the times. Renewable energy development first saw the light as a countermeasure for an energy crisis in the period immediately before the first oil crisis. However, its meaning as a countermeasure for environmental problems, such as global warming, was emphasized later when oil prices declined. In the subsequent period, renewable energy moved into the center, or everyone's hopes as one of the new industries that would support Japan's economic development. Furthermore, renewable energy has been viewed recently as an important component of the best energy mix in the era after a period when exclusive reliance on atomic power generation became difficult. In Case Study 3, the new meanings of renewable energy development were raised in accordance with the context of different times. As a result, the survival of the project over the years became clear. Who, and for what, caused such changes in meanings? In what specific ways are new realities constructed socially, with or without intentions? Perspectives turn to these points when we adopt social construction models.

According to Vivien Burr, social constructionism refers to approaches that contain at least one of the following four assumptions: (1) a critical stance on self-evident knowledge, (2) historical and cultural distinctiveness, (3) knowledge

---

[9]Twain (1876).

supported by social processes, and (4) knowledge and social actions going together.[10] Of these four, the assertions in (1) that the presence of self-evident knowledge is criticized and (2) that social phenomena vary according to histories and cultures are relatively easy to understand. However, the assertions in (3) and (4) may require some explanation. The assertion in (3) refers to the fact that people build knowledge about the world in cooperation with each other. We can rephrase this point and say that knowledge is a socially shared collective idea that acts as a base for undertaking social activities. In other words, knowledge loses its legitimacy unless activities are undertaken with itself as their base. The assertion in (4) shows that constructed knowledge supports certain social actions and rejects others. Actions that should be taken in society based on knowledge are determined. Others are rejected when certain actions are chosen. For example, according to Burr, drunks were once considered to be offenders, and were sent to prisons under the National Prohibition Act. Today, however, they are sent to hospitals, instead of prisons, as alcoholics. Where drunks are accommodated depends on the socially shared knowledge of each period and place. Knowledge of all types is born as a result of observation of the world from various perspectives when the assumptions mentioned above are adopted as foundations. A truth does not exist as an objective reality. It is we who construct a truth socially ourselves.

In that case, academic efforts to clarify social phenomena are compelled to turn from the innocent direction of trying to find the true characteristics of human psychology and social structures. This is the case because, based on the perception stated above, simple and true characteristics that are able to explain all social phenomena do not exist. Instead of such characteristics, academic efforts are made on historical studies on the emergence of the various forms of social structures, and social practices and systems that give rise to those various forms. In making these efforts, we must pay attention to the fact that human actions produce a society, in addition to the fact that a society affects human actions.[11] The descriptions of histories and cultures will be emphasized more strongly as a way of gaining an understanding of a society when we stand on the perception that a society is the whole where microscopic (individual) actions and macroscopic (social) structures affect each other. When that happens, the approaches move closer to those of humanities. Based on the perception explained above, in this chapter I choose to refer to research aimed at explaining social phenomena from the perspective of social reconstruction processes built by the interactions of individuals influenced by society as studies based on social construction models. I examine studies based on this way of thinking in the following.

The way we understand the world is said to originally come from other people in the past and today, instead of objective realities, when we adopt these models as our base. Based on this view, all values in the world, including truth, goodness, and

[10]Burr (1995).

[11]Kato (2011) can be cited as a business administration study that pursued this point comprehensively using the innovation of technologies and organizations as a resource.

beauty, are nothing but values that we have acquired in society. For example, we are all born into our respective worlds. The conceptual frameworks and categories that people use in our cultures already exist in these worlds. People acquire these concepts and categories by developing their ability to use language through everyday life and education. Even economic sense, which appears to be viewed as too self-evident, has an aspect that is acquired as a product of education through the cultural prescription of actions suited to living. For example, the assessment that donations to other members of society leads to greater benefits than personal consumption in the long run is not economic sense alone. Such an assessment is impossible without the cultural understanding that defines who the other members of society related to us are. In the meantime, such social systems are not permanent. The world is constructed through practice each time people communicate with each other and take performative actions using concepts and categories. Social construction models do not try to explain social phenomena from the perspectives of human psychology and external social structures. They reject all those positions and find explanations for phenomena in the process of interactions that take place among people on a daily basis. The process of interactions causes social phenomena to exist ex post facto. In other words, it is relationships that create realities.

This way of thinking originated from *The Social Construction of Reality* written by Peter L. Berger and Thomas Luckmann. It later became the forerunner to numerous subsequent studies as sociology of social problems through *Constructing Social Problems* written by Malcolm Spector and John I. Kitsuse.[12] Why do we call a given social phenomenon a social problem? How can we distinguish matters that become the substance of formal objections? What do people use to persuade others? What is the effective rhetoric for lodging claims? How do counterclaims come into existence? A large number of empirical studies based on this approach have been produced in connection with points like these. These studies ask for our common sense again and sound an alarm to the easygoing and unreflective reproduction of social actions in our everyday lives. In these studies, the solutions to social problems are discussed from the perspective of correcting negative problems, and the realization of social ideas is discussed from the perspective of proposing positive policies. The processes through which actions for such objectives succeed or fail become the subjects of those studies. Research in which social construction models are used includes the following.

Wilbur J. Scott clarified how post-traumatic stress disorder (PTSD) was recognized as an illness and became a formal mental disability. In 1980, psychiatrists added various types of suffering caused by mental trauma, such as nightmares, declined responses to external conditions, and sleep disorders, to their list of illnesses. There were numerous political struggles along the way. The neurosis of soldiers exposed to extreme stress on the battlefield has been known from the beginning, but soldiers who responded in this manner were often assumed to be cowards. According to Scott (1990), cooperating psychiatrists argued that mental

[12]Berger and Luckmann (1967), Spector and Kitsuse (1977).

trauma and numbness on the battlefield caused acts of cruelty in Vietnam and crimes and social maladjustment after returning home when a group of Vietnam War veterans started the antiwar movement in 1967. They tried to understand the suffering experienced by the veterans in terms of the very nature of war, instead of as individual psychological problems. A category of illness known as PTSD was established in this way as a result of many years of study undertaken by antiwar veterans and psychiatrists.[13]

Dana Y. Takagi analyzed the disputes between student organizations and universities regarding ways of making university admission fair and impartial, with the admission of Asian Americans to these institutions in the 1980s as a topic. At the beginning, this issue was discussed in the context of discrimination against Asian Americans, leading to public disputes over the question of inequality. However, proposals for treating Caucasian admission candidates favorably to secure racial diversity and a desirable student balance appeared subsequently as Asian American students grew in number. Later, the assertion that the high hurdles set for Asian American students resulted from affirmative actions for members of other ethnic minorities, such as African Americans and Hispanics, also appeared. In making these assertions, facts presented as numerical values and statistical figures were used in entirely opposite ways depending on the contexts asserted by the respective parties regarding the question of what socially proper university admission means. This was the situation that emerged.[14] This development shows that the meanings of facts that are considered to be objective change according to the standpoints.

With those studies as the start, a large number of subsequent studies were conducted energetically on new topics that began to attract public attention. Topics such as drug-induced suffering, child abuse, feminism, social withdrawal, and passive smoking were, for example, studied in Japan.[15] All these studies focused on the processes through which a certain social concept gains a broad social consensus.

Studies based on social construction models force us to reconsider the way we perceive our society. Things that we consider to be a matter of course may not have been that way at all until a little while ago. Furthermore, things viewed as a matter of course today may not be regarded that way tomorrow. Observation of objective realities becomes impossible when this position is adopted. The absence of differences between researchers in the position of observers and other members of society means that researchers also affect how outcome and assessment will occur by observing this method of social perception. The descriptions of phenomena from outside the facts become theoretically impossible for that reason. Researchers cannot escape from being parties that observe phenomena from the inside while participating in them. This limitation means that we cannot eliminate the possibility for forecasts to prove otherwise, and even more so if people engage in reflection. It

[13]W. J. Scott (1990): 294–310.

[14]Takagi (1990): 578–592.

[15]Nakagawa et al. (2001), Akagawa (2006), Nakagawa and Akagawa (2013).

also means that researchers cannot predict or control the future of social phenomena in a privileged manner after all.

Accordingly, what can we do based on the way of thinking described above? The pursuit of a completely objective and external existence that guides human actions is meaningless when we adopt this way of thinking. Therein lies the principle that the selection and execution of personal actions, making reflective judgment case by case, conceiving the future of the society, and believing their legitimacy, at least on a subjective level, produces social phenomena ex post facto as we build relationships in society. This idea is not necessarily based on the economic sense described in Case Study 1 or the adaptation to social systems explained in Case Study 2. For this reason, the idea works as a railroad switch for social reproduction based on this sense or adaptation. It becomes a base for public nature that also underpins democracy.

However, thought patterns that are strongly influenced by the economic sense and social systems that people who live in contemporary society share as their common sense are already implanted in our own thought frameworks. Unfortunately, these patterns also affect our descriptions of the social phenomena based on personal actions in Case Study 3. For this reason, the exclusive characteristics of Case Study 3 are difficult to isolate. This is a problem. Case Study 3 explained the occasional presence of actions that differ from the assumptions in Case Studies 1 and 2 (such as entrepreneurship), which take the unrealized future into consideration. However, overlaps of a considerable degree with the principles explained in other models inevitably emerge when such descriptions are attempted in studies of contemporary society. The characteristics may be isolated as different behavioral principles, such as myths, when we study primitive cultures. In fact, however, the pure separation of actions for the future that can be called their remainder is extremely difficult in contemporary society where capitalist economic principles and social systems, such as bureaucracy, are functioning as if they are myths that cannot be questioned. Furthermore, people try to apply the benefits of existing social systems strategically when they take actions, bearing the future in mind. This tendency invites a situation in which Case Study 2 and Case Study 3 are particularly difficult to distinguish in writing.

Ontological gerrymandering (arbitrary demarcation in ontology) discussed by Steve Woolgar and Dorothy Pawluch is another fundamental criticism of social constructionism in general. Ontological gerrymandering criticizes studies based on social constructionism that seek to relativize the definitions of social phenomena by arguing that they are not absolute on the one hand, but that insist that their conditions and actions are stated objectively on the other.[16] This criticism denounces the arbitrariness of the division of definitions by researchers in the first place. The argument that such definitions equal the arbitrary division of constituencies (gerrymandering) works to the advantage of specific candidates.

[16]Woolgar and Pawluch (1985): 214–227. Taira and Nakagawa (2000), pp. 21–22.

As an example of gerrymandering, Woolgar and Pawluch criticize explanations of changes in the social definition of marijuana discussed by Spector and Kitsuse in *Constructing Social Problems*. In the 1930s, marijuana was considered to be dangerous and addictive. By the 1960s, however, it was no longer categorized as an addictive substance. Spector and Kitsuse maintained that the nature of marijuana cannot explain this change in the definition if its nature remained constant during that time. They argued that we must direct our attention to other factors including the methods of thinking adopted by various groups, the concept of addiction, evidence supporting this view, political strategies and tactics, and support provided by governmental organizations to analyze the change in the definition.[17]

In response to this argument, Woolgar and Pawluch criticize Spector and Kitsuse for overlooking the fact that the very acts of identifying the nature of marijuana or of emphasizing changes in conditions and actions can be interpreted as objections to the definition of marijuana. They maintain that researchers are defining assumed actions and conditions under their discussions through their own undertakings for naming, identifying, and describing the actions and conditions, but this point is concealed.

Certainly, we must say that researchers are unintentionally practicing social movements by themselves if they are, as Woolgar and Pawluch point out, discovering social problems using definitions that they have laid down based on their subjective views as grounds. However, this is a problem that cannot be eliminated completely, in theory, as long as researchers have opportunities to affect social phenomena under their observation. To begin with, researchers are assumed to have lodged one claim already at the point where they selected marijuana as their research topic. Their claim is that marijuana should be treated as a problem. For that reason, researchers are suspected of making arguments based on their subjective views, pretending they are performing objective studies. To cite a classic example, such criticism is sometimes expressed against Case Study 3 in this book in forms including the suspicion that the method used for stating the facts in the study is only the result of a certain method of value judgment by the author and the doubt that the differences between the study and nonfiction are paper-thin.

If we carry out a thorough investigation of the position taken by Woolgar and Pawluch, however, researchers are undertaking their studies based on value judgements of some kind in the first place. This condition could mean that definitions that suit the convenience of researchers are mixed into all human attempts to explain society using concepts of some kind. For those reasons, an appropriate position on how matters, including academic concepts, are understood in society is that flexible agreements that are understandable to a certain degree and connectible to practice by others exist in society. In other words, how those matters are can be described subjectively in ways readers can understand while such agreements remain in force. As a matter of course, such descriptions are not exclusive and objective. Multiple explanations that insist on their respective validity can exist.

[17]Woolgar and Pawluch (1985): 214–227, Taira and Nakagawa 2000, p. 23.

The relative merits of those explanations depend on their respective persuasiveness. In that sense, Case Study 3 examined in this book is one of the cases that could be written from many other perspectives.

Making an astute observation of this way of thinking, we can also say that the destination for social construction models is moving closer to the perception that the most important thing in realizing our personal intentions is whether or not we can win a battle for superior labeling. The analysis of claims and counterclaims certainly shows that studies based on social construction models are anticipated to produce practical and beneficial knowledge by generalizing this approach. However, it is impossible for specific individuals who have mastered the techniques of claims to define social phenomena completely and arbitrarily. There is also no guarantee that other people will accept arbitrary definitions as they are. In dealing with such definitions, people turn to factors such as economic sense and social systems as clues for determining proper understanding from time to time. For this reason, my discussions at this point again return to Case Studies 1 and 2.

What are the methods for analyzing society that exist under these conditions, after all? I would like to consider this point more deeply in the next and following sections, reviewing the overall picture of ways of thinking in the background of the case studies examined in this book. I would like to emphasize the significance of analyzing organizations from multiple perspectives in the sections to follow.

8.3 Significance of Analysis Using Three Models

In the sections of this book through the previous chapter, I described the history of the Sunshine Project through three case studies, gradually increasing the resolution along the way. My observations focused on the overall management of the project in Case Study 1, routines and their systematization at the respective organizations in Case Study 2, and individual intentions and actions in those routines in Case Study 3. As a matter of course, the field of vision narrows as the resolution increases, as in the case of looking at something through a microscope. In exchange for the change, the information obtained becomes detailed.

As the conclusion of the book, I explain its motive and significance in this chapter; in other words, the reasons why I made such an attempt in this book. In *Essence of Decision*, Allison had the objective of unraveling the mystery of the Cuban missile crisis that occurred as a historic incident in 1962. In unraveling this mystery, as assumptions for advancing the work, Allison presented three models as the methodological viewpoints that we tacitly adopt when we observe and analyze social phenomena. They are the first model (the rational model), the second model (the organizational model), and the third model (the political model). The different aspects that the same incident appears to present in different models used for perception, in addition to the analysis of the Cuban crisis itself, are what makes Allison's book interesting.

Allison's first model explains that the series of events known as the Cuban crisis was entirely the outcome of rational judgment made by the governments of the United States and the now defunct Soviet Union. However, his second model stresses the point that government orders have not actually reached the lower levels of organizations completely and that the process of the Cuban crisis resulted from daily routines simply performed during the crisis, rather than rational judgment made by the governments. According to his third model, the analysis of the existing state of the Cuban crisis and the course of responses resulted from political conflicts within the governments. In this way, one incident was explained differently from the perspectives of rational calculation, organizational routines, and political conflicts.

The first rational action-taker model used the analogy of appropriate actions as a way of understanding government policies. In this model, government policies are regarded as appropriate standardized actions taken by governments. Explaining and predicting such actions equals assuming one decision-maker takes actions rationally based on calculations and calculating the rational actions this individual should take under certain conditions in view of specific objectives and targets. This model explains policies from the perspective of how rational it was for the Soviet Union to decide to construct a missile base in Cuba. This model is also known as the classic model. Some analysts have preferred using it as a method that has existed for a long time.

The second organizational process model takes the view that what the first model calls actions and choices are in fact the output of a huge organization functioning in accordance with regular behavioral patterns. This model assumes that government policies are not on the level of choices by a single individual, and that they are ultimately actualized as organizational output. There is no guarantee that organizations always work as ordered by decision-makers if organizational characteristics, procedures, and repertoire produce this output. This is the case because the condition means that organizations function by referring to themselves. Accordingly, in this analysis, policies are explained by presenting the patterns of organizational actions that have produced actions after distinguishing the government organizations that were involved in decision-making. The clarification of the patterns of organizational actions is essential for this analysis. This is the case because the forecast of future policy results will ideally become possible by applying such patterns to the future if their clarification makes organizational actions predictable. In this model, the policies were explained from the perspective of the type of organizational output generated by the construction of a Soviet missile base in Cuba.

The third governmental politics model is an attempt to explain policy-based phenomena, which are viewed as rational choices in the first model and organizational output in the second model, as the derivative results of political bargaining. In other words, for this third model, a government explanation of a given policy is synonymous with finding out who took the action concerned and who did what to this particular individual. The future of policies is predicted by distinguishing the games in which problems arise, the corresponding players, and the relative strengths and abilities of the players. Factors such as mutual persuasion by players through discussions, rhetoric used in such discussions and appeals to third parties, including alliances with other key players and public opinion, naturally come under the spotlight when these points are examined. In this model, policies were explained from the perspective of the political outcome produced by the U.S. economic blockade of Cuba as a result of such exchanges.

The relationships between these three models are concisely explained as follows. In the first model, the best action that rational decision-makers should choose against external conditions that require a response is decided. However, rational actions are not always self-evident because decision-makers cannot obtain information about all the external conditions in advance. Under this condition, the first model assumes that decision-makers adopt the best action after obtaining as much accessible information about the external conditions as possible. In such a case, two important aspects for decision-makers are (1) to secure the information necessary for understanding the external conditions accurately and making rational decisions and (2) to rationally choose the best action within the scope of the information that could be obtained. In other words, rational calculation is important. In that respect, we can say that the first model corresponds to rational models that have been used in the history of organizational theories.

There was a tacit assumption for the first model, however. In this model, subordinate organizations are assumed to execute judgments by decision-makers as they are. In reality, however, subordinate organizations do not always share the judgments of decision-makers as they are according to the latter's intentions. This is the case because subordinate organizations attempt to execute orders received from decision-makers in an organized manner after translating them into procedures that they can practice. On the way, subordinate organizations reinterpret orders as familiar procedures that they can practice easily in some cases where the orders cannot be translated into practicable procedures or where the orders from superiors are in opposition to the procedures based on their own organizational rules. Similar actions also arise regarding the types of information that should be conveyed in cases where subordinate organizations provide information to decision-makers. Subordinate organizations study the matters that they are supposed to examine based only on their daily routines. For this reason, there is a possibility that important information outside the daily routines may be eliminated before it reaches the decision-makers. In that sense, we can say that decision-makers are actually judging matters based on information supplied by subordinate organizations only, and that subordinate organizations are executing only the practicable orders. If so, the cynical view that routines at subordinate organizations may be making decisions, in effect, also becomes possible. One might refer to it as a phenomenon called the power of individuals in the control of procedures. The second model pays attention to the various functions that the existence of such organizational routines brings to phenomena. These functions include the problem of the dysfunction of formal organizations. In that sense, we can say that this model, too, corresponds to natural system models that have been used in the history of organizational theories.

The third model also throws doubt on the judgment made in the first model. The first model assumes an accurate understanding of the conditions, rational judgments, and the execution of actions according to the judgements. The understanding of conditions by decision-makers and rationality are important in this model. In the meantime, the third model questions cases where two or more parties take an active part in decision-making. The understanding of conditions varies from one individual to another. For this reason, the understanding may not agree even if the respective parties judge matters rationally. Furthermore, individuals attempt to guide decisions into directions that lead to the expansion of their own influence, the guarantee of their positions, economic gains and the like in some cases. In such cases, political conflicts arise and, as a result, organizations end up producing output based on bargaining and compromises that will never arise when decisions are made by a single decision-maker acting rationally based on calculations. What is more, such output can be unintended consequences, such as those that are rational for limited individual self-interest but that are irrational for a government as a whole. Decisions made at the top will also be distorted if and when the active participants in decision-making intentionally order subordinate organizations to implement work slowdowns or disobedience for their own interests, not only at the judgment stage but also at the execution stage. As explained above, the second and third models view as problems events that occur in cases where the three conditions assumed in the first model, namely the understanding of conditions, rational judgments, and the execution of actions according to orders, are not satisfied. This can be viewed as a perspective that is indispensable for clarifying actual organizational phenomena in our everyday lives. In this book, I performed this analysis from multiple perspectives, using the Sunshine Project as an example.

8.4 Application of the Third Model—From Political Conflicts to Social Construction Based on Agreements

These three models advocated by Allison are potentially useful; we can apply these models to a broad range of cases. This is the case because the knowledge and implications produced through a given case study present entirely different aspects even if the policy is one and the same when analysis is performed from the different perspectives of rational choices, organizational output, and political results. In the first model, policies are the result of rational choices by governments. For this reason, governments should collect as much useful external information as possible and decide on their policies rationally based on an accurate assessment of the situations. The information necessary for making rational judgements is assumed to have been in short supply or situations accidental enough to invalidate advance rational judgments are predicted to have emerged in cases where policies fail to achieve results.

For the second model, it is ideally important for governments to enable organizations to play their intended roles by making the structure of their internal organizations as rational and suited to the conditions as possible, and by arranging the internal flow of information systematically. Policies require these conditions to achieve the desirable results. Enabling organizations to play their intended roles is not easy, though. In the first model, the cause of policy failures is explained as a lack of information or the emergence of unexpected situations to the end. However, the second model can explain situations where the autonomy of various organizations subordinate to governments sometimes prevents the smooth execution of policies, even when sufficient external information is provided. Conditions such as sectionalism, visual field constriction, short sight, imperfect mutual communication, a lack of adjustments, incompetence learned through exhaustive routine tasks, and reduced capacity to cope with changes in conditions may arise at subordinate organizations. These are phenomena widely known as the dysfunctions of a bureaucracy. In the meantime, in some cases, knowledge development through the creative aspect of organizations and coordination through the supply of places have a positive impact on policies as a whole. These cases show that the autonomy of organizational components works positively for entire organizations in some cases and negatively for them in others. In that sense, the capacity of organizations is a double-edged sword. A key mystery that organizational management theories have addressed for a long time is found precisely at this point.[18]

In the third model, the units for analysis become smaller and move down to microscopic individual levels. In this model, various players with individual interests are assumed to try guiding decision-making from a policy point of view to a direction that is advantageous for them. There were tacit assumptions in both the first model and the second model. The tacit assumption in the first model was that organizations are supposed to take rational actions for achieving their targets. The same assumption in the second model was that organizations are supposed to take greater measures for maintaining themselves than those for achieving their targets. However, these unconditional assumptions are taken away in the world of individual intentions and actions where resolution is increased by one level. Only the individual concerned knows his or her own inner world. The application of the third model offers the advantage of enabling the confirmation of the true picture that becomes visible after assumptions, such as those mentioned above, are removed. However, it becomes extremely important for us as researchers to understand the realm of meanings; in other words, how to find out the individual actions in ways that are understandable and position them in the overall context of facts, because such external standards that are available for reference do not exist any longer.

Ideally, meaningless bargaining must be prevented and cooperative relationships among players must be guided into a direction that is rational for governments as a

[18]For the dysfunction of bureaucracy, refer to Merton (1949), Selznick (1949), Gouldner (1954a, b), Blau (1963), Blau and Meyer (1971). For theories on knowledge building, refer to Nonaka (1990), Nonaka and Takeuchi (1995). For theories on places, refer to Itami (1999, 2005).

whole to produce the desirable results with plans and policies, based on the assumption that individuals engage in political activities in cooperation with others. However, doing these things is difficult in reality, and that is a problem. Like the second model, this third model has an advantage in terms of the fact that it can explain policy failures. The results do not necessarily become rational for entire governments, as assumed in the first model, when the actions of players who engage in political bargaining are composed. Involuntary functions and dysfunctions that occur as a result of attempts to achieve results at the respective organizational levels, such as those focused on in the second model, do not subsume the outcome either. Actions based on mutual readings by multiple players sometimes reach conclusions that none of them ever intended through their composition. In such cases, we must return to the level of the intentions and motives of individuals who took an active part in these actions in order to understand and explain the causes of the overall results. This need shows that the third model has strong realities for the explanations of past phenomena and persuasiveness based on such realities. At the same time, however, forecasting the future based on this model is extremely difficult in the sense that the mechanical application of the model does not allow such a forecast because the ability to read the actions of others is necessary for prediction itself. Allison specifies the examples of methods for explaining results that are derived from various games of bargaining played by the parties concerned as follows: (1) Ask which result from what type of bargaining among the players produced important decisions and actions. (2) Set the units for analysis, in other words, politically derived results. (3) Focus on certain concepts—the awareness of players, motives, positions, power, and tactics. (4) Use certain reasoning patterns. A certain action taken by a government becomes a result derived from bargaining among game players when the action is based on this model. Accordingly, analysts in the third model explain an incident where they discovered who produced the action concerned and who did what to that particular individual. Predictions are made by distinguishing the games in which problems emerge, the corresponding players, and the relative strengths and abilities of those players.[19] This method will prove useful when we attempt to understand organizational phenomena at a deeper level. More importantly, we must know how the individuals concerned interpreted these phenomena to understand them at such levels. For this reason, we inevitably direct our attention to the understanding of the realm of meanings for action-takers.

[19]Allison (1971), p. 7.

8.5 Organizational Analysis from Multiple Perspectives: Multiple Conceptual Lenses

In the following section, I would like to discuss the issue of building a bridge between theories and history, which is a central theme of this book, from broader perspectives. Allison called the multiple frameworks used for analyzing facts conceptual lenses. His explanations are as follows. These three lenses differ in terms of what they magnify, what they assume to be important, what they clarify, what they obscure, and what they play down. Different conceptual lenses produce totally different pictures regarding the aspects of historic incidents that are stressed and viewed as important. In the historic event known as the Cuban Crisis, as well, different cognitive frameworks used for reference attract close attention to the different aspects of the crisis that caused its aggravation toward a nuclear war. Accordingly, there cannot be a single prescription for social phenomena. In the following section, I would like to explain the positive meanings of pluralism that permit such diverse interpretations in the analysis of social phenomena.

In the preface of his book, Allison provides the following interesting explanations for the motives for attempting the substantial case analysis and theoretical description of the history of the Cuban crisis. Foreign policy researchers belong to two schools: the traditional school that focuses on case studies and the scientific school that attaches importance to theories. The former has subtlety and art in the descriptions of cases. The latter has the strictness required by systems and science. Allison states in his book, *Essence of Decision*, that the case studies he wrote may be criticized for lacking art on the one hand and attacked for not having the strictness required by science for theoretical chapters on the other hand.

As a matter of course, traditional and scientific schools share the explanation of something as their objective. Allison also admits this point. Researchers affiliated with the traditional school attempt to understand the uniqueness of events. In the meantime, members of the scientific school work to grasp events as cases of propositions that are more general. To borrow the words of Allison, from the perspective of the scientific school, researchers who belong to the traditional school excessively emphasize nuances and accidental factors in surroundings that may be better treated as matters unrelated to the essence, instead of common and frequent factors. This tendency applies to historical studies that attach importance to dilettante episodes, for example. To state it the other way around, from the viewpoint of the traditional school, members of the scientific school appear to analyze specific suitable aspects only because of their eagerness to seek universality.[20] In many cases, universal knowledge remains trivial discoveries that are already known, even if the knowledge is accurate. Attempts to study phenomena within the reach of light called theories are like searching for a key dropped during a night walk only in the

[20]Allison (1971), pp. ii–iii (Introduction).

areas under the streetlights due to the available light. There is no guarantee that the key will always be within the range of the streetlights.

In a second approach, interesting historic phenomena are events that are exceptional in some way in the first place. We must know that. In what way do those exceptional events have universality? Or, in what way are these events exceptional phenomena? These questions can be seen as two sides of the same coin. In that sense, researchers in both schools are working to understand and explain why events took place. They are acting in the same way regarding this point. To begin with, they cannot prescribe exceptions unless they are aware of the horizon of universality. However, they cannot understand why exceptional phenomena arise in this world in a manner that is guided by accidental factors by simply observing the universal side.[21] Trying to explain a phenomenon that has occurred only once in history using multiple theoretical analysis frameworks is the same as building multiple alternative hypotheses in person and causing them to compete with each other. As a matter of fact, it is precisely such an attempt that becomes an effective way of building a bridge between studies of history and those of theories.

It is an awareness of the problems we assume that decides on the useful matters that should be stated, the matters that should not be stated, the matters that are important, and the matters that are unimportant among various facts and types of information. We should instead direct our attention to the advantages of multi-faceted approaches, if that is unavoidable. Historical facts show diverse nuances when light is shed on them from different angles. The multifaceted descriptions of history can provide a greater volume of knowledge.

Following Allison, I attempted to pursue these themes simultaneously in this book based on the problem awareness described above. For that reason, I adopted a slightly different structure from the ordinary descriptions of history. The subject of descriptions in this book is a series of processes called the formation, continuation, and dissolution of the Sunshine Project. The descriptions mainly consisted of a series of developments from the beginning to the end of this Project. Those processes are historical facts that occurred only once. However, I described their history three times in this book as stories from three different perspectives. Readers must have viewed the series of events in the Sunshine Project from a different perspective on the problems in each case study, because each study placed emphasis on different aspects. The three rays of light threw light on different aspects of the same phenomenon. This book was a work for comparing three homemade, middle-range theories as well, so to speak.

However, I must be prepared to be subject to criticisms for this approach from two different camps. These criticisms resemble the attacks faced by Allison. I expect one of them to come from researchers in the so-called scientific school who approve the methodology of natural science. Theories are generally assumed to be good when they are simple, as Occam's razor shows. Explanations are worthwhile

[21]Refer to Yonekura (1998): 678–692 for an attempt at laying a foundation for business history studies from this perspective.

because they can illustrate complex phenomena simply. An approach is considered to be superior when it can control the noises around a given phenomenon, reduce the number of variables as much as possible, and explain or predict the phenomenon with a small number of variables. From this perspective, writing new facts in a case study in the name of the discovery of historical facts just because they are found makes the causal relationships among phenomena in the case concerned unclear. The criticism that the juxtaposition of multiple case studies for a single fact shows resignation may also be possible.

To state it another way, selective case descriptions in a manner that can be converted to simple principles are already underhand in one way or another from the viewpoint of researchers who belong to the traditional school. Historical descriptions should state the distinct aspects of diverse facts by referring to objective evidence. This is the case because refusing lighthearted understanding by discrediting the simple explanations of theoretical research with actual discoveries, discovering new variables that prior studies have overlooked and emphasizing the complexity of historical facts become an issue when we write about cases. Historical studies are important for finding the variables necessary for building new theories.

8.6 Levels of Systems

I would like to examine further what caused the different views on research behind these two approaches in the following section by looking at the system types advocated by Kenneth Boulding. As Numagami Tsuyoshi already explained in his book *Kōi no keieigaku* [Toward An Action System Theory of Management], Boulding's system complexity puts various phenomena around the world in order based on an idea called systems. Boulding points to a hierarchy that exists for them. As shown in Table 8.1, the hierarchy in his model has nine levels. From bottom to top, they are called (1) frameworks, (2) clockworks, (3) control systems, (4) open systems, (5) self-replication systems for growth, (6) systems with inner images, (7) symbol processing systems, (8) many-headed systems, and (9) complex and unspecifiable systems.

I believe many studies in the past that tried to explain social and organizational phenomena with models of some kind tacitly assumed one of these systems, if not self-consciously, for their subjects.

There appear to be many researchers who believe that organizations are control systems (on the third level) in the fields of business administration and organizational theories. Studies of social science and those of natural science are basically not distinguished from each other on this level. There is an engineering view of social phenomena behind such thinking. The advocacy of practical styles of research that, for example, view organizations as tools and think about how to use those tools effectively for achieving targets set voluntarily in person characterizes this viewpoint. Humans take the same actions under the same conditions. Humans

Table 8.1 Levels of Systems

Levels	Characteristics	Examples	Classifications
(9) Transcendental: complex and unspecifiable systems	Unavoidable inability to know	Metaphysics and aesthetics	Understanding the meanings of social phenomena
(8) Social organizations: many-headed systems	Value systems Meanings	Companies Governments	
(7) Humans: symbol processing systems	Self-consciousness Production, absorption and ability to interpret symbols Sense of the passage of time	You I	
(6) Animals: systems with inner images	Mobility Self-awareness Specialized sensory receptors Highly developed nervous systems Structure of knowledge (images)	Dogs Cats Elephants Whales and dolphins	Functional reproduction of organic systems
(5) Ontogenic: self-replication systems for growth	Division of labor (cells) Distinguished parts that interact Growth according to blueprints	Plants	
(4) Open systems	Self-maintenance Material throughput Energy input Reproduction	Cells Rivers Flames	
(3) Control systems	Self-control Feedback Information transmission	Thermostats Homeostasis Autopilots	Rational causal relationships among physical phenomena
(2) Clockworks	Cyclical events Simple and regular (regulated) movements Equation or the state of balance	Solar system Simple mechanisms (clocks and pulleys) Systems of balance in economics	
(1) Frameworks	Labels and terminology Classification systems	Anatomical charts Geographical summaries Indexes Catalogs	

Source Material prepared by the author using Boulding (1956) as a base and Pondy and Mitroff (1979), Numagami (2000), Hatch and Cunliffe (2006), Blaschke (2008), as supplementary explanations. The classifications in the far right column are those produced by the author

are regarded as beings similar to materials that react in the same way in all situations in explanatory frameworks on lower levels, including the first, the second, and the third. For example, homo economicus as assumed in economics is presumed to react in the same way to the same incentive. Human actions become predictable precisely for that reason. As stated above, humans can be treated in the same way as materials in this approach. Modern economics have been able to enjoy the advantages of this methodology by achieving their development with physics as a model.[22]

In the field of organizational theories, however, informal organizations were discovered after the Hawthorne Works experiments. The view that organizations are not completely controlled by business managers was born, and the approach subsequently led to the theory that organizations are autonomous organisms. For example, people who view organizations as systems with inner images (on the sixth level) must assume that organizations have images called norms and cultures and that their members follow collective ideas unconsciously within themselves. However, as on the seventh and higher levels, understanding the semantic content of those images is not emphasized on this level.[23] This is the case because people on the following side are not viewed as main actors who have a wish to change these images intentionally. Studies that assume emergent organizational characteristics of some kind and aim to specify these characteristics are located on this level.

Organizations are no longer tools that can be used freely when this viewpoint is adopted. They are understood to adapt to external conditions and maintain and expand themselves while performing their functions. Organizations that became systems with the introduction of values advocated by Philip Selznick are perfect examples. We find in them organizations that are the aggregates of relationships that are not reductionistic. We cannot divide organisms into components and return them to their original conditions by reassembling those components. Likewise, organizations that are the holistic aggregates of those relationships may become different entities once they are disassembled and reassembled. Unlike the actions of machines, the actions of living animals are hard to control. In the same way, living organizations are hard to control completely from the outside. We may be able to call this principle, which is unique to living organizations, organizational capacity. The capacity is a collective characteristic that is hard to control from the outside. The capacity acts functionally for the targets set by outsiders in some cases and works dysfunctionally for them in others. Many researchers point out the conceptual incompleteness of functionalism when it comes to using it for the research of organizational theories. However, there are ideas called rules and routines that adapt to external conditions and keep expanding themselves through self-reproduction

[22] Arakawa (1999).

[23] As Kagono Tadao pointed out, properly, we must question the situation where "systems do not produce effects unless they are justified" because systems have been "justified as matters that are at least subjectively rational" by people within themselves. Kagono (1988), p. 6.

and selection in the background of the way of thinking based on social system theory in a broad sense.[24] I consider that studies that look for the new possibilities of organizational theories in collective attributes that are hard to grasp with methodological, individualistic research and that are present in areas such as organizational capacity, networks, places, industrial accumulation, and teamwork fall under this category. However, I would like to call attention to the point that the individuals concerned are not always aware of the attributes, such as those stated above, in this approach. There is a definition that states that norms and cultures as collective realities or systems exist there because they are practiced in reality. At the same time, there is another definition that states that those things exist because people within systems share the perception that they are there. In other words, there could be systems that are not practiced as actions, but that exist because people within systems recognize their existence. There may be phenomena that are repeated as actions, but they are not understood as systems by the people within them either. That is to say, systems do not indicate external and observable rules and patterns for actions only. They are also ideological realities whose normative and binding characteristics are supported by the intersubjectivity of people. Systems feature this aspect.[25] We cannot objectively recognize the existence of systems in the latter sense unless we understand the realm of meanings for them. We need to be aware of this point.

For the reasons explained above, people who attempt to apply functional analysis on the fourth to sixth levels to social phenomena are criticized by people on the lower levels for not being able to return its collective elements to methodological individualism from the perspective of natural science, and are attacked by critics on the higher levels for processing the substance of its realm of meanings in the name of functions and not treating the substance as a subject of understanding from a humanities point of view.[26] Functional explanations in social science exhibit this weakness.

Let us move to the higher levels. There is an important reason why system views are required on the seventh and higher levels. The reason is that humans can change their own actions reflectively based on their interpretations of the meanings of external conditions. This makes us, as humans, different from materials and simple organisms. There is no guarantee that we will take the same action in the same conditions when their meanings differ, because we grasp the conditions around us through meanings. How do humans understand the meanings of the conditions around them? How can humans change their own actions reflectively? How can humans improve their society based on such reflections? These are long-ranging issues in social science that also overlap with social philosophy and social thought.

---

[24]Elster (1983).

[25]Seiyama (1995), p. 164.

[26]However, there are also functionalists like Niklas Luhmann who position communication as a system component and raise theories that social systems exist as a result of its linkage based on meanings. Refer to Luhmann (1968). This work was written before Luhmann began to use the concept of "autopoiesis" (a system capable of reproducing and maintaining itself).

Understanding the realm of meanings in social phenomena is an indispensable first step toward grasping this point. However, understanding this realm of meanings does not immediately make a society, such as the one assumed on lower levels, easy to control. If anything, the understanding can be seen as useful for gaining an insight into the reasons why such control is theoretically difficult. However, questions such as what it means to plan a society based on this condition and how we can live without yielding to the various lighthearted forces of social control can materialize precisely because such control is difficult. Studies based on such system views on higher levels show us this point.

Surveyed from this position, the scope of applicability for rational models is a portion of all social phenomena. Rational actions are a special case among intended actions. In particular, reasons for taking strategic actions must be clear to everyone. Jürgen Habermas expressed this point as follows in his book, *Zur Logik der Sozialwissenschaften*:

> In the case of strategic actions, there is no need to absorb meanings thought subjectively based on cultural traditions in the first place …
>
> Meanings intended by strategic actions can be defined to allow just one interpretation at all times. In other words, they are nothing less than rules that say that variables are able to be measured or at least compared should be maximized or optimized …
>
> Marginal cases called strategic actions are convenient because they can monologically [*monologisch*] fix meanings to thought subjectively.[27]

There is no problem explaining social phenomena using rational models if we can limit the research subjects to strategic actions. However, not all social phenomena observed in reality are as easy to understand as those actions.

I think that more researchers in the field of sociology have studied organizations from the position of higher system levels based on their view that organizations consist of actions that are hard to understand and are not self-evident than those in the field of business administration. Sociologists have taken the lead in those studies that stepped into the realm of meanings and relationships for people within various social organizations. Ethnography heads the list of such approaches. The meanings of norms and cultures that are socially woven by people who deviate from the life of ordinary and sensible citizens, such as motorcycle gang and street gang members, were mainly emphasized in this field. Ethnography based on careful observations and superb insights is anthropology that focuses on contemporary society. It radiates the light of attractiveness peculiar to itself as a work in itself.[28] This is the case because such anthropology allows us to understand, with fresh surprising impressions, the meanings of social actions taken by specific unknown groups. Like excellent historical studies, outstanding ethnography shows the characteristics of the realm of meanings for individual once-in-history research

[27] J. Habermas (1970).

[28] Outstanding literature on organizational ethnography includes the following; Kanai et al. (2010), Van Maanen (1988), Satō (1984), Whyte (1943), Willis (1977).

subjects and the universality behind that realm. However, it may be difficult to expect studies of this type to produce forecasts for social phenomena in ways that resemble simple demonstrations on lower levels.

We can understand the point stated below when we arrange various studies about organizations in order, assuming the system levels described above. The understanding of the realm of meanings for a given phenomenon deepens as we explain the phenomenon in greater depth using a framework on a higher level. However, forecasting the future of the phenomenon from the perspective of natural science in a simple manner becomes difficult in exchange for the deepened understanding. This is the case because humans change their actions based on reflections.

In the meantime, forecasts for a given phenomenon become easier as we explain the phenomenon in greater depth using a framework on a lower level. This is the case because such explanations assume that humans repeat the same response in a given environment if the other conditions are the same. The derivation of managerial implications favored by business administration experts means that demonstrating possible explanations on levels lower than those of prior studies enhances the sure predictability of phenomena. The clear demonstration that a phenomenon assumed to be a system with inner images on the sixth level may be in fact explained with a control system on the lower third level will be a major contribution. However, we should be aware that such a demonstration has a tacit tendency somewhere to encourage the idea that understanding the realm of meanings for humans and society is unnecessary.

From this perspective, I have attempted to clarify the differences that become visible through the cross-comparison of multiple middle-range theories for a single historical case in this book.

8.7 Controllability in Uncontrollable Matters: Agreements on Images of the Future

Innovations are one of the central themes for business administration today. What should we do to achieve innovations with new combinations? There is a limit to just observing the patterns found in past social phenomena when we seek an answer to this question. Regarding approaches in social science based on the interpretations of meanings, Imada Takatoshi states, "awareness of reality based on the interpretations of meanings is more effective than such awareness based on hypothetico-deductive approaches in many cases where the potential future universality of matters that are not universal now needs to be discussed".[29]

Just repeating past practices is not effective for studying ways of designing systems for the future. That is the case because humans can change their own actions based on reflections. If so, we can say that clarifying a mechanism that

[29] Imada (1986), p. 9.

produces, from within systems, factors that cause social changes with external shocks as mere catalysts is a challenge that we confront today. If systems are understood as ideological realities consisting of intersubjectivity, the reasons for their maintenance are sought in structures that mutually make predictions interdependently and circularly, which cause people within systems to think that people within other systems will follow the same systems.[30] In that case, actions adapted to external conditions cannot be assumed in a unified and conclusive manner because there is no absolute, neutral, and objective standard of any kind that all people must follow. Accordingly, we cannot specify certain external conditions as the main causes of changes. For this reason, we must question a mechanism through which new systems are produced from within such interdependent and circular prediction structures for explaining the production and reproduction of systems. Ideas that act as reference points for the actions of others and the creation of new meanings are two of the things that break those circles. Clarifying the very mechanism that causes prediction structures that exist as ideological realities to change leads to the explanation of the birth of not only economic and business administration systems but also entrepreneurs and innovations in a broad sense.

What prescribes relationships among organizations is the mutual reading of the patterns of interests that result from the commitments made by various entities. We can call the control of such patterns an act that is aimed at the further mutual reading of the interest patterns. We can say that what is truly important in matters taught at business and other schools is not mere knowledge, but the enhancement of reading ability for that purpose. Artificial system design is a process for producing agreements on the images of the future of systems that restrict ourselves. Accordingly, plans formed as anchoring points through the process can become both restrictions and possibilities for participating bodies.

We are living in the midst of the formation process for a society in which multiple entities participate, whether we are aware of it or not. To realize the society we want, we must understand and predict the actions of others. I believe that the significance of social science lies in the formation of agreements for the future, while understanding the realm of meanings for each of those people, grasping society as a whole, and always reflecting on the lighthearted understanding of society supported by preconceived ideas.

References

Akagawa, M. (2006). *Kōchikushugi o saikōchiku suru* [*Reconstructing constructionism*]. Tokyo: Keiso Shobo.

Allison, G. T. (1971). *Essence of decision: Explaining the Cuban missile crisis*. Boston: Little.

Almond, G. A. (1950). *The American people and foreign policy* (1st ed.). New York: Harcourt.

[30]Seiyama (1995), p. 110.

Arakawa, A. (1999). *Shisōshi no nakano kindai keizaigaku: Sono shisōteki keishikiteki kiban* [*Modern economics in the history of ideas: Their ideological and formal foundation*]. Tokyo: Chuōkōronsha.

Berger, P. L., & Luckmann, T. (1967). *The social construction of reality: A treatise in the sociology of knowledge*. New York: Anchor Books.

Blaschke, S. (2008). *Structures and dynamics of autopoietic organizations: Theory and simulation*. Wiesbaden: Gabler.

Blau, P. M. (1963). *The dynamics of bureaucracy: A study of interpersonal relations in two government agencies* (2nd ed.). Chicago: University of Chicago Press.

Blau, P. M., & Meyer, M. W. (1971). *Bureaucracy in modern society* (2nd ed.). New York: Random House.

Boulding, K. E. (1956). General system theory: The skeleton of science. *Management Science 2* (3).

Burr, V. (1995). *An introduction to social constructionism*. London; New York: Routledge.

Elster, J. (1983). *Explaining technical change: A case study in the philosophy of science*. Cambridge Cambridgeshire, New York, Oslo: Cambridge University Press.

Gouldner, A. W. (1954a). *Wildcat strike*. Yellow Springs, Ohio: Antioch Press.

Gouldner, A. W. (1954b). *Patterns of industrial bureaucracy*. Glencoe, Ill.: Free Press.

Habermas, J. (1970). *Zur Logik der Sozialwissenschaften*. Frankfurt a. M: Suhrkamp.

Hatch, M. J., & Cunliffe, A. L. (2006). *Organization theory: Modern, symbolic, and postmodern perspectives* (2nd ed.). New York: Oxford University Press.

Hilsman, R. (1967). *To move a nation: The politics of foreign policy in the administration of John F. Kennedy* (1st edn.). Garden City, New York: Doubleday.

Huntington, S. P. (1961). *The common defense: Strategic programs in national politics*. New York: Columbia University Press.

Satō, I. (1984). *Bōsōzoku no esunogurafī: Mōdo no hanran to bunka no jubaku [Ethnography of motorcycle gangs: A revolt of mode and the spells of a culture]*. Tokyo: Shinyōsha.

Imada, T. (1986). *Jiko soshikisei: Shakai riron-no fukkatsu [Self-organization: Revival of social theories]*. Tokyo: Sōbunsha.

Itami, H. (1999). *Ba no manejimento: Keiei no shin paradaimu [Management of places: New paradigm for business administration]*. Tokyo: NTT Publishing.

Itami, H. (2005). *Ba no ronri to manejimento* [*The logics and management of places*]. Tokyo: Tōyō Keizai Inc.

Kagono, T. (1988). *Soshiki ninshiki ron: Kigyō ni okeru sōzō to kakushin no kenkyū* [*Theory on organizational perception: Research on development and innovation at companies*]. Tokyo: Chikura-Shobo.

Kanai, T., Satō, I., Kunda, G. , & Van Maanen, J. (2010). *Sosiki esunogurafī* [*Organizational ethnography*]. Tokyo: Yuhikaku.

Kato, T. (2011). *Gijutsu shisutemu no kōzō to kakushin: Hōhōronteki shiza ni motozuku keieigaku no tankyū [The structure and innovation of technological systems: The pursuit of business administration based on a methodological viewpoint]*. Tokyo: Hakuto-Shobo.

Kimura, O. (2007). *Ondanka bōshi seisakuron (9): Kokka gijutsu kaihatsu purojekuto* [*Discussion of policies for preventing global warming (9): National project for developing technologies*]. Tokyo: Central Research Institute of Electric Power Industry.

Lindblom, C. E. (1965). *The intelligence of democracy: Decision making through mutual adjustment*. New York: Free Press.

Luhmann, N. (1968). *Zweckbegriff und Systemrationalität: Über die Funktion von Zwecken in sozialen Systemen*. J.C.B. Mohr: Tübingen.

Merton, R. K. (1949). *Social theory and social structure: Toward the codification of theory and research*. Glencoe, Ill.: Free Press.

Nakagawa, N., & Akagawa, M. (Eds.). (2013). *Hōhōron toshiteno kōchikushugi [Constructionism as methodology]*. Keiso Shobo: Tokyo.

Nakagawa, N., Kitazawa, T., & Doi, T. (Eds.). (2001). *Shakai kōchikushugi no supekutoramu: Pāsupekutibu no genzai to kanōsei* [*The spectrum of social constructionism: The present and potentials of perspectives*]. Kyoto: Nakanishiya Shuppan.
Neustadt, R. E. (1960). *Presidential power: the politics of leadership*. New York: Wiley.
Nonaka, I. (1990). *Chishiki sōzō no keiei: Nihon kigyō no episutemorojī [A theory of organizational knowledge creation: Epistemology of Japanese companies]*. Tokyo: Nihon Keizai Shimbunsha.
Nonaka, I., & Takeuchi, H. (1995). *The knowledge-creating company: How Japanese companies create the dynamics of innovation*. New York: Oxford University Press.
Numagami, T. (2000). *Kōi no keieigaku: Keieigaku ni okeru itosezaru kekka no tankyū* [*Toward an action system theory of management: The pursuit of unintended consequences in business administration*]. Tokyo: Hakuto-Shobo Publishing.
Pondy, I. R., & Mitroff, I. I. (1979). Beyond open system models of organization. In B. M. Staw (Ed.), *Research in organizational behavior: An annual series of analytical essays and critical reviews* (vol. 1). Greenwich, CT: JAI Press.
Schilling, W. R. (1962). The politics of national defense: Fiscal 1950. In W. R. Schilling, P. Y. Hammond, & G. H. Snyder (Eds.), *Strategy, politics, and defense budgets*. New York: Columbia University Press.
Scott, W. J. (1990). PTSD in DSM-III: A case in the politics of diagnosis and disease. *Social Problems, 37*(3), 294–310.
Seiyama, K. (1995). *Seidoron no kozu* [*The structure of institutional theory*]. Tokyo: Sōbunsha.
Selznick, Ph. (1949). *TVA and the grass roots: A study in the sociology of formal organization*. Berkeley and Los Angeles: University of California Press.
Spector, M., & Kitsuse, J. I. (1977). *Constructing social problems*. Menlo Park, Calif.: Cummings Publishing.
Taira, H., & Nakagawa, N. (2000). *Kōchikushugi no shakaigaku: Ronsō to giron no esunogurafī* [*The sociology of constructionism: Ethnography of disputes and discussions*]. Tokyo: Sekaishisōsha.
Takagi, D. Y. (1990). From discrimination to affirmative action: Facts in the Asian American admissions controversy. *Social Problems, 37*(4), 578–592.
Twain, M. (1876). *The adventures of Tom Sawyer*. Hartford: American Publishing.
Van Maanen, J. (1988). *Tales of the field: On writing ethnography*. Chicago: University of Chicago Press.
Whyte, W. F. (1943). *Street corner society: The social structure of an Italian slum* (4th ed.). Chicago: University of Chicago Press.
Willis, P. E. (1977). *Learning to labour: How working class kids get working class jobs*. Farnborough, Eng.: Saxon House.
Woolgar, S., & Pawluch, D. (1985). Ontological gerrymandering: The anatomy of a social problems explanation. *Social Problems, 32*(3), 214–227.
Yonekura, S. (1998). Keieishigaku no hōhōron: Itsudatsu, fukisokusei, shūkansei [Methodology of business history: Deviation, irregularity and subjectivity]. *Ikkyo ronsō* [*The Hitotsubashi Review*] *120*(5), 678–692.

Chapter 9
Developments After the Project

9.1 Sudden Changes After the Completion of the Sunshine Project

Japan has made concerted public and private sector efforts to develop the solar power generation industry for many years. In fact, companies in Japan had continued making efforts to develop and commercialize solar power generation as an industry for a long time since the oil crisis in the 1970s, supported by a national project called the Sunshine Project. Under these conditions, the sunlight conversion efficiency of solar cells improved and a shift toward lower cell and module prices progressed. Japan rose to the top of the world in terms of the volumes of solar cells produced and introduced at the end of the 1990s. Japan went on to lead the world in terms of the cumulative volume of solar power generation introduced in 1997. Japan then overtook the United States and became the largest producer of solar cells in 1999. The rapid progress achieved by the Japanese solar power generation industry up to that point was a happy story of the fruits borne by the Sunshine Project at long last (Fig. 9.1).[1]

However, this situation began to change swiftly in the middle of the 2000s. The rapid change occurred because the introduction of the feed-in tariff system (a fixed price purchase system, hereinafter referred to as the FIT system) by Germany in 2004 caused demand for solar cells to expand rapidly worldwide. The parties concerned in Japan had not assumed such rapid market expansion until that point. In a way, Japanese companies were not prepared or ready to deal with the situation. Competition commenced to take a share of the rapidly expanding European market.

With the introduction of the FIT system by Germany (and Spain), the volume of solar power introduced began to increase explosively in Europe. Japan had been the global leader in terms of the volume of solar power produced until then, but that

---

[1]NEDO BOOKS Editorial Committee, ed., *Naze nihon ga taiyōkō hatsuden de sekaiichi ni naretanoka* [Why Japan Was Able to Become the World Number One at Photovoltaic Power Generation] (Kawasaki: Shin'enerugī Sangyō Gijutsu Sōgō Kaihatsu Kikō, 2007).

M. Shimamoto, *National Project Management*,
Advances in Japanese Business and Economics 25,
https://doi.org/10.1007/978-981-15-3180-4_9

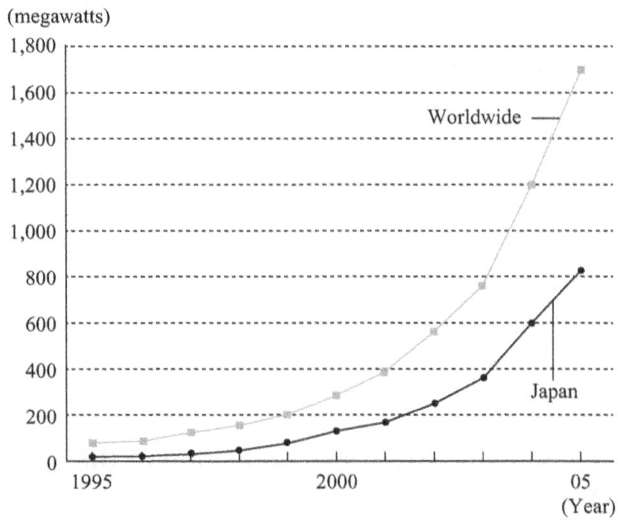

Fig. 9.1 Solar cell production volume in Japan (1). *Source PV News*

was during the period when the market was primarily limited to the country's small territory. Solar cell shipments from Japan to overseas markets surpassed those within Japan in 2004.[2] Rising companies in a variety of countries that had been in business for less than ten years expanded their production volumes in a geometrical progression year after year by making huge capital investments to deal instantly with the increasing demand. The volume of solar cells produced worldwide grew in leaps and bounds, from 1,700 MW in 2005, 2,500 MW in 2006, 3,733 MW in 2007, 6,941 MW in 2008, 10.9 GW in 2009, and 23.9 GW in 2010. In this environment, Japanese companies failed to keep up with the trend toward rapid output expansion (Fig. 9.2). China and Germany overtook Japan in terms of solar cell production volume by 2008. Figure 9.3 shows the countries that are making their presence felt in the production of solar cells. In the meantime, Germany and Spain outstripped Japan, the previous global leader, in 2005 and 2008 respectively in terms of the cumulative volume of solar power generation introduced. Figure 9.4 shows the countries with large solar power generation markets. As it turned out, it was only for a period of five years from 1999 to 2004 when Japan led the world in

[2]More solar cells manufactured in Japan were shipped to domestic destinations than to locations overseas. Those shipped to destinations in Japan and overseas were 186.2 MW and 83.1 MW in 2002 and 225.0 MW and 82.7 MW in 2003, respectively. However, those shipped to destinations overseas surpassed those shipped domestically in 2004, with the former amounting to 383.9 MW and the latter totaling 274.2 MW. In 2005, the difference widened, with solar cells shipped overseas reaching 578.7 MW against those shipped in Japan at 305.1 MW (Tawada (2011), p. 52.

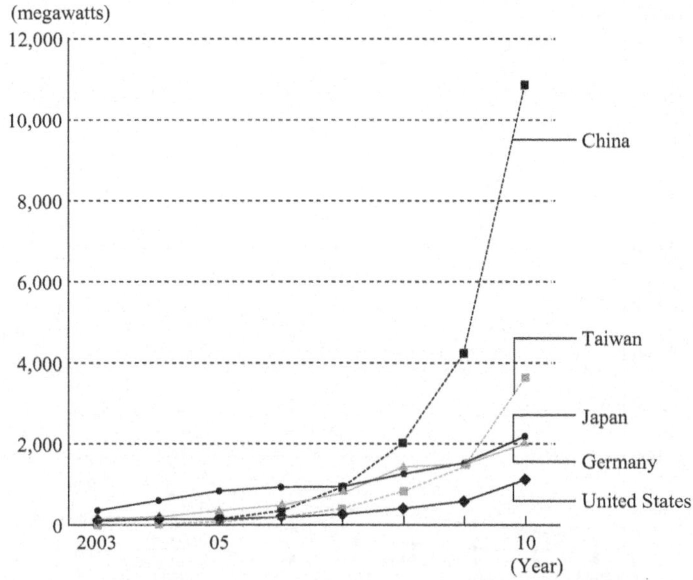

Fig. 9.2 Annual solar cell production volumes in Each Country. *Note* Japan had led the world in terms of production volume until 2007, but Germany and China overtook it in 2008. Later, Japan passed Germany in terms of volume in 2010, but Taiwan outstripped it in the same year. *Sources* EPI from Worldwatch; Prometheus Institute; Greentech Media

terms of the volume of solar cell production and the volume of solar power generation introduced. Those five years corresponded with a period when Japan was a main market. In the international structure that was established subsequently, solar cells produced in China and Taiwan rapidly flew to markets in Europe, such as Germany and Spain.

The changes described above mean a simultaneous decline in the relative presence of Japanese companies. From 2000 to 2006 Sharp led the world in terms of the volume of solar cells produced. In addition to Sharp, three other Japanese companies, Kyocera, Sanyo Electric, and Mitsubishi Electric, were on the list of top ten solar cell manufacturers in the world in 2006. However, these Japanese companies fell in rank as new companies specializing in solar cells, such as Q Cells, Suntech Power, and First Solar, made rapid strides in the subsequent period. Sharp and Kyocera were the only Japanese companies left in the top ten list in 2009. Sharp then became the only Japanese survivor on the list in 2010 (Table 9.1). What had happened during those years?

In this chapter, I would like to clarify the reasons why Japanese companies rapidly lost their relative superiority and the measures that leading manufacturers in other countries are planningto take by analyzing the trends in the solar power generation industry, which has undergone significant changes in recent years.

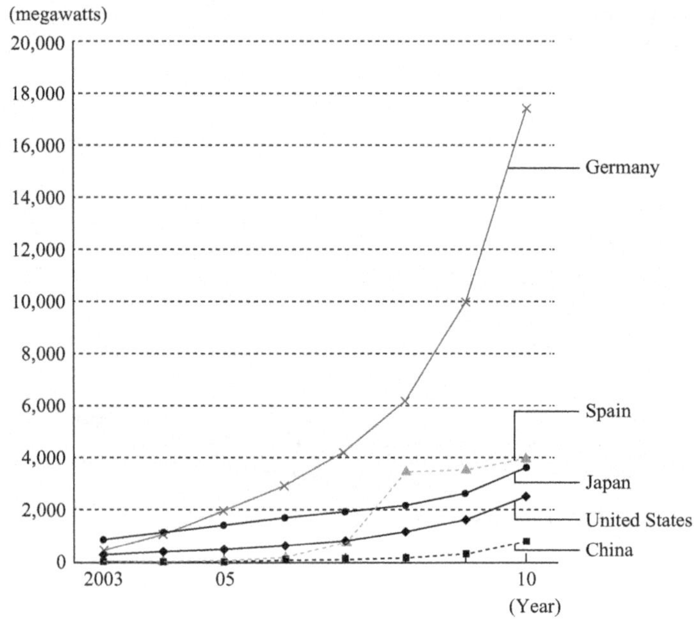

Fig. 9.3 Cumulative volume of solar power generation introduced by Each Country. *Note* Japan had led the world in terms of the cumulative volume of solar power generation introduced until 2004, but Germany, Spain and China overtook the country in 2005, 2008 and 2010, respectively. *Source* IEA, "Trends in photovoltaic applications: Survey report of selected IEA countries between 1992 and 2010" (Report IEA-PVPS T1-20:2011), 2011

Fig. 9.4 Japanese share of the volume of solar cells produced Worldwide. *Source* Material prepared by the author based on *PV News*

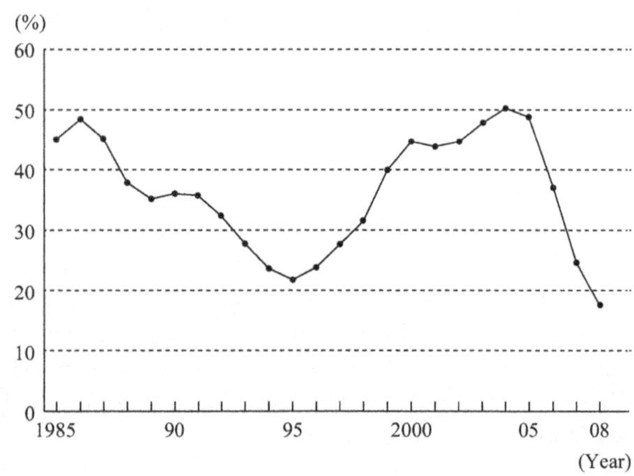

Table 9.1 Global ranking of solar cell manufacturers in terms of production volume (based on cells)

Ranking	2006	2007	2008	2009	2010
1	**Sharp (Japan)**	Q Cells	Q Cells	First Solar	Suntech Power
2	Q Cells (Germany)	**Sharp**	First Solar	Suntech Power Sharp	JA Solar
3	**Kyocera (Japan)**	Suntech Power	Suntech Power	**Sharp**	First Solar
4	Suntech Power (China)	**Kyocera**	**Sharp**	Q Cells	Yingli Solar
5	**Sanyo Electric (Japan)**	First Solar (United States)	**Kyocera**	Yingli Solar JA Solar	Trina Solar
6	**Mitsubishi Electric (Japan)**	Motech	Yingli Solar	JA Solar	Motech
7	Motech (Taiwan)	Solar World (Germany)	JA Solar	**Kyocera**	Q Cells
8	Schotte Solar (Germany)	**Sanyo Electric**	Motech	Trina Solar (China)	Gintech
9	Shell Solar (Netherlands)	Yingli Solar (China)	SunPower (United States)	SunPower	**Sharp**
10	BP Solar (United States)	JA Solar (China)	**Sanyo Electric**	Gintech (Taiwan)	Canadian Solar (Canada)

Source PV News; Photon International for 2009; and PV Insights for 2010

9.2 Changes in the Competitive Environment from the Middle of the 2000s

After the completion of the Sunshine Project, the Japanese solar power generation industry appeared to sail smoothly until the middle of the 2000s. However, unexpected major changes occurred in the period that followed. These were the rapid market expansion with the introduction of the FIT system by Germany in 2004, simultaneous market entries by new companies in China, Germany and the United States, and the explosive market expansion that accompanied their market participation mentioned above (Fig. 9.5). The FIT system is a system that guarantees the purchase of electric power for 20 years at a price several times higher than the price at the time the solar power generation facility was introduced. Under the system, the charge is reduced at a fixed rate each year. This system is designed to offer greater advantages to parties that introduce solar power generation facilities earlier. As a result, all the parties competed to introduce the facilities, causing a boom. The purpose of their introduction was not limited to making a contribution to

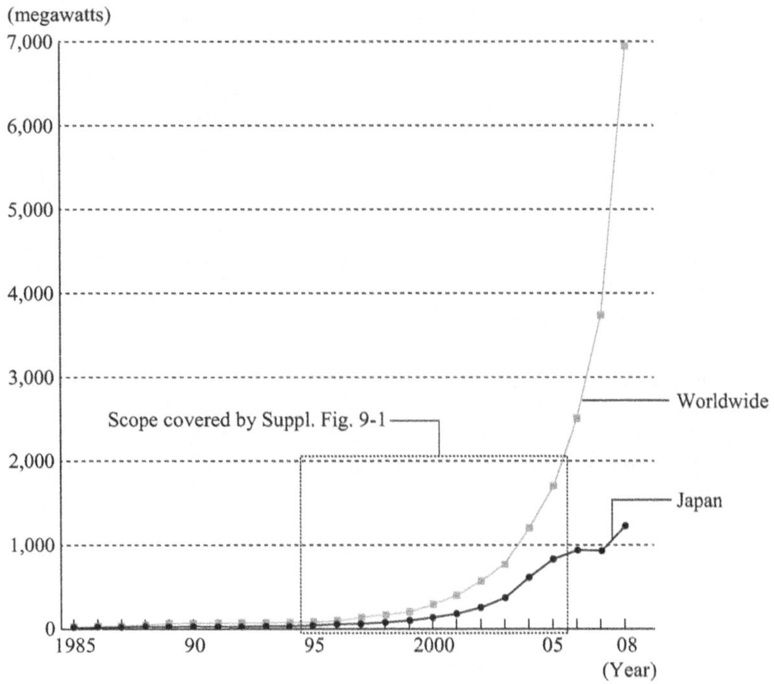

Fig. 9.5 Solar cell production volume in Japan (2). *Source PV News*

environmental and energy issues. Their introduction actually meant investment as well. It is natural for the establishment of solar power generation systems for the purpose of investment to advance at a rapid pace with the emergence of a situation where the purchase of solar cells and the sale of electric power offer a higher return on investment than deposits at banks or the purchase of government bonds.[3] The FIT system changed solar power generation into not only a solution for environmental and energy issues, but also a product for investment that is expected to yield a high return. The FIT system had irreversible and material effects on the development of the solar power generation industry in that a policy goal was achieved with the practical use of an economic incentive.

In the meantime, the Japanese government discontinued subsidies for solar power generation around the same time. The subsidies were terminated in fiscal year 2005. The timing for this step was misguided when compared with German and Spanish policy support for solar power generation that made strides around the same time. For this reason, demand for solar power generation for housing use declined in Japan (Fig. 9.6), causing the volume of solar cells produced by Japanese companies to stagnate during the same period. As a result, demand rose in Germany

[3]Wadaki (2008), p. 20.

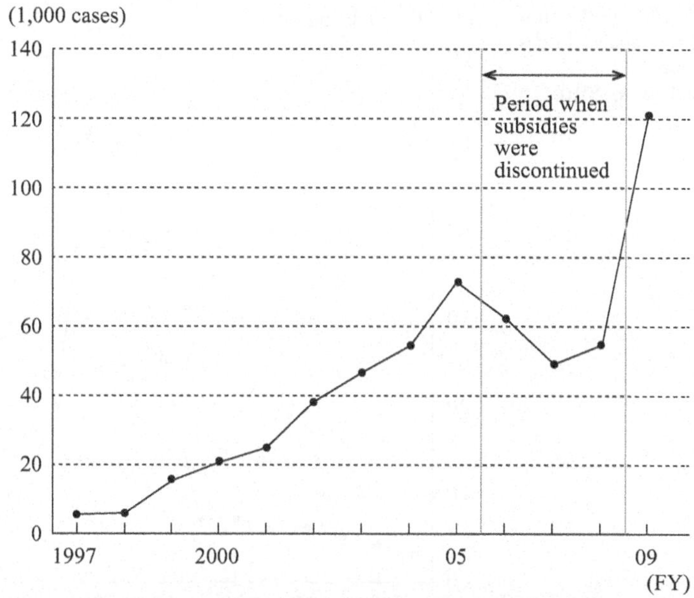

Fig. 9.6 Cases of solar cell introduction to Japanese Houses. *Sources* New Energy Promotion Council, "Survey on the state of solar power generation systems introduced for housing use in fiscal 2008," July 2009; *Nikkei Electronics*, Feb. 8, 2010, p. 38

but fell in Japan, leading to a major difference in the volumes of solar power generation introduced in the two countries. The difference in introduction volumes also affected the manufacturing costs in each company, because the difference has effects on volume efficiency. The discontinuation of subsidies and prices stuck at high levels cooled the Japanese market for solar power generation for housing use. As a result, the operating results for solar cells began falling into the deficit range at some of the companies in Japan.

Observing the rapid market expansion caused by the introduction of the FIT system, many companies in other countries attempted to expand their solar cell output. Naturally, supplies of silicon began to run short and its price to increase when the production of solar power generation systems increased all at once in many countries as crystalline silicon solar cells began to comprise the mainstream. The spot price of crystalline silicon soared about ten times and reached US$400–500 per kilogram in the period from 2007 to 2008 (Fig. 9.7).[4]

Japanese companies worked to develop technologies until the first half of the 2000s, waiting for the day when the solar power generation industry would take off. However, the solar cell market expanded at a more rapid pace than they had anticipated in the second half of the 2000s. Japanese companies misread the speed of

[4]Yamaka (2009), p. 30.

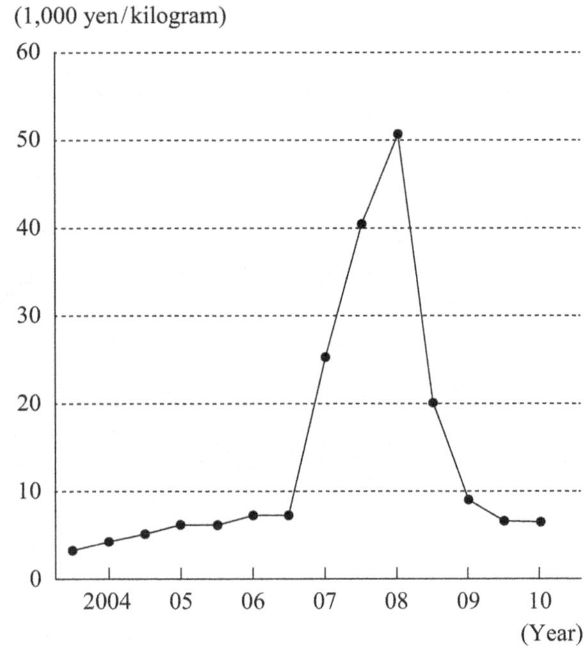

Fig. 9.7 Changes in the spot price of silicon. *Source* Prices are taken from *Nikkei Electronics*, Feb. 8, 2010, p. 35

the market expansion and expanded their production late. This delay in the end invited a decline in their relative standing. As a matter of fact, First Solar, Q Cells, and Suntech Power expanded their production volumes rapidly from 2006 to 2007 when Japanese companies took time to deal with the unexpected situation. These three newcomers obtained funds amounting to tens of billions of yen by listing their shares based on their forecasts for demand expansion triggered by the environmental and energy policies adopted in each country. Using these funds, they made bold investments in manufacturing facilities and signed long-term contracts for procuring raw materials. All three of the companies invested their management resources intensively in solar cells, making the most of their advantages as specialized solar business operators. They worked to bolster their production capability using the funds they procured. Manufacturing cost per product unit falls as the production volume rises. So, the three companies were able to supply relatively inexpensive products to markets that expanded rapidly under the effects of government policies in this way, relying on their cost competitiveness based on the economies of scale.

This competition for rapid, bold investments in solar cells may significantly affect corporate business performance in the future as well. As a matter of course, this possibility does not eliminate competition for cutting down development and manufacturing costs and competition for improving performance. However, to win shares of markets that keep expanding rapidly, investment strategies that make the most of mobility for manufacturing optimal products with technologies at the present time will be required more than technological development that looks ahead solely to the remote future.

9.3 Strategies Adopted by Each Company

As explained above, solar power generation is an industry that has been growing remarkably on the whole. At the individual company level, however, their strategies are not necessarily uniform. The types of solar cells manufactured by each company differ too. They are also attaching importance to different markets. In the following section, I would like to explain the strategies adopted by the leading manufacturers in Japan, namely Sharp, Kyocera, and Sanyo Electric, and their overseas counterparts, including Suntech Power (China) , Q Cells (Germany), and First Solar (the United States). Sharp, Kyocera, Suntech Power, and Q Cells mainly sell polycrystalline silicon solar cells. Among these crystalline silicon solar cell manufacturers, Sharp and Q Cells are working to launch the production of thin-film crystalline silicon solar cells on a full scale to prepare for the tightening of the silicon supply–demand balance, like the situation that occurred from 2007 to 2008. In the meantime, Sanyo Electric is dealing with the heterojunction with intrinsic thin layer system (HIT system), and First Solar is working on compound semiconductors using cadmium telluride (CdTe). The former and the latter offer advantages in the areas of performance and price, respectively.

The key points for comparing the strategies adopted by each company include (1) development (the solar cell type chosen; in other words, crystalline silicon, thin-film silicon or a new type offering higher performance), (2) purchase (the method for procuring silicon used as a raw material; in other words, internal production or purchase from other companies), (3) production (the determination of factory sites and investment scales overseas), and (4) sales (the target market chosen and the development of a distribution network). We can find the course of the solar power generation business strategies adopted by each company by paying attention to the combination of decisions made in the areas of development, purchase, production, and sales.[5] These four areas are interrelated. The procurement of silicon becomes a matter of commercial life or death when crystalline silicon solar cells are positioned in the center. In that case, solar power generation systems for housing use under heavy installation site restrictions become the main targets, because crystalline solar cells are small in terms of surface area, but they offer a high sunlight conversion efficiency. Door-to-door visits, construction companies, sales agents at shopping malls and the like are suited to being their distribution channels. In the meantime, silicon procurement becomes relatively unimportant when thin-film silicon solar cells are positioned in the center because the volume of silicon used in them is small. Conversely, however, the required surface area becomes inevitably large because their sunlight conversion efficiency is lower than that of crystalline solar cells. For that reason, applications such as large power

[5]The descriptions in the remaining portion of Sect. 9.3 are based on *Taiyō denchi sangyō sōran 2007* [Complete Guide to the Solar Cell Industry—2007 Edition] (Tokyo: Sangyō Times Inc., 2007) and *Taiyō denchi sangyō sōran 2010* [Complete Guide to the Solar Cell Industry—2010 Edition] (Tokyo: Sangyō Times Inc., 2009) unless stated otherwise.

plants that are relatively free from area restrictions and are hoping to lower the cost of solar cells have advantages. Accordingly, markets are sought in that direction. Choices, such as the course chosen for focusing operations or the combined use of multiple approaches for diversifying risks, are left up to the strategic judgment of each company.

There are also thin-film solar cells using compounds such as CdTe, in addition to those based on silicon. The thin-film solar cells require no silicon. They offer the advantage of a considerably low manufacturing cost. At the same time, however, their sunlight conversion efficiency is not high. Consumers also have doubts about the harmful substances, such as cadmium, used in some of those thin-film solar cells. As these points suggest, the various technologies each have their own advantages and disadvantages. Finding appropriate markets for them is a way of achieving profits in this industry.

9.3.1 Sharp: Hedging Risks with Multiple Development Approaches and Advancing into the Upstream and Downstream Sections of Value Chains

As explained in Chap. 5, Sharp launched its basic research on solar cells long before the oil crisis of 1959. The company was a solar power generation pioneer in Japan. It remains a top runner today. Sharp began the mass production of modules in 1963 and supplied solar cells to lighthouses in 1966. The company then had its solar cells mounted on a man-made satellite in 1976 and marketed solar pocket calculators for general use in 1980.[6] However, the dramatic growth of Sharp's business began in 1994 with the launch of a subsidy system for solar cells for housing use. Sharp started selling solar cells for housing use that year. The company's solar power generation business began growing rapidly from that point (Fig. 9.8). Sharp had a share of less than 10% of the Japanese market for solar cells until then. However, the company increased the volume of solar cells produced smoothly year after year by making capital investments ahead of its competitors. Sharp finally overtook Kyocera and became the largest solar cell producer in the world in 2000. The company maintained this top position until 2005. Its solar cell segment achieved net sales of 158 billion yen and an operating profit margin of about 10% (equaling its operating income of approximately 15 billion yen) in fiscal year 2005.[7]

However, the markets for solar power generation changed significantly in the period that followed. Demand expanded globally, but Sharp was unable to secure

[6]Sharp (1996), p. 36 and *Taiyō denchi sangyō sōran 2007* [Complete Guide to the Solar Cell Industry—2007 Edition] (Tokyo: Sangyō Times, Inc., 2007), p. 22.

[7]*Taiyō denchi sangyō sōran 2007* [Complete Guide to the Solar Cell Industry—2007 Edition] (Tokyo: Sangyō Times, Inc., 2007), p. 23.

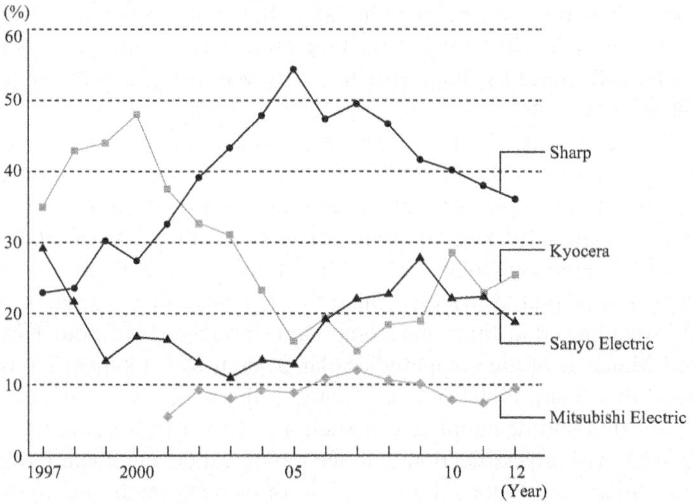

Fig. 9.8 Market shares for leading solar cell manufacturers in Japan (in value terms). *Source* Material prepared by the author based on Nikkei Sangyo Shimbun, ed., *Nikkei Market Shares*, editions for the respective years, (Nikkei Inc.); Yano Research Institute Ltd., ed., *Dictionary of Japanese Market Shares*, editions for the respective years, (Yano Research Institute Ltd.)

sufficient silicon for use as a raw material. As a result, the company failed to achieve its target net sales for fiscal year 2006. Net sales for Sharp's solar cells amounted to 151 billion yen in the following year, fiscal year 2007, but an operating loss of 3.6 billion yen resulted for the company in that year. In fiscal year 2008, net sales grew again, to 157.1 billion yen, but operating loss swelled to 16.1 billion yen, reflecting the contraction of the European market due to FIT reviews and declining panel prices.[8] The delay in raw material procurement and the operating loss for two years in a row prompted the company to take immediate countermeasures.

Sharp had already changed course from the manufacture of crystalline silicon solar cells only and started producing thin-film silicon solar cells in 2005. In other words, the company launched the full-scale manufacture of thin-film solar cells that can be made with a small volume of silicon. Sharp was able to apply thin-film technologies for TFT liquid crystals used for LCD panels at that point. This was an advantage that the company had. Sharp was able to enjoy the economies of scope by establishing thin-film solar cell plants at the same sites as LCD panel plants, sharing infrastructure installations and accumulating the plants of component manufacturers. In 2010, a manufacturing line for thin-film solar cells was completed

[8]*Taiyō denchi sangyō sōran 2010* [Complete Guide to the Solar Cell Industry—2010 Edition] (Tokyo: Sangyō Times, Inc., 2000), p. 44.

at the company's manufacturing plant in Sakai. Including the volume manufactured at an existing plant in Katsuragi, Nara Prefecture, the volume of crystalline and thin-film solar cells made by Sharp rose to 1,210 MW per year (including 550 MW for crystalline solar cells).[9]

Taking a further step, Sharp is advancing the development of compound solar cells that are extremely high in sunlight conversion efficiency as next-generation products, in addition to their crystalline and thin-film counterparts. The company achieved the world's highest sunlight conversion efficiency of 36.9% with a triple-junction compound solar cell using indium and gallium in 2011. This achievement was reported widely. Sharp is handling solar cells of all types, including crystalline, thin-film, and compound varieties. Morimoto Hiroshi, then the General Manager of the company's Solar System Development Division, said, "I feel proud that Sharp is the only company in the world that is dealing with the production, design and development of such a large variety of solar cells."[10]

In addition to risk diversification with technological development using multiple approaches, Sharp has adopted the policy of seizing both the upstream and downstream segments of value chains, not only in the manufacture of cells and modules but also in their purchase and sales. In the upstream operation of raw material procurement, the company began manufacturing silicon on a scale of approximately 1,000 tons per year in 2007 while advancing cooperation with polycrystalline silicon manufacturers in the United States.[11] The idea of stabilizing the silicon supply through vertical integration and preparing for demand fluctuations lies in the background of these steps. In the meantime, in downstream operations, Sharp is undertaking the independent power producer (IPP) project with Enel Green Power in Italy. This is a project for building solar power plants in Italy and undertaking power generation businesses in the country using those plants. Sharp and Enel Green Power signed a two-company joint-venture contract and established a new company for the businesses in 2010.[12]

---

[9]Heavy & Chemical Industries News Agency Editorial Department, ed., *Sekai no Taiyō denchi sangyō* [Solar Cell Industries Around the World]. (Tokyo: The Heavy & Chemical Industries News Agency, 2010), p. 30.

[10]Nikkei Trendy Net, November 8, 2011, http://trendy.nikkeibp.co.jp/article/pickup/20222207/1038545/.

[11]Development Bank of Japan Inc., "Taiyōkō hatsuden o meguru saikin no dōkō: hageshisa o masu kokusai kyōsō to kaihatsu ga susumu hakumaku kei [Recent trends in solar power generation: intensifying global competition and advancing development of thin-film solar cells]," Kongetsu no topikku, no. 122-1, April 23, 2008.

[12]Heavy & Chemical Industries News Agency Editorial Department, ed., *Sekai no Taiyō denchi sangyō* [Solar Cell Industries Around the World]. (Tokyo: The Heavy & Chemical Industries News Agency, 2010), p. 30.

Sharp is aiming to make profits from the businesses associated with solar cells, in addition to the manufacture of cells and modules it has engaged in up to this point, by expanding its operations to the manufacture of silicon used as a raw material and the management of solar power plants, two areas that are highly profitable. The company is finding business opportunities not only in production but also in the upstream raw material segment and the downstream power generation business. The return on sales for Sharp's solar cell segment could not be called high in fiscal year 2009 (the fiscal year ended March 2010), but the segment achieved net sales of 208.7 billion yen in the same year, recovering its operating income to 3.3 billion yen.[13] The solar cell segment posted net sales of 265.5 billion yen and operating income of 2.1 billion yen again in the following year, fiscal year 2010.

The strategies adopted by Sharp can be summarized as follows. First, the company is at the moment focusing its cell development on crystalline solar cells, but is increasing the production of thin-film silicon solar cells that use small volumes of silicon. In addition to handling both crystalline and thin-film products, Sharp is achieving results in the development of compound solar cells that are high in sunlight conversion efficiency. Second, the company is aiming to secure wafers based on long-term contracts and save on silicon in the area of raw material purchases while preparing a system of stable supply in cooperation with polycrystalline silicon manufacturers in the United States. At the same time, the company is advancing the internal production of silicon used in solar cells. The move represents its advance into upstream operations. Third, in the area of production, Sharp is focusing on substantial output expansion in anticipation of increased global demand by taking steps including an increase in thin-film solar cells manufactured at its plant in Sakai. Fourth, Sharp is focusing on solar cells used for public works, in addition to those used for housing, regarding the sales of its products and the cultivation of new demand. The company even made inroads into solar power generation businesses in Europe as well. These moves represent Sharp's advance into downstream operations.

As the author explained above, Sharp is characterized by the strategies of hedging the risk of technological selection with advances in all directions, including crystalline, thin-film and compound solar cells, of securing a wide range in value chains from their upstream to downstream segments, instead of limiting its operation to cell production only, and of turning those segments into revenue sources.

Regrettably, however, the price advantages of thin-film solar cells were weakened because of developments in 2009 and subsequently, including sudden drops in silicon prices and FITsystem revisions in Europe. The conditions surrounding solar power plants in Italy grew severer as well. It appears to be difficult to say at this point that the business environment is shifting in a direction that Sharp wants.

[13]*The Nikkei/Nihon Keizai Shimbun*, May 7, 2010.

9.3.2 *Kyocera: Concentration on Polycrystalline Silicon and Consumer Products*

As described in Chap. 7, Kyocera commenced solar cell research and development in 1975 and started mass-producing solar cells using the polycrystalline silicon cast method as the first company in the world to do so in 1982. The cast method is a method for manufacturing wafers by casting raw silicon and slicing its blocks. The method enables users to reduce the volume of silicon used. In the subsequent period, Kyocera placed polycrystalline silicon solar cells at the center of its technological development and accumulated technologies related to this method. Kyocera launched a solar power generation system for housing use onto the market in 1993. The company became the world leader, ahead of Sharp, in terms of the volume of solar cells manufactured in 1998 and 1999.[14]

In terms of production, Kyocera has prepared a consistent internal production system that covers all processes ranging from silicon casting to wafer slicing to module manufacturing in order to raise production efficiency. We can improve the performance of modules by increasing their sunlight conversion efficiency or by making them larger. We can reduce the number of modules required for the same output by enhancing their performance as well. Kyocera's policy was to improve performance and expand production.

In the purchase of raw silicon, Kyocera had a hard time securing silicon until 2006 as well. Silicon procurement was essential for Kyocera, which targeted crystalline silicon solar cells. In these circumstances, the company made silicon procurement certain and without delay, even when silicon was expensive. Kyocera began expanding its solar cell production rapidly about one year ahead of its rivals, in fiscal year 2007.[15]

In the meantime, the equipping of Toyota's Prius model with Kyocera's solar cells in 2009 deserves attention as a measure for increasing product sales. Kyocera Vice President Tatsumi Maeda made the following statement: "Our solar cells will be mounted on the roof of the Prius after 25 years of research, during which they produced consecutive losses. Our solar cells have moved from the shade to a place in the sun."[16] The 25 years mentioned by Maeda suggests that his company involved itself in the solar cell business following the first oil crisis. Toyota began to rate the solar cells mounted on its car models in 2006. The company came to the conclusion that Kyocera's products were the finest in terms of durability and reliability after comparing the performance of solar cells produced by multiple manufacturers and itself. Kyocera also began around the same time preparing an aggressive sales network, in addition to cultivating new demand. Door-to-door sales

[14]Heavy & Chemical Industries News Agency Editorial Department, ed., *Sekai no Taiyō denchi sangyō* [Solar Cell Industries Around the World]. (Tokyo: The Heavy & Chemical Industries News Agency, 2010), p. 26.

[15]*Nikkei Micro Device*, August 2008, p. 37.

[16]*Nikkei Business*, June 8, 2009, p. 18.

had been a major marketing approach, but the company announced a sales contract with Aeon Co., Ltd. in May 2009, established sales agents (Kyocera Solar Corporation franchisees) at Aeon-operated shopping malls, and stepped up the sales of solar power generation systems for housing use because the number of inquiries about suppliers grew. The number of these sales agents surpassed 100 in April 2010.[17] Kyocera's solar cells achieved operating income of more than 10 billion yen in fiscal year 2009.[18]

We can summarize Kyocera's strategies as follows. First, Kyocera focuses on the cast method for polycrystalline silicon and does not deal with thin-film products, as Sharp does in the area of cell development. The company has viewed polycrystalline silicon solar cells as prospective winners since placing its focus on them. Second, Kyocera has prepared a system for raw material procurement ahead of its competitors and has built a structure that enables it to obtain silicon stably at low prices in exchange for bulk orders. Production concentration on polycrystalline silicon solar cells inevitably leads to this policy. Third, Kyocera has substantially increased the production of cells and modules since 2007 in an attempt to meet expanding demand. Fourth, in the area of product sales, Kyocera bolstered its sales network in cooperation with Aeon and came up with new approaches, including the equipment of the Prius car with its solar cells in partnership with Toyota.

The policies stated above show that Kyocera's strategies are aimed at narrowing its focus on polycrystalline silicon, making the utmost efforts to procure silicon stably by mass-producing products and cultivating markets for new applications, such as automobiles, instead of limiting the market to solar cells for housing use. Kyocera's focus on solar cells for general use centered on polycrystalline products to the end is working favorably in the declining stage for silicon prices. It is a factor behind the company's strong business performance.

9.3.3 Sanyo Electric: Focusing on High-Performance HIT Solar Cells

Like Kyocera, Sanyo Electric commenced solar cell studies in 1975. The company differed from Sharp and Kyocera in that its research began with amorphous semiconductors. Sanyo Electric developed its product in the subsequent period by applying amorphous semiconductor studies to solar cells. For that reason, the company's business began with thin-film solar cells using amorphous silicon. Sanyo Electric became the first company to mass-produce solar cells for pocket calculators in 1980 because its approach was advantageous for manufacturing small solar cells used indoors. Later, the company engaged in research on single crystal solar cells and polycrystalline silicon solar cells in addition to amorphous silicon

[17] Kyocera Website, http://www.kyocera.co.jp/news/2010/0401_ariz.html.

[18] *The Nikkei/Nihon Keizai Shimbun*, May 7, 2010.

solar cells. Sanyo Electric productized HIT solar cells that it had developed in 1990 by combining those technologies in 1997.[19] In this method, amorphous silicon, a Sanyo Electric strong point, is connected to single crystal silicon wafer substrates that excel in sunlight conversion efficiency. Sunlight conversion efficiency at the highest level in the industry characterizes HIT solar cells. In fact, Sanyo Electric achieved the highest sunlight conversion efficiency of 23% in 2009, using a crystalline silicon solar cell of a practical size measuring more than 100 square centimeters. These HIT solar cells are staple products for Sanyo Electric today. The company's solar cell operations posted net sales of 81.1 billion yen in fiscal year 2008. Sanyo Electric has invested a total of 100 billion yen in production buildups in the period since fiscal year 2009. [20]

For procuring silicon used as a raw material, Sanyo Electric established Sanyo Solar of Oregon LLC., a company that manufactures silicon ingots and wafers, in 2009. The establishment of the company was aimed at supplying silicon ingots stably in future where demand for solar cells increases. [21]

Sanyo Electric adopted the policy of selecting technologies to differentiate its products from those of its competitors by making the most of the advantages of high efficiency HIT method solar cells. However, the company went through a lot in dealing with the situation when the silicon supply–demand balance tightened. At that point, Sanyo Electric taxed its ingenuity and used polycrystalline silicon in HIT solar cells that had been based on single crystal silicon because the former used a smaller volume of its raw material. Furthermore, the company started developing thin-film solar cells through a joint project with another company. In 2009, Sanyo Electric established Sanyo Eneos Solar Co., Ltd. a joint corporation with Nippon Oil Corporation (now JX Nikko Nisseki Energy Corporation) to attempt the development and practical application of thin-film microcrystal silicon tandem solar cells, making the most of the two companies' strengths. However, the situation changed as a result of a sharp fall in silicon prices that caused thin-film solar cells to rapidly lose their advantages over their polycrystalline counterparts. The joint corporation was dissolved in effect by the summer of 2010 with the withdrawal of all employees who had been dispatched there from the two parent companies. Sanyo Electric is now waiting for the right time to develop thin-film solar cells.[22] In these circumstances, the company is at the moment focused on high-performance HIT solar cells.

---

[19]*Taiyō denchi sangyō sōran 2007* [Complete Guide to the Solar Cell Industry—2007 Edition] (Sangyō Times, Inc., 2007), p. 36.

[20]Heavy & Chemical Industries News Agency Editorial Department, ed., *Sekai no Taiyō denchi sangyō* [Solar Cell Industries Around the World]. (Tokyo: The Heavy & Chemical Industries News Agency, 2010), p. 28.

[21]Heavy & Chemical Industries News Agency Editorial Department, ed., *Sekai no Taiyō denchi sangyō* [Solar Cell Industries Around the World]. (Tokyo: The Heavy & Chemical Industries News Agency, 2010), p. 29.

[22]*The Nikkei/Nihon Keizai Shimbun*, July 24, 2010.

We can summarize Sanyo Electric's strategies as follows. First, Sanyo Electric positions high-performance HIT solar cells at the center of its cell development. The company is expanding its solar cell production by concentrating its efforts on products made using this HIT method. Sanyo Electric proceeded with a project for developing thin-film solar cells at one point, but it stopped their development and production when the prices of silicon plunged. Second, in the area of raw material purchases, the company built an ingot production system in the United States to prepare for the future tight supply–demand balance while waiting for silicon prices to drop without hurrying procurement. Furthermore, the company worked to reduce the volume of silicon consumed through measures including the use of polycrystalline silicon in HIT solar cells.[23] Third, Sanyo Electric positioned HIT solar cells at the center of production and advanced a plan based on large investments for rapidly doubling their output in order to meet growing demand worldwide. Fourth, to sell products, the company is promoting their high performance, taking advantage of the characteristic large volume of power generated in a small installation space. Sanyo Electric successfully differentiated its products using HIT solar cells when newly established companies entered this market one after another and the market declined temporarily due to the FIT system reviews and a business slump.

9.4 Trends Among Overseas Companies: How Long Will Their Rapid Advance Continue?

In the following section, I would like to summarize the trends among specialist solar cell companies in other countries that have been making their presence felt rapidly in the course of the last ten years or so.

9.4.1 Q Cells

Established as a company specializing in solar cells in Germany in 1999, Q Cells started producing crystalline silicon solar cells on a commercial basis with 19 employees in 2001. The production volume was just 0.4 MW in the initial year, but the company rapidly expanded its production capacity in the subsequent period, taking over from Kyocera as the world's second largest manufacturer in 2005. Q Cells outproduced Sharp and became the world's largest producer in 2007. The company retained this position in the following year.

[23] *Nikkei Micro Device,* August 2008, p. 38.

During this time, Q Cells listed its shares on the Frankfurt and other stock exchanges in 2005 and raised funds totaling 313 million euro (approximately 42.5 billion yen). Polycrystalline silicon solar cells comprise key products for the company. To deal with the tightening the silicon supply–demand balance, Q Cells established a subsidiary through which it is developing thin-film solar cells using silicon and compounds, respectively. Q Cells achieved its remarkable business growth in this way and become the world's largest solar cell manufacturer within a short period of time. Increased demand in Germany, the country of the company's origin, supported by the FIT system, was in the background of this achievement.

However, the revision of the FIT system that began in 2009 unexpectedly forced Q Cells into a difficult situation. This situation occurred because Germany lowered the purchase prices. Net sales for Q Cells decreased 36% to minus 362.5 million euro (around 52 billion yen) due to changes in other external conditions for the solar power generation business during this period, including the worsening of the global economy, rises in solar cell prices and offensives mounted by companies in emerging nations. As a result, the company stopped operating manufacturing lines in Germany by the end of 2009 and reduced its workforce by 500 people. Q Cells has continued to carry out business restructuring in the subsequent period, working to cut costs with screened investments and raise sunlight conversion efficiency. [24]

9.4.2 Suntech Power

Founded by Shi Zhengrong, a solar power generation researcher, Suntech Power is the largest solar cell manufacturer in China. At present, the company has manufacturing plants for cells and modules in Luoyang, Qinghai, and Nagano in Japan, in addition to Wuxi where it is headquartered. Suntech Power also owns a manufacturing plant for thin-film solar cells in Shanghai. Established in China in 2001, the company launched a 15 MW production line in 2003 and expanded its production smoothly in the subsequent period. Suntech Power raised US$400 million (approximately 43 billion yen) by performing an initial public offering (IPO) on the New York Stock Exchange in 2005, the first Chinese company to do so. This is a point about the company that deserves special mention. In the upstream area of raw material procurement, Suntech Power signed a long-term supply contract for polycrystalline silicon with a silicon manufacturer in 2005 to realize a system of stable procurement at low cost. However, the company was unable to produce up to its capacity in fiscal year 2006 owing to a shortage of silicon supplied as a raw material. For that reason, Suntech Power signed a ten-year supply contract for polycrystalline silicon with Monsanto Electronic Materials Company (MEMC) in

[24]Heavy & Chemical Industries News Agency Editorial Department, ed., *Sekai no Taiyō denchi sangyō* [Solar Cell Industries Around the World]. (Tokyo: The Heavy & Chemical Industries News Agency, 2010), p. 145.

the United States in 2007. Under the long-term contract, the company is seeking to secure the raw material stably for a long period of time. In the meantime, for downstream sales operations, Suntech Power acquired MSK (now Suntech Power Japan Corporation), a solar cell module manufacturer in Japan, in 2006. The company's production capacity surpassed 1 GW in 2008, making the company the first solar cell manufacturer in the world to reach that level.

However, Suntech Power experienced a slowdown of demand in the period that followed. The company dismissed 400 employees, or 30% of its workforce, at one point when its utilization rate plunged to 50%.[25] The shipment volume subsequently recovered, but net sales for Suntech Power dropped year on year because product prices fell during the same period. In these circumstances, the company reviewed the structure of its shipments that placed disproportionate emphasis on Europe and worked on expanding sales in the United States and Japan. In the Japanese market, Suntech Power has been selling its products through stores operated by Yamada-Denki Co. Ltd. since 2008, using Suntech Power Japan as a foothold. The company later expanded this approach to sell its solar cells in all stores under the direct management of Yamada-Denki. Taking further steps, Suntech Power signed contracts for the installation of large-scale solar power generation systems with the state government and local governments in China, anticipating a future increase in demand in the Chinese market. Demand for solar power generation in China is low at present, and China is exporting 98% of solar cells produced in the country. However, the day when the domestic market in China starts growing is still in the future. Suntech Power is taking steps in anticipation of the arrival of that day.

9.4.3 First Solar

First Solar was founded in 1999 with Solar Cells, a company that had been established in the first half of the 1990s, as its predecessor. First Solar developed specialized technologies for the vapor deposition of CdTe and CdS on glass with support received from the National Renewable Energy Laboratory (NREL). The company launched commercial production in 2002. After listing its shares on the NASDAQ market in 2006, First Solar expanded its production rapidly, overtaking Q Cells and achieving the world's largest production in 2009. The sunlight conversion efficiency for the company's products is not necessarily high, at around 10%. However, First Solar's products stand out with their strong prices at the lowest level among solar cell modules in the world. The company can complete a module in the extremely short period of two and a half hours by performing work

[25]Heavy & Chemical Industries News Agency Editorial Department, ed., *Sekai no Taiyō denchi sangyō* [Solar Cell Industries Around the World]. (Tokyo: The Heavy & Chemical Industries News Agency, 2010), p. 54.

such as the delivery of glass substrates, thin-film deposition, and evaluation in sequential processes. No company in Japan has attempted to increase the production of this type of solar cells because Cd (cadmium) is a substance that is highly toxic in an isolated state and it had an image of past pollution. However, First Solar did not hesitate. The company says that its solar cells use compounds instead of cadmium in an isolated state, and the compounds do not spread into the environment unless the modules are damaged, as they are sealed with glass at both ends. First Solar promises to recover used modules through its own efforts as well. For this operation, the company built a system for securing a fund at the point of module sales and ensuring recovery and recycling by outsourcing the administration of the fund to a third party. First Solar rapidly gained the position of the world's largest solar cell manufacturer with this system of guaranteed module recovery and the inexpensive nature of its products.

First Solar, a company capitalizing on no silicon use and low prices, increased its net sales and profits further in 2009, when other manufacturers faced tough conditions. In a further step, the company stepped up its production capacity by establishing module manufacturing bases in Malaysia and Germany. In the meantime, First Solar is building large solar power plants in China, the United States, and Canada to expand its power supply.

These three companies boosted their net sales and operating income sharply in the second half of the 2000s. However, subsequent developments produced different business results from fiscal year 2009 (Fig. 9.9). Net sales for Suntech Power and Q Cells fell, with financial crises after the global financial crisis in 2008 and the revision of the FIT system as triggers. In particular, Q Cells fell into a state of operating loss. The rapid loss of competitiveness due to relatively high module prices attributable to comparatively expensive silicon based on long-term contracts caused the loss at Q Cells. [26] In this environment, only First Solar achieved growth in revenues and profits. The company has swept the board. The reason for this is that First Solar is the only company in the industry that positions CdTe solar cells using compound semiconductors as its main business. CdTe solar cells compare unfavorably with their crystalline counterparts in terms of sunlight conversion efficiency. However, they increased net sales under a business slump with their overwhelmingly low manufacturing cost, high manufacturing speed and associated low module prices as their advantages. Nonetheless, we can no longer expect the sunlight conversion efficiency of CdTe solar cells to improve. We cannot be sure how long First Solar will remain in the global leader's position. In any case, the solar cell companies that maintained a competitive advantage amid violent silicon price fluctuations were First Solar, which was characterized by low prices, and Sanyo Electric, which was characterized by high performance. That was an interesting result.[27]

[26] *Shūkan Ekonomisuto* [Weekly Economist], August 3, 2010, p. 30.
[27] *Nikkei Business*, February 8, 2010, p. 43.

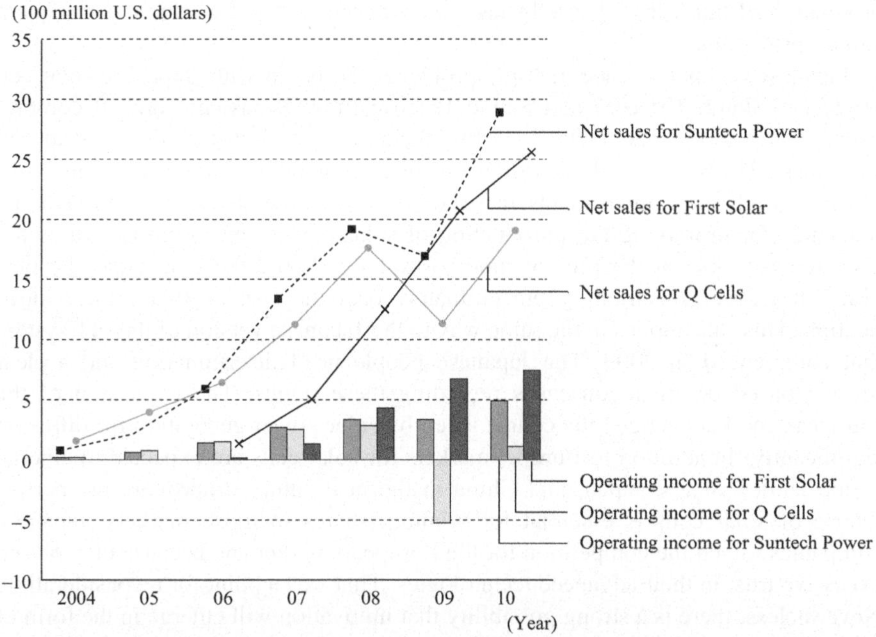

Fig. 9.9 Changes in Business Results for Leading Solar Cell Manufacturers Overseas. *Source* Material prepared by the author based on the annual reports of each company

9.5 Conclusion

I would like to answer the question I raised in Chap. 1 in the following way, based on the outcome of my analysis of the trends in the solar power generation industry, which has undergone significant changes in recent years. Why did Japan lose its position as the global leader so rapidly? The solar power generation industry left the previous stage of step-by-step commercialization through repeated research and development in the middle of the 2000s. The industry has transformed itself rapidly into a profitable business. The FIT system in Germany made it possible for companies to turn the installation of solar power generation systems into a product for investment. This change caused demand to increase rapidly in Europe. In response to this golden opportunity, new companies specializing in solar cells in each country, such as Q Cells, Suntech Power, and First Solar, increased their output substantially using funds they raised by means including share flotation, and leaped rapidly into the position of the world's leading solar cell producers. In the meantime, Japanese manufacturers took their time to ride the wave because of delays in their silicon procurement. As a result, their production share dropped. At the same time, a threat known as the tight silicon supply–demand balance encouraged the companies to plan the large-scale production of thin-film solar cells, which are low in terms of sunlight conversion efficiency and manufacturing cost. However, the production of thin-film solar cells has not advanced as expected due to subsequent silicon price falls.

Japanese companies face multiple problems. To begin with, Japanese solar cell producers Sharp, Kyocera and Sanyo Electric (now Panasonic) are all comprehensive electronics manufacturers, while their major overseas rivals are solar cell specialists. The presence of other divisions in need of similar large investments may prevent rapid, large investments in solar cells by comprehensive manufacturers. Subsidies for promoting the introduction of solar power generation to houses are also indispensable at the present time. How long should these subsidies be provided, however? Electricity users ultimately bear the cost of solar power introduction. This rule applies in the same way to the Japanese version of the FIT system that commenced in 2009. The Japanese people need determination and a clear resolve based on their consensus regarding these points. The expansion of the European market changed the course taken by solar power generation for diffusion significantly. In addition to Europe, markets for solar cells are expected to emerge in the United States, Japan, and China in the near future. Improving the performance of solar cells is essential for winning shares in those markets. Japanese companies joined the competition for the European market late because they placed excessive trust in their advanced technologies. That was a point for reconsideration. Nevertheless, there is a strong possibility that innovation will emerge in the form of a new type of solar cell that offer higher sunlight conversion efficiency than current silicon-based products. It goes without saying that the studies and development of advanced technologies are important in that respect. In the first place, solar power generation systems that take a long time before they pay off, such as 20 years, require durability. Inexpensive solar power generation systems with low performance abound and compete with each other at present. This is a situation that should be improved urgently.

In a solar power generation business for which the market is expanding and competition is global, the future of Japanese companies depends on the business models they build. In these circumstances, each company must sustain aggressive investments to ensure that solar cells will not follow in the steps of semiconductors and liquid crystal displays before them. At the same time, the Japanese government must continue its policy assistance in order to enable the country's solar power generation industry to maintain and improve its international competitiveness from the viewpoint of making contributions to Japan's environmental and energy policies. A green innovation has just begun.[28]

---

[28]This chapter has been prepared by refining and correcting two articles written by the author. They are Shimamoto (2010, 2012).

References

Tawada, Y. (Ed.). (2011). *Puro ga oshieru taiyō denchi no subete ga wakaru hon: Saishin no taiyō denchi gijutsu o kiso kara manandemiyō: Shijō saikyō karā zukan* [*A book by a professional that tells you everything about solar cells: Let's learn solar cell technologies from the basics: The best illustrated dictionary in color*]. Tokyo: Natsumesha.

Sharp. (1996). *Taiyō denchi no sekai kaiteiban* [*The world of solar cells revised edition*]. Self-pub.

Shimamoto, M. (2010). Taiyōkō hatsuden no hanseiki: sankangaku niyoru gijutsu kaihatsu to shijō kaitaku no chōsen [A half century of solar power generation: Joint academy, industry and government attempts to develop technologies and markets]. *Hitotsubashi Business Review 58*(1).

Shimamoto, M. (2012). Nihon no taiyōkō hatsuden sangyō wa fukkatsu suruka [Will the Japanese solar power generation industry come to life again?]. *Hitotsubashi Business Review 59*(4).

Wadaki, T. (2008). *Bakuhatsu suru taiyō denchi sangyō: Nijūgo-okuen shijō no genjō to mirai* [*The exploding solar cell industry: The current state and future of a 25-trillion yen market*]. Tokyo: Tōyō Keizai Inc.

Yamaka, K. (2009). *Sorā uōzu: Gekidō-suru Taiyō denchi bijinesu no zenbō* [*Solar wars: The complete story of the rapidly changing solar cell businesses*]. Tokyo: Energy Forum.

Bibliography

Publications

Akagawa, M. (2006). *Kōchikushugi o saikōchiku suru* [*Reconstructing constructionism*]. Tokyo: Keiso Shobo.

Allison, G. T. (1971). *Essence of decision: Explaining the Cuban missile crisis*. Boston: Little.

Almond, G. A. (1950). *The American people and foreign policy* (1st ed.). New York: Harcourt.

Anchordoguy, M. (1989). *Computers Inc.: Japan's challenge to IBM*. Cambridge, Mass.: Council on East Asian Studies, Harvard University Press.

Arakawa, A. (1999). *Shisōshi no nakano kindai keizaigaku: Sono shisōteki keishikiteki kiban* [*Modern economics in the history of ideas: Their ideological and formal foundation*]. Tokyo: Chuōkōronsha.

Berger, P. L., & Luckmann, T. (1967). *The social construction of reality: A treatise in the sociology of knowledge*. New York: Anchor Books.

Blau, P. M., & Meyer, M. W. (1971). *Bureaucracy in modern society* (2nd ed.). New York: Random House.

Blau, P. M. (1963). *The dynamics of bureaucracy: A study of interpersonal relations in two government agencies* (2nd ed.). Chicago: University of Chicago Press.

Blaschke, S. (2008). *Structures and dynamics of autopoietic organizations: Theory and simulation*. Wiesbaden: Gabler.

Boulding, K. E. (1956). General system theory: The skeleton of science. *Management Science, 2*(3).

Burr, V. (1995). *An introduction to social constructionism*. London, New York: Routledge.

Callon, S. (1995). *Divided sun: MITI and the breakdown of Japanese high-tech industrial policy, 1975–1993 (Stanford*. California: Stanford University Press.

Clark, B. R. (1960). *The open door college: A case study*. New York: McGraw Hill.

Denki Shimbun. (Ed.). (1991). *Shōgen daiichiji sekiyu kiki: Kiki wa sairaisuruka?* [*Testimony from the first oil crisis: Will there be another crisis?*]. Tokyo: Nihon Denki Kyōkai Shimbunbu [The Newspaper Division of the Japan Electric Association].

Denpatsu. (1984). *Denpatsu 30-nen shi* [*30 year history of Denpatsu*] (30th edn.). Year History Editorial Committee. Tokyo: Denpatsu.

Den'yūkai (A social gathering at the electrotechnical laboratory under the AIST). (1995). In (Ed.), *Enerugī kenkyūsha eno messēji senpai kara kōhai e: Atarashii mirai no kensetsu no tameni* [*A message from senior energy researchers to their juniors: For building the new future*]. Tokyo: Powersha.

M. Shimamoto, *National Project Management*,
Advances in Japanese Business and Economics 25,
https://doi.org/10.1007/978-981-15-3180-4

Electrotechnical Laboratory 100th Anniversary Commemorative Projects Executive Committee 100 Year History Committee. (Ed.). (1995). *Denshi gijutsu sōgō kenkyūjo hyakunenshi* [*100 year history of the electrotechnical laboratory*]. Tokyo: Electrotechnical Laboratory 100th Anniversary Commemorative Projects Supporter's Association.

Electrotechnical Laboratory of the Agency of Industrial Science and Technology. (1991). *Denshi gijutsu sōgō kenkyūjo saikin no junen shi soritsu hyakushūnen kinen shi* [*The last 10 years at the electrotechnical laboratory in the journal commemorating the 100th anniversary of the laboratory's establishment*]. Self-pub.

Elster, J. (1983). *Explaining technical change: A case study in the philosophy of science*. Cambridge Cambridgeshire, New York, Oslo: Cambridge University Press.

Evan, W. M. (1976). *Organization theory: Structures, systems, and environments*. New York: Wiley.

Fransman, M. (1999). *Visions of innovation: The firm and Japan*. New York: Clarendon Press.

Fuji Keizai. (1997). *'97 Taiyō denchi kanren shijō no zenbō* [*An overall picture of markets related to solar cells in 1997*]. Tokyo: Fuji Keizai.

Geological Survey of Japan Editorial Committee. (Ed.). (1982). *Chishitsu chōsasho hyakunenshi* [*Hundred year history of the geological survey of Japan*]. Tokyo: Kōgyō gijutsuin Chishitsu Kenkyūjo Soritsu Hyakushūnen Kinen Kyōsan Kai [Geological Survey of Japan 100th Anniversary Commemoration Supporter's Association].

Gouldner, A. W. (1954). *Patterns of Industrial Bureaucracy*. Glencoe, Ill.: Free Press.

Gouldner, A. W. (1954). *Wildcat strike*. Yellow Springs, Ohio: Antioch Press.

Gusfield, J. R. (1955). Social structure and moral reform: A study of woman's christian temperance union. *American Journal of Sociology, 61*(3).

Hatch, M. J., & Cunliffe, A. L. (2006). *Organization theory: Modern, symbolic, and postmodern perspectives* (2nd edn). New York: Oxford University Press.

Habermas, J. (1970). *Zur Logik der Sozialwissenschaften*. Frankfurt a. M: Suhrkamp.

Heavy & Chemical Industries News Agency Editorial Department. (Ed.). (2010). *Sekai no Taiyō denchi sangyō* [*Solar cell industries around the world*]. Tokyo: The Heavy & Chemical Industries News Agency.

Hilsman, R. (1967). *To move a nation: The politics of foreign policy in the administration of John F. Kennedy* (1st edn.). Garden City, New York: Doubleday.

Ishizaki, S., & Hirono, T. (Eds.). (1985). *Nijūisseiki eno enerugī: gijutsu ga hiraku shin enerugī* [*Energy for the 21st century: New energy developed by technologies*] (p. 69). Tokyo: Tsūshō Sangyō Chōsakai.

Hoffmann, S. (1965). Restraints and choices in American foreign policy. In S. Hoffmann (Ed.), *Daedalus* 91, no. 4 (1962). Reproduced in *The state of war: Essays on the theory and practice of international politics*. New York: Praeger.

Horigome, et al. (1972). Taiyō hatsuden shisutemu no teian [A proposal for a solar power generation system]. *Denshi gijutsu sōgō kenkyūjo ihō* [*Electrotechnical Laboratory Journal*], *36*(10).

Huntington, S. P. (1961). *The common defense: Strategic programs in national politics*. New York: Columbia University Press.

Imada, T. (1986). *Jiko soshikisei: Shakai riron-no fukkatsu* [*Self-organization: Revival of social theories*]. Tokyo: Sōbunsha.

Imai, K. (1982). Shin'enerugī Kaihatsu no senryaku to soshikiron. Institute of Business Research, Hitotsubashi University, Discussion Paper no. 103.

Imai, K. (1985). Sentan gijutsu bun'ya ni okeru sangyōseisaku. In I. Miyazaki & T. Usui (Eds.), *Sentan gijutsu to Nihon Keizai* [*Advanced technology and Japanese economy*]. Tokyo: Nippon Hyōron sha.

Investigative Committee on Industrial Technologies. (Ed.). (1974). *Gijutsu kaihatsu josei seido to tsūsanshō no ōgata purojekuto* [*System for assisting technological development and large projects by the ministry of international trade and industry*]. Tokyo: Sangyō Kagaku Kyōkai.

Itami, H., et al. (Eds.), (1998). *Kēsubukku Nihon kigyō no keiei kōdō 4: Kigyōka no gunzō to jidai no ibuki* [*Casebook 4 on administrative actions taken by Japanese companies: A group of entrepreneurs and the signs of the times*. Tokyo: Yuhikaku.

Itami, H. (1999). *Ba no manejimento: Keiei no shin paradaimu* [*Management of places: New paradigm for business administration*]. Tokyo: NTT Publishing.

Itami, H. (2005). *Ba no ronri to manejimento* [*The logics and management of places*]. Tokyo: Tōyō Keizai Inc.

Itō, T. (1998). *Tekunoporisu seisaku no kenkyū* [*The study of technopolis policy in Japan*]. Tokyo: Nippon Hyōron sha.

Janowitz, M. (1960). *The professional soldier: A social and political portrait*. Glencoe, Ill.: Free Press.

Kagono, T. (1988). *Soshiki ninshiki ron: Kigyō ni okeru sōzō to kakushin no kenkyū* [*Theory on organizational perception: Research on development and innovation at companies*]. Tokyo: Chikura-Shobo.

Kahn, H. (1965). *On escalation: Metaphors and scenarios*. New York: Praeger.

Kanai, T., Satō, I., Kunda, G. , & Van Maanen, J. (2010). *Sosiki esunogurafī* [*Organizational ethnography*]. Tokyo: Yuhikaku.

Kanamori, H., Kenjirō, A., & Moriguch, C. (Eds.), (1990). *Keizai jiten* [*Yuhikaku dictionary of economic terms*]. Tokyo: Yuhikaku.

Kato, T. (2011). *Gijutsu shisutemu no kōzō to kakushin: Hōhōronteki shiza ni motozuku keieigaku no tankyū* [*The structure and innovation of technological systems: The pursuit of business administration based on a methodological viewpoint*]. Tokyo: Hakuto-Shobo.

Kikkawa, T., & Tsūshō sangyō seisakushi hensan iinkai [Editorial Committee on the History of Japan's Trade and Industry Policy]. (Eds.), (2011). *Tsūshō sangyō seisakushi 1980–2000* [*History of Japan's Trade and Industry Policy 1980–2000*] (Vol. 10) Natural Resources and Energy Policy. Tokyo: Keizai Sangyō Chōsakai.

Kikkawa, T. (2008). Nihon no genshiryoku hatsuden: sono rekishi to kadai. *Hitotsubashi Review of Commerce and Management, 3*(1).

Kikkawa, T., Shimamoto, M., Suzuki, K., Tsuboyama, Y., & Hirano, S. (Eds.), (2012). *Idemitsu Kōsan no jiko kakushin* [*The renewal of Idemitsu Kōsan*]. Tokyo: Yuhikaku.

Kikuchi, M. (Ed.). (1982). *Amorufasu handōtai no kiso* [*The basics of amorphous semiconductor*]. Tokyo: Ohmsha.

Kimura, O. (2007). *Ondanka bōshi seisakuron (9): Kokka gijutsu kaihatsu purojekuto* [*Discussion of policies for preventing global warming (9): National project for developing technologies*]. Tokyo: Central Research Institute of Electric Power Industry.

Kitano, T. (1996). *Keieigaku genron: Atarasii kachi taikei no sozō* [*Principles of business management: Creating a new value system*]. Tokyo: Tōyō Keizai Inc.

Komiya, R., Okuno, M., & Suzumura, K. (Eds.). (1984). *Nihon no sangyō seisaku* [*Industrial policy of Japan*]. Tokyo: University of Tokyo Press.

Kuwano, Y. (1985). *Amorufasu: Fukashigi na hishōshitsu busshitsu* [*Amorphous: A mysterious amorphous material*]. Tokyo: Kōdansha.

Kuwano, Y. (1992). *Taiyō denchi o tsukaikonasu: Taiyō denchi ga hiraku shinjidai* [*Using solar cells efficiently: A new era of solar cells will unfold*]. Tokyo: Kodansha.

Kuwano, Y. (2011). *Taiyo denchi wa donoyōni hatsumei sare seichōshitanoka: Taiyō denchi kaihatsu no rekishi* [*How solar cells were invented and how they grew: The history of solar cell development*. Tokyo: Ohmsha.

Kyocera Corporation Solar Energy Division. (Ed.). (1994). *Taiyō enerugī eno chōsen: Taiyō denchi no jidai ga yattekita* [*A challenge to solar energy: The era of solar cells began*]. Tokyo: Seibunsha.

Lindblom, C. E. (1965). *The intelligence of democracy: Decision making through mutual adjustment*. New York: Free Press.

Little, D. (1991). *Varieties of social explanation: An introduction to the philosophy of social science*. Boulder: Westview Press.

Luhmann, N. (1968). *Zweckbegriff und Systemrationalität: Über die Funktion von Zwecken in sozialen Systemen*. Tübingen: J.C.B. Mohr.

Meadows, D. H., Meadows, D. L., Randers, J., & Behrens, W. W., III. (1972). *The limits to growth: A report for the club of Rome's project on the predicament of mankind*. New York: Universe Books.

Merton, R. K. (1949). *Social theory and social structure: Toward the codification of theory and research*. Glencoe, Ill.: Free Press.

Messinger, S. L. (1955). Organizational transformation: A case study of declining social movement. *American Sociological Review, 20*(1).

Michels, R. (1911). *Zur Soziologie des Parteiwesens in der modernen Demokratie: Untersuchungen über die oligarchischen Tendenzen des Gruppenlebens*. Leipzig: Klinkhardt.

Miller, R., Lessard, D. R., & IMEC Research Group. (2000). *The strategic management of large engineering projects: Shaping institutions, risks, and governance*. Cambridge, Mass.: MIT Press.

MITI Editorial Committee on the History of Japan's Trade and Industry Policy. (Ed.). (1991). *History of Japan's trade and industry policy, vol. 13, (IV) the age of diversification (1971–1979) Part 2*. Tokyo: Tsūshō Sangyō Chōsakai.

MITI Kōgyō gijutsuin Kenkyū kaihatsukan shitsu [MITI AIST, Office for Research and Development]. (Ed.). (1987). *Ōgata Purojekuto nijūnen no ayumi: Wagakuni sangyō gijutsu no ishizue o kizuku* [*The course taken by large projects in the last 20 years: Laying foundations for industrial technologies in Japan*]. Tokyo: Tsūshō Sangyō Chōsakai.

MITI Kōgyō gijutsuin Kenkyū kaihatsukan shitsu. (1973, July). *Atarashii kurīn enerugī gijutsu no kaihatsu keikaku: Sanshain keikaku* [*Development of new clean energy technology: The sunshine project*]. Self-pub.

MITI Kōgyō gijutsuin Sanshain keikaku suishin honbu [MITI AIST Sunshine Project Promotion Office] (Ed.). (1980). *Kagayakeru taiyō enerugī: Sanshain Keikaku – taiyō enerugī kenkyū kaihatsu no genjō* (rev. edn.). Tokyo: Ōkurashō Insatsukyoku.

MITI Kōgyō gijutsuin Sanshain keikaku suishin honbu. (Ed.). (1984). *Sanshain keikaku jūnen no ayumi* [*Ten year history of the sunshine project*]. Tokyo: Sanshain keikaku jusshūnen kinen jigyō suisin konwakai [Sunshine Project 10th Anniversary Commemorative Projects Promotion Committee].

MITI Kōgyō gijutsuin Sanshain keikaku suishin honbu. (Ed.). (1980). *Sanshain keikaku no kasokuteki suishin senryaku: Sangyō gijutsu shingikai shin enerugī gijutsu kaihatsu bukai chūkan hōkoku o chūshin toshite* [*Acceleratory promotion strategies for the sunshine project: Based primarily on an interim report by the industrial technology advisory committee's new energy technology development subcommittee*]. Tokyo: Tsūsan Seisaku Kōhōsha.

MITI Kōgyō gijutsuin Sanshain keikaku suishin honbu. (Ed.). (1974). *Sanshain keikaku: Shin'enerugī gijutsu eno chōsen* [*The sunshine project: A challenge for new energy technologies*]. Tokyo: Daiichi Hōki Shuppan.

MITI Kōgyō gijutsuin Sanshain keikaku suishin honbu. (1976). *FY1975 research outcomes report*. MITI Kōgyō gijutsuin: Tokyo.

MITI Kōgyō gijutsuin Sanshain keikaku suishin honbu. (1997). *Nyū Sanshain Keikaku*. Tokyo: Nihon sabgyō shinkō kyōkai [Japan Industrial Technology Association].

MITI Kōgyō gijutsuin Sanshain keikaku suishin honbu. (1981). *Shōwa gōjugo-nendo seika hōkokusho* [*FY1980 Annual Report*]. Self-pub.

MITI Kōgyō gijutsuin Sanshain keikaku suishin honbu. (1985). *Showa gojūkyū-nendo seika hōkokusho taiyōhatsuden no kachi hyōka ni kansuru chōsa kenkyū* [*FY1984 Annual Report*]. Self-pub.

MITI Kōgyō gijutsuin Sanshain keikaku suishin honbu. (1986). *Shōwa rokujū nendo seika hōkokushotaiyōhatsuden no kachi hyōka ni kansuru chōsa kenkyū (2)* [*FY1985 Annual Report*]. Self-pub.

MITI Kōgyō gijutsuin Sanshain keikaku suishin honbu. (1982). *Shōwa gōjuroku-nendo seika hōkokusho* [*FY1981 Annual Report*]. Self-pub.

MITI Kōgyō gijutsuin Sanshain keikaku suishin honbu. (1987). *Shōwa rokujūichi-nendo seika hōkokusho* [*FY1986 Annual Report*]. Self-pub.

MITI Kōgyō gijutsuin Sanshain keikaku Suishinhonbu. (Ed.). (1980). *Kagayakeru taiyō enerugī: Sanshain Keikaku – taiyō enerugī kenkyū kaihatsu no genjō* (rev. edn.). Tokyo: Ōkurashō Insatsukyoku.

MITI Kōgyō gijutsuin Sanshain keikaku Suishinhonbu (Ed.). (1984). *Sanshain keikaku jūnen no ayumi* [*Ten year history of the sunshine project*]. Tokyo: Sanshain keikaku jusshūnen kinen jigyō suisin konwakai.

MITI Kōgyō gijutsuin. (Ed.). (1974). *Shin'enerugī gijutsu kenkyū kaihatsu keikaku (Sanshain keikaku)* [*New energy technology development project (The Sunshine Project)*]. Tokyo: Nihon Sangyō Gijutsu Shinkō Kyōkai.

MITI Kōgyō gijutsuin. (Ed.). (1985). *Kōgyō gijutsuin nenpō*. Tokyo: MITI Kōgyō Gijutsuin.

MITI Kōgyō gijutsuin. (Ed.). (n.d.). *Kōgyō gijutsuin shōkai* [*Introducing the AIST*] (FY1973 Edition). Self-pub.

MITI Kōgyō gijutsuin. (Ed.). (n.d.). *Kōgyō gijutsuin shōkai* [*Introducing the AIST*] (FY1997 Edition). Self-pub.

MITI Kōgyō gijutsuin. (Ed.). (n.d.). *Wagakuni no sangyō gijutsu seisaku no enkaku* [*History of the industrial technology policy of Japan*]. Self-pub.

MITI Kōgyō gijutsuin, Sanshain keikaku Suishinhonbu. (Ed.). (1984). *Sanshain keikaku jūnen no ayumi* [*Ten year history of the sunshine project*]. Tokyo: Sanshain keikaku jusshūnen kinen jigyo suisin konwakai.

MITI Kōgyō gijutsuin. (1973, June). *Shin'enerugī gijutsu kaihatsu keikaku (Sanshain keikaku)* [*New energy technology development project (The Sunshine Project)*]. Memorandum.

MITI Sōgō enerugī taisaku suishinhonbu jimukyoku [Energy Measures Promotion Division Secretariat]. (Ed.). (1980). *Sekiyudaitai enerugīhō no kaisetsu* [*An explanation of the law concerning promotion of the development and introduction of alternative energy*]. Tokyo: Tsūshō Sangyō Chōsakai.

MITI. (Ed.). (1979). *Nijūisseiki eno enerugī senryaku*. Tokyo: Tsusho Sangyo Chōsakai.

MITI. (Ed.). (1993). *Nyū sanshain keikaku handobukku* [*The handbook of new sunshine project*]. Tokyo: Tsūshō Sangyō Chōsakai.

MITI. (Eds.). (1973). *Nihon no enerugī mondai*. Tokyo: Tsūshō Sangyō Chōsakai.

Miyazaki, I., & Usui, T. (1985). *Sentan gijutsu to Nihon keizai* [*Advanced technology and Japanese economy*]. Tokyo: Nippon Hyōron sha.

Morgenthau, H. J. (1972). *Politics among nations: The struggle for power and peace* (5th ed.). New York: Knopf.

Morris, P. W. G., & Hough, G. H. (1987). *The anatomy of major projects: A study of the reality of project management*. Chichester; New York: Wiley.

Nakagawa, N., & Akagawa, M. (Eds.). (2013). *Hōhōron toshiteno kōchikushugi* [*Constructionism as methodology*]. Tokyo: Keiso Shobo.

Nakagawa, N., Kitazawa, T., & Doi, T. (Eds.). (2001). *Shakai kōchikushugi no supekutoramu: Pāsupekutibu no genzai to kanōsei* [*The spectrum of social constructionism: The present and potentials of perspectives*]. Kyoto: Nakanishiya Shuppan.

NEDO BOOKS Editorial Committee. (Ed.). (2007). *Naze nihon ga taiyōkō hatsuden de sekaiichi ni naretanoka* [*Why Japan was able to become the world number one at photovoltaic power generation*]. Kawasaki: Shin'enerugī Sangyō Gijutsu Sōgō Kaihatsu Kikō.

NEDO. (Ed.) (1990). *NEDO jūnen no ayumi* [*Ten year history of NEDO*]. Tokyo: NEDO.

NEDO. (1996). *Shin enerugī gijutsu kaihatsu kankei dētashū sakusei chōsa: Taiyōkō hatsuden* [*Research for preparing collections of data related to the development of new energy technologies: Solar power generation*]. Self-pub.

Neustadt, R. E. (1960). *Presidential power: The politics of leadership*. New York: Wiley.

Nishida, M. (2000). *Inobēshon to keizai seisaku* [*Innovation and economic policy*]. Tokyo: Yachiyo Shuppan.

Nonaka, I., & Takeuchi, H. (1995). *The knowledge-creating company: How Japanese Companies create the dynamics of innovation*. New York: Oxford University Press.

Nonaka, I. (1990). *Chishiki sōzō no keiei: Nihon kigyō no episutemorojī* [*A theory of organizational knowledge creation: Epistemology of Japanese companies*]. Tokyo: Nihon Keizai Shimbunsha.

Numagami, T. (1992). Ninchi moderu to shiteno gijutsu: soshiki to nettowāku no keiei kōsōryoku [Technologies as cognitive models: The powers of management conception of organizations and networks]. *Business Review*, *40*(2), 50–65.

Numagami, T. (2000). *Kōi no keieigaku: Keieigaku ni okeru itosezaru kekka no tankyū* [*Toward an action system theory of management: The pursuit of unintended consequences in business administration*]. Tokyo: Hakuto-Shobo Publishing.

Ono, G. (1999). *Gendai Nihon no sangyō seisaku: Dankaibetsu seisaku kettei no mekanizumu* [*Industrial policy of contemporary Japan*]. Tokyo: Nihon Keizai Shimbunsha.

Ohtaki, S. (1981), Daikibo kenkyū kaihatsu purojekuto no soshiki kan bunseki – 'Genshiryoku seitetsu projekuto' no jirei ni kansuru yobiteki kōsatsu. *Senshu Keieigaku ronsyu*, *32*.

Tawada, Y. (Ed.). (2011). *Puro ga oshieru taiyō denchi no subete ga wakaru hon: Saishin no taiyō denchi gijutsu o kiso kara manandemiyō: Shijō saikyō karā zukan* [*A book by a professional that tells you everything about solar cells: Let's learn solar cell technologies from the basics: The best illustrated dictionary in color*]. Tokyo: Natsumesha.

Perrow, C. (1972). *Complex organizations: A critical essay*. Glenview, Ill.: Scott, Foresman.

Pondy, I. R., & Mitroff, I. I. (1979). Beyond open system models of organization. In B. M. Staw (Ed.), *Research in organizational behavior: An annual series of analytical essays and critical reviews* (vol. 1). Greenwich, CT: JAI Press.

Popper, K. R. (1960). *The poverty of historicism*. New York: Basic Books.

PVTEC. (1996). *PVTEC gonen no ayumi* [*Five year history of PVTEC*]. Tokyo: PVTEC.

Quade, E. S. (Ed.). (1964). *Analysis for military decisions: The Rand lectures on systems analysis*. Chicago: Rand McNally.

Safety and Environment Center (SEC) for Petroleum Development, Engineering Advancement Association of Japan, *SEC Nyūsu*, No. 80, January 2012.

Sakaiya, T. (1975). *Yudan!* [*Negligence!*]. Nihon Keizai Shimbun: Tokyo.

Sakaiya, T. (1976). *Dankai no sedai* [*Baby boomers*]. Tokyo: Kōdansha.

Sakakibara, K. (1986). Kyōdō kenkyū kaihatsu project no soshiki to management: Chō LSI gijutsu kenkyū kumiai no kēsu. In K. Imai (Ed.), *Inobēshon to soshiki*. Tokyo: Tōyō Keizai Inc.

Sangyō kōzō shingikai. (1971). *70-nendai no tsūshō sangyō seisaku: Sangyō Kōzō Shingikai chūkantōshin* [*International trade and industry policy in the 1970s*]. Tokyo: Ōkurashō Insatsukyoku.

Sangyō Times Inc. (2007). *Taiyō denchi sangyō sōran 2007* [*Complete guide to the solar cell industry – 2007 edition*]. Tokyo: Sangyō Times Inc.

Sangyō Times Inc. (2009). *Taiyō denchi sangyō sōran 2010* [*Complete guide to the solar cell industry – 2010 edition*]. Tokyo: Sangyō Times Inc.

Satō, I. (1984). *Bōsōzoku no esunogurafī: Mōdo no hanran to bunka no jubaku* [*Ethnography of motorcycle gangs: A revolt of mode and the spells of a culture*]. Tokyo: Shinyōsha.

Sawada, S. (1998). Horigome Takashi Hakushi no Ayundekita Michi [The course taken by Dr. Horigome Takashi]. *Kōgyō gijutsu* [*Industrial Technologies*], *39*(4).

Sawai, M., & Tsushō sangyō seisakushi hensan iinkai [Editorial Committee on the History of Japan's Trade and Industry Policy] (Eds.). (2011). *Tsūshōsangyō seisakushi 1980–2000* [*History of Japan's trade and industry policy 1980–2000*] (vol. 9) Industrial Technology Policy. Tokyo: Keizai Sangyō Chōsakai.

Schelling, T. C. (1960). *The strategy of conflict*. Cambridge, MA: Harvard University Press.

Schilling, W. R. (1962). The politics of national defense: Fiscal 1950. In W. R. Schilling, P. Y. Hammond, & G. H. Snyder (Eds.), *Strategy, politics, and defense budgets*. New York: Columbia University Press.

Schwenk, C. R. (1988). *The essence of strategic decision making*. Lexington, MA: Lexington Books.
Scott, R. A. (1967). The selection of clients by social welfare agencies: The case of the blind. *Social Problems, 14*(3).
Scott, W. J. (1990). PTSD in DSM-III: A case in the politics of diagnosis and disease. *Social Problems, 37*(3), 294–310.
Scott, W. R. (1998). *Organizations: Rational, natural, and open systems* (4th ed.). Upper Saddle River, N.J.: Prentice Hall.
Seiyama, K. (1995). *Seidoron no kozu* [*The structure of institutional theory*]. Tokyo: Sōbunsha.
Selznick, Ph. (1948). Foundations of the theory of organization. *American Sociological Review, 13* (1).
Selznick, Ph. (1957). *Leadership in administration: A sociological interpretation*. Evanston, Ill: Row, Peterson.
Selznick, Ph. (1960). *The organizational weapon: A study of bolshevik strategy and tactics*. Glencoe, Ill.: Free Press.
Selznick, Ph. (1949). *TVA and the grass roots: A study in the sociology of formal organization*. Berkeley and Los Angeles: University of California Press.
Selznick, Ph. (1996). Institutionalism 'old' and 'new'. *Administrative Science Quarterly*, *41*(2) (40th anniversary issue).
Sharp. (1996). *Taiyō denchi no sekai kaiteiban* [*The world of solar cells revised edition*]. Self-pub.
Shigen ererugīchō. (Eds.). (1993). *Enerugī seisaku no ayumi to shinten*. Tokyo: Tsūshō Sangyō Chōsakai.
Shigen enerugīchō. (Ed.). (1995). *Shin enerugī binran heisei shichi-nendo ban* [*New energy handbook, FY1995 edition*]. Tokyo: Tsūshō Sangyō Chōsakai.
Shigen enerugīchō. (2013). *FY2012 annual report on energy (Energy White Paper 2013)*. Tokyo: Nikkei Printing.
Shigen enerugīchō. (Ed.). (1987). *Nijūisseiki e no enerugī tenbō*. Tokyo: Denryoku Shinpōsha.
Shigen enerugīchō. (Ed.). (1992 and 1995). *Sōgō enerugī tōkei*. Tokyo: Tsushō Sangyō Chōsakai.
Shigen enerugīchō. (1995). *Shin-enerugī binran Heisei shichi-nendo ban*. Tokyo: Tsūshō Sangyō Chōsakai.
Shigen ererugīchō. (2014). *FY2013 annual report on energy (Energy White Paper 2014)*. Tokyo: Nikkei Printing.
Shigen ererugīchō. (Ed.). (1985). *Shigen enerugi dētā shū*. Tokyo: Denryokushinpōsha.
Shigen ererugīchō. (Ed.). (1986). *Shigen enerugi dētā shū*. Tokyo: Denryokushinpōsha.
Shigen ererugīchō. (Ed.). (1987). *Shigen enerugi dētā shū*. Tokyo: Denryokushinpōsha.
Shigen ererugīchō. (Ed.). (1988). *Shigen enerugi dētā shū*. Tokyo: Denryokushinpōsha.
Shigen ererugīchō. (Ed.). (1989). *Shigen enerugi dētā shū*. Tokyo: Denryokushinpōsha.
Shigen ererugīchō. (Ed.). (1990). *Shigen enerugi dētā shū*. Tokyo: Denryokushinpōsha.
Shigen ererugīchō. (Ed.). (1991). *Shigen enerugi dētā shū*. Tokyo: Denryokushinpōsha.
Shigen ererugīchō. (Ed.). (1992). *Shigen enerugi dētā shū*. Tokyo: Denryokushinpōsha.
Shigen ererugīchō. (Ed.). (1993). *Shigen enerugi dētā shū*. Tokyo: Denryokushinpōsha.
Shimamoto, M. (1999). YS-11 Purojekuto no sosiki dezain: Yōranki kōkūki sangyō to tsūsan kanryō (Akawaza Shōichi) [Oranizational design of the YS-11 project: The rise of the Japanese aircraft industry and a MITI bureaucrat (Akawaza Shōichi)]. *Ikkyo ronsō* [*The Hitotsubashi Review*], *121*(5), 685–701.
Shimamoto, M. (2012). Nihon no taiyōkō hatsuden sangyō wa fukkatsu suruka [Will the Japanese solar power generation industry come to life again?]. *Hitotsubashi Business Review, 59*(4).
Shimamoto, M. (2010). Taiyōkō hatsuden no hanseiki: sankangaku niyoru gijutsu kaihatsu to shijō kaitaku no chōsen [A half century of solar power generation: Joint academy, industry and government attempts to develop technologies and markets]. *Hitotsubashi Business Review, 58* (1).
Shin'enerugī zaidan. (Ed.). (1997). *Shin enerugī kaihatsu riyō jittai chōsa hōkokusho sōgōhen*. Tokyo: Shin'enerugī Zaidan Keikaku Honbu.

Spector, M., & Kitsuse, J. I. (1977). *Constructing social problems*. Menlo Park, CA: Cummings Publishing.

Subcommittee on the Development of Amorphous Solar Cells. (1979). *Report by the subcommittee on the development of amorphous solar cells (draft)*. Unpublished manuscript.

Taira, H., & Nakagawa, N. (2000). *Kōchikushugi no shakaigaku: Ronsō to giron no esunogurafī* [*The sociology of constructionism: Ethnography of disputes and discussions*]. Tokyo: Sekaishisōsha.

Taiyōkō Hatsuden Gijutsu Kenkyū Kumiai [Photovoltaic Power Generation Technology Research Association]. (Ed.). (1998). *Taiyōkō hatsuden: Sono hatten to tenbō* [*Solar power generation: Its development and prospects*]. Tokyo: Art Studio, 76.

Takagi, D. Y. (1990). From discrimination to affirmative action: Facts in the Asian American admissions controversy. *Social Problems, 37*(4), 578–592.

Takeuchi, H., et al. (1986). *Kigyō no jiko kaikaku: Kaosu to sōzō no manējimento* [*Corporate self-reforms: Management of chaos and creation*]. Tokyo: Chuōkōron Shinsha.

Tanaka, K., & Shimizu, T. (1976). Hishōshitsu handōtai ni kansuru futatsu no kokusai kaigi [Two international conferences on amorphous semiconductors]. *Journal of the Physical Society of Japan, 31*(9).

Tekunorisāchi Kenkyūjo [Technoresearch Co., Ltd.]. (1997). *Nyū sanshain keikaku purojekuto taishō gijutsu no dōnyū jōkyō chōsa hōkokusho*. Shizuoka: Tekunorisāchi Kenkyūjo.

Tsuruta, T. (1982). *Sengo nihon no sangyō seisaku* [*Industrial policies in postwar Japan*]. Tokyo: Nihon Keizai Shimbunsha.

Twain, M. (1876). *The adventures of Tom Sawyer*. Hartford: American Publishing.

Van Maanen, J. (1988). *Tales of the field: On writing ethnography*. Chicago: University of Chicago Press.

Wadaki, T. (2008). *Bakuhatsu suru taiyō denchi sangyō: Nijūgo-okuen shijō no genjō to mirai* [*The exploding solar cell industry: The current state and future of a 25-trillion yen market*]. Tokyo: Tōyō Keizai Inc.

Wakasugi, R. (1986). *Gijutsu kakushin to kenkyū kaihatsu no keizai bunseki*. Tokyo: Tōyō Keizai Inc.

Watanabe, C. (1994). Japanese industrial science & technology policy at a turning point. In *International Conference on Understanding Government R&D Investment Decisions*.

Watanabe, C. (Ed.). (2001). *Gijutsu kakushin no keiryō bunseki: Kenkyū kaihatsu no seisansei shūekisei no bunseki to hyōka* [*Numerical analysis of technological innovation*]. Tokyo: Nikka Giren Shuppansha.

Watanabe, C. (2007). *Gijutsu keizai shisutemu* [*Techno-economic systems*]. Tokyo: Soseisha.

Watanabe, C. (1995). Identification of the role of renewable energy: A view from Japan's challenge, the new sunshine program. *Renewable Energy, 6*(3).

Watanabe, C., Miyazaki, K., & Katsumoto, M. (Eds.). (1998). *Gijutsu keizairon* [*Theory of techno-economics*]. Tokyo: Nikka Giren Shuppansha.

Watanabe, C. (1998). Systems option for sustainable development. In OECD IEA CERT 18th Meeting.

Watanabe, S. (2007). *Soshiki shakaigaku* [*Organization sociology*]. Kyoto: Minerva Shobo.

Weber, M. (1947). *The theory of social and economic organization* (A. M. Henderson & T. Parsons, Trans.). New York: Oxford University Press.

Whyte, W. F. (1943). *Street corner society: The social structure of an Italian slum* (4th ed.). Chicago: University of Chicago Press.

Willis, P. E. (1977). *Learning to labour: How working class kids get working class jobs*. Farnborough, Eng.: Saxon House.

Wohstetter, R. (1962). *Pearl Harbor: Warning and decision*. Stanford, Calif.: Stanford University Press.

Woolgar, S., & Pawluch, D. (1985). Ontological gerrymandering: The anatomy of social problems explanations. *Social Problems, 32*(3), 214–227.

Yamaka, K. (2009). *Sorā uōzu: Gekidō-suru Taiyō denchi bijinesu no zenbō* [*Solar wars: The complete story of the rapidly changing solar cell businesses*. Tokyo: Energy Forum.

Yonekura, S. (1998). Keieishigaku no hōhōron: Itsudatsu, fukisokusei, shūkansei [*Methodology of business history: Deviation, irregularity and subjectivity*]. *Ikkyo ronsō* [*The Hitotsubashi Review*], *120*(5), 678–692.

Yonekura, S., & Shimamoto, M. (1998). Kyōsō to keikaku no chōsei: Yōranki konpyūtā sangyō to tsūsan kanryō (Hiramatsu Morihiko) [Coordination between competition and planning: The rise of the Japanese computer industry and a MITI bureaucrat (Hiramatsu Morihiko)]. In H. Itami, et al. (Eds.), *Kēsubukku Nihon kigyō no keiei kōdō 1: Nihonteki keiei no seisei to hatten* [*Casebook 1 on administrative actions taken by Japanese companies: Generation and development of the Japanese style of management*]. Tokyo: Yuhikaku.

Yoshikai, M. (1985). *Nihon no sangyō gijutsu seisaku* [*Industrial technology policy of Japan*]. Tokyo: Tōyō Keizai Inc.

Documents

Development Bank of Japan Inc. (2008, April 23). Taiyōkō hatsuden o meguru saikin no dōkō: hageshisa o masu kokusai kyōsō to kaihatsu ga susumu hakumaku kei [Recent trends in solar power generation: Intensifying global competition and advancing development of thin-film solar cells]. Kongetsu no topikku, no. 122-1.

Energy Issue Study Group at the Electric Power Division of Denki Shikenjo. (1970). Denki shikenjo ni okeru enerugī gijutsu no kenkyū kadai [Themes for energy technologies research at Denki Shikenjo]. Unpublished manuscript.

Energy Issue Study Group at the Electric Power Division of Denki Shikenjo. (1969). Enerugīmondai ni kansuru burēn sutōmingu [Brainstorming on energy issues]. Unpublished manuscript.

Kurokawa, K. (1977). Taiyōkō hatsuden shisutemu no kenkyū kaihatsu [Research and development of solar power generation systems]. Unpublished manuscript.

Material provided by Ishikawa, F. (1979, December). Shin enerugī sōgō kaihatsu kikō (kashō) no setsuritsu ni tsuite [Regarding the establishment of a general development organization for new energy (Provisional Name)]. Memorandum.

Material provided by Ishikawa, F. (1979, December). Arukōru senbai jigyō no toriatsukai ni tsuite [Regarding the handling of the exclusive sale of alcohol]. Memorandum.

Material provided by Tanaka, K. (1977). Pamphlet for the first seminar on the properties and application of amorphous materials.

Material provided by Tanaka, K. (1980, February 14). Amorufasu hakumaku taiyō denchi no kiso kenkyū [*Fundamental research on amorphous thin-film solar cells*]. Unpublished manuscript.

Material provided by Tanaka, K. (1979). *The development of amorphous photovoltaic cells in Japan*, by M. H. Cohen.

MITI AIST Minister's Secretariat, General Affairs Division. (1974). *Shōwa gojū-nendo Shin Seisaku* [*FY1975 New Policies*]. Memorandum.

Sakudo, K. (1977, August 4). Shōchō ryūchihi no tokubetsu sochi no irai ni tsuite [Request for a special measure for manager retention expenses]. Unpublished manuscript.

Suzuki, K. (1973, May). Shin'enerugī gijutsu kaihatsu seido no seitei oyobi shin'enerugī gijutsu kaihatsubu no secchi ni tsuite [Regarding the establishment of a new energy technology development scheme and new energy technology development department]. Memorandum.

Tanaka, K. (1980, February 14). Amorufasu hakumaku taiyō denchi no kiso kenkyū [Basic research on amorphous thin-film solar cells]. Unpublished manuscript.

Tanaka Kazunobu's note on the Agency of Industrial Science and Technology's hearings on the Sunshine Project, 1977.

Tanaka Kazunobu's pocket notebook for fiscal year 1977, May 23, 1977.

The Japan Society of Applied Physics. (1980). Daiikkai handōtai no hikari kidenryoku kōka no kiso to ōyō ni kansuru shinpojium [1st Symposium on the Basics and Application of the Photoelectromotive Force of Semiconductors].
Tomae, H. (1996). Kyōdō kenkyū ni okeru ritateki shinrai [Altruistic trust in joint research and development](PhD dissertation). Hitotsubashi University.

Newspapers and Magazines

Asahi Shimbun, August 23, 1973.
Kyōto Shimbun, October 12, 1985.
Mainichi Shimbun, January 17, 1974.
NEDO News, January 1981.
NEDO News, March 1981.
NEDO News, October 1981.
NEDO News, January 1983.
NEDO News, March 1983.
NEDO News, September–October 1983.
NEDO News, January 1984.
NEDO News, July 1984.
NEDO News, September–October 1984.
NEDO News, March 1986.
NEDO News, May 1987.
NEDO News, December 1987.
NEDO News, March 1988.
NEDO News, October–November 1988.
NEDO News, October–November, 1988.
Nikkei Business, March 27, 1978.
Nikkei Business, June 8, 2009.
Nikkei Business, February 8, 2010.
Nikkei Micro Device, August 2008.
Nikkei Sangyō Shimbun, September 10, 1975.
Nikkei Sangyō Shimbun, March 18, 1976.
Nikkei Sangyō Shimbun, January 11, 1977.
Nikkei Sangyō Shimbun, February 4, 1977.
Nikkei Sangyō Shimbun, June 2, 1977.
Nikkei Sangyō Shimbun, May 8, 1979.
Nikkei Sangyō Shimbun, May 9, 1979.
Nikkei Sangyō Shimbun, May 30, 1979.
Nikkei Sangyō Shimbun, August 15, 1979.
Nikkei Sangyō Shimbun, September 1, 1979.
Nikkei Sangyō Shimbun, October 2, 1980.
Nikkei Sangyō Shimbun, March 26, 1981.
Nikkei Sangyō Shimbun, April 2, 1981.
Nikkei Sangyō Shimbun, June 3, 1981.
Nikkei Sangyō Shimbun, August 22, 1982.
Nikkei Sangyō Shimbun, June 22, 1983.
Nikkei Sangyō Shimbun, July 20, 1983.
Nikkei Sangyō Shimbun, October 25, 1984.
Nikkei Sangyō Shimbun, December 3, 1986.
Nikkei Sangyō Shimbun, December 28, 1987.
OHM, November issue, 1984.
President, September issue, 1984.

The Nikkei/Nihon Keizai Shimbun, October 16, 1973.
The Nikkei/Nihon Keizai Shimbun, January 25, 1974.
The Nikkei/Nihon Keizai Shimbun, September 6, 1975.
The Nikkei/Nihon Keizai Shimbun, December 20, 1975.
The Nikkei/Nihon Keizai Shimbun, December 31, 1975.
The Nikkei/Nihon Keizai Shimbun, January 8, 1976.
The Nikkei/Nihon Keizai Shimbun, January 20, 1977.
The Nikkei/Nihon Keizai Shimbun, August 24, 1977.
The Nillei/Nihon Keizai Shimbun, January 30, 1978.
The Nkkei/Nihon Keizai Shimbun, September 14, 1981.
The Nikkei/Nihon Keizai Shimbun, October 22, 1981.
The Nkkei/Nihon Keizai Shimbun, December 21, 1981.
The Nikkei/Nihon Keizai Shimbun, July 4, 1990.
The Nikkei/Nihon Keizai Shimbun, May 7, 2010.
The Nikkei/Nihon Keizai Shimbun, July 24, 2010.
Regional Economy pages, *The Nikkei/Nihon Keizai Shimbun*, August 21, 1982.
Regional Economy pages, *The Nikkei/Nihon Keizai Shimbun*, July 23, 1983.
Shūkan Ekonomisuto [Weekly Economist], August 3, 2010.
WiLL, October issue, 1984.

Websites, Databases, Etc.

Nikkei Trendy Net. (2011, November 8). http://trendy.nikkeibp.co.jp/article/pickup/20222207/1038545/.
Kyocera Website. http://www.kyocera.co.jp/news/2010/0401_ariz.html.

Interviews Conducted by the Author

Horigome, T., interview by author, June 13, 1998.
Horigome, T., interview by author, September 16, 1998.
Horigome, T., telephone conversation with author, November 19, 1998.
Ishikawa, F., interview by author, June 4, 1998.
Kimura, K., interview by author, September 9, 1998.
Kimura, K., telephone conversation with author, November 10, 1998.
Konagai, M., interview by author, October 12, 1998.
Konno, K., interview by author, March 11, 1998.
Kurokawa, K., interview by author, April 20, 1998.
Kurokawa, K., interview by author, April 29, 1998.
Kuwano, Y., interview by author, October 29, 1998.
Mr. X at company S (name withheld by request), interview by author, September 10, 1998.
Murozono, M., interview by author, September 11, 1998.
Nebashi, M., & Ken, S., interview by author, May 8, 1998.
Saito, T., interview by author, September 7, 1998.
Sawada, S., interview by author, June 22, 1998.
Suzuki, N., interview by author, September 2, 1998.
Tanaka, K., interview by author, October 15, 1998.
Tani, T., interview by author, July 9, 1998.
Wakimura, K., interview by author, September 7, 1998.

Names of People

M. Shimamoto, *National Project Management*,
Advances in Japanese Business and Economics 25,
https://doi.org/10.1007/978-981-15-3180-4

Index

M. Shimamoto, *National Project Management*,
Advances in Japanese Business and Economics 25,
https://doi.org/10.1007/978-981-15-3180-4

Subject Index

M. Shimamoto, *National Project Management*,
Advances in Japanese Business and Economics 25,
https://doi.org/10.1007/978-981-15-3180-4

The manufacturer's authorised representative in the EU is Springer Nature Customer Service Centre GmbH, Europaplatz 3, 69115 Heidelberg, Germany. If you have any concerns regarding our products, please contact ProductSafety@springernature.com

Printed and bound by CPI Group (UK) Ltd, Croydon, CR0 4YY
30/09/2026
02225254-0014